MATHS IN FOCUS

MATHEMATICS EXTENSION 2

YEAR **12**

Jim Green
Janet Hunter
Series editor: Robert Yen

NELSON
A Cengage Company

Maths in Focus 12 Mathematics Extension 2
1st Edition
Jim Green
Janet Hunter
ISBN 9780170413435

Publisher: Robert Yen and Alan Stewart
Project editor: Anna Pang
Permissions researcher: Catherine Kerstjens
Editor: Elaine Cochrane
Cover design: Chris Starr (MakeWork)
Cover image: iStock.com/cmart7327
Text design: Sarah Anderson
Project designer: Justin Lim
Project manager: Jem Wolfenden
Production controller: Alice Kane

Any URLs contained in this publication were checked for currency during the production process. Note, however, that the publisher cannot vouch for the ongoing currency of URLs.

For product information and technology assistance,
in Australia call **1300 790 853**;
in New Zealand call **0800 449 725**

For permission to use material from this text or product, please email **aust.permissions@cengage.com**

ISBN 978 0 17 041343 5

Cengage Learning Australia
Level 7, 80 Dorcas Street
South Melbourne, Victoria Australia 3205

Cengage Learning New Zealand
Unit 4B Rosedale Office Park
331 Rosedale Road, Albany, North Shore 0632, NZ

For learning solutions, visit **cengage.com.au**

Printed in China by 1010 Printing International Limited.
2 3 4 5 6 7 23 22 21 20

PREFACE

Maths in Focus 12 Mathematics Extension 2 is written for the new Mathematics Extension 2 syllabus (2017). Although this is a new book, students and teachers will find that it contains the familiar features that have made *Maths in Focus* a leading senior mathematics series, such as clear and abundant worked examples in plain English, comprehensive sets of graded exercises, chapter *Test yourself* exercises and practice sets of mixed revision and exam-style questions.

The Mathematics Extension 2 course is designed for students who intend to study mathematics at university, possibly majoring in the subject.

The theory presented in this book follows a logical order, although some topics may be learned in any order. We have endeavoured to produce a practical text that captures the spirit of the course, providing relevant and meaningful applications of mathematics.

The *NelsonNet* teacher website contains additional resources such as worksheets, *ExamView* quizzes and questionbank, topic tests and worked solutions (see page viii). We wish all teachers and students using this book every success in embracing the new Mathematics Extension 2 course.

ABOUT THE AUTHORS

Jim Green is Head of Mathematics at Trinity Catholic College, Lismore, where he has spent most of his teaching career of over 35 years. He has taught pre-service teachers at the Southern Cross University, written HSC examinations and syllabus writing drafts, composed questions for the Australian Mathematics Competition and recently co-authored *Nelson Senior Maths 11-12 Specialist Mathematics for the Australian Curriculum*.

Janet Hunter is Head of Mathematics at Ascham School, Edgecliff, and has worked in the finance sector and in tertiary/adult education as a lecturer. She has been a senior HSC examiner and judge, an HSC Advice Line adviser, served on the editorial team for the MANSW journal *Reflections*, and recently co-authored *Nelson Senior Maths 11-12 Specialist Mathematics for the Australian Curriculum*.

CONTRIBUTING AUTHORS

Jim Green and **Janet Hunter** also wrote the topic tests.

Roger Walter wrote the *ExamView* questions.

Shane Scott wrote the worked solutions to all exercise sets.

CONTENTS

1

2

3

4

SYLLABUS REFERENCE GRID

Topic and subtopic	Maths in Focus 12 Mathematics Extension 2 chapter
PROOF	
MEX-P1 The nature of proof	2 Mathematical proof 5 Further mathematical induction
MEX-P2 Further proof by mathematical induction	5 Further mathematical induction
VECTORS	
MEX-V1 Further work with vectors	3 3D vectors
V1.1 Introduction to three-dimensional vectors V1.2 Further operations with three-dimensional vectors V1.3 Vectors and vector equations of lines	
COMPLEX NUMBERS	
MEX-N1 Introduction to complex numbers	1 Complex numbers
N1.1 Arithmetic of complex numbers N1.2 Geometric representation of a complex number N1.3 Other representations of complex numbers	
MEX-N2 Using complex numbers	4 Applying complex numbers
N2.1 Solving equations with complex numbers N2.2 Geometric implications of complex numbers	
CALCULUS	
MEX-C1 Further integration	6 Further integration
MECHANICS	
MEX-M1 Applications of calculus to mechanics	7 Mechanics
M1.1 Simple harmonic motion M1.2 Modelling motion without resistance M1.3 Resisted motion M1.4 Projectiles and resisted motion	

MATHS IN FOCUS AND NEW CENTURY MATHS 11–12

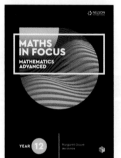

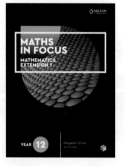

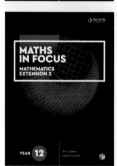

ABOUT THIS BOOK

AT THE BEGINNING OF EACH CHAPTER

- Each chapter begins on a double-page spread showing the **Chapter contents** and a list of chapter outcomes

- **Terminology** is a chapter glossary that previews the key words and phrases from within the chapter

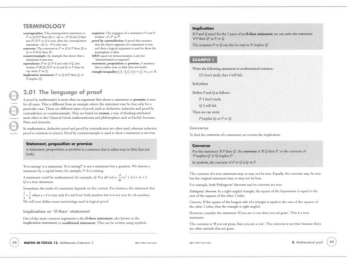

IN EACH CHAPTER

- Important facts and formulas are highlighted in a shaded box.

- Important words and phrases are printed in red and listed in the Terminology chapter glossary.

- Graded exercises include exam-style problems and realistic applications.

- Worked solutions to all exercise questions are provided on the *NelsonNet* teacher website.

- **Investigations** explore the syllabus in more detail, providing ideas for modelling activities and assessment tasks.

- **Did you know**? contains interesting facts and applications of the mathematics learned in the chapter.

2.04 Proofs involving numbers

In this section we will develop some techniques for proving common properties of numbers.

Properties of positive integers

An **even number** can be described by the formula $2n$ where $n \in \mathbb{N}$.

An **odd number** can be described by the formula $2n - 1$ where $n \in \mathbb{N}$.

A **square number** can be described by the formula n^2 where $n \in \mathbb{N}$.

A number, X ($X \in \mathbb{N}$), is **divisible** by another number, p ($p \in \mathbb{N}$), if $\exists\, Y$ ($Y \in \mathbb{N}$) such that $X = pY$.

Note that in general, properties such as even, odd, multiples, factors, refer to the **positive integers** only. For instance, we would not usually say −7 is an odd number or that $\frac{4}{9}$ is a square number.

EXAMPLE 11

Determine whether this statement is true:

> All camels have one hump.

Solution

Most camels have one hump, but the Bactrian camel has two humps.
Therefore the statement is false.

EXAMPLE 12

Explain what is wrong with the following argument.
Teacher: Smoking is bad for you.
Student: I have a counterexample: social media is bad for you.

Solution

The student has found another habit that is bad for you rather than a counterexample to show that smoking is not bad for you.

Exercise 2.03 Proof by counterexample

1 Find a counterexample to prove that each statement is not true.
   a $\forall n \in \mathbb{R}, n^2 \geq n$
   b $\forall n \in \mathbb{R}, n^2 + n \geq 0$
   c All prime numbers are odd.
   d $\forall x, y \in \mathbb{Z}$ and $n \in \mathbb{N}$, $x^n + y^n = (x + y)(x^{n-1} - x^{n-2}y + x^{n-3}y^2 - \ldots - x^2y^{n-2} + y^{n-1})$
   e $\forall x \in \mathbb{Z}, x + \frac{1}{x} \geq 2$

2 Find a counterexample for each statement to demonstrate it is false.
   a If $n^2 = 100$, then $n = 10$.
   b The statement $x^3 - 6x^2 + 11x - 6 = 0$ is true for $x = 1, 2, 3, \ldots$
   c All lines that never meet are parallel.
   d If an animal lays eggs, then it is a bird.

3 Decide whether each statement is true or false. If it is false, provide a counterexample.
   a If a quadrilateral has diagonals that are perpendicular, then it is a square.
   b If $p \leq 3$, then $\frac{1}{p} \geq \frac{1}{3}$
   c $(x + y)^2 \geq x^2 + y^2$ for all $x, y \in \mathbb{R}$
   d If $pq = rq$ then $r = p$
   e All rectangles are similar.

4 In each case, determine whether the counterexample shows that the statement is false.
   a *Statement*: $x^2 = 3x - 2$ for $x \in \mathbb{N}$
      Counterexample: For $x = 3$, LHS $= 3^2 = 9$
      RHS $= 3(3) - 2 = 7$ $\therefore$ LHS $\neq$ RHS
      The statement is not true.
   b *Statement*: All dogs are domesticated.
      Counterexample: Cats are domesticated.
      The statement is not true.

5 Decide whether the statement below is true or false. If it is false, find a counterexample.
   If $a > b$ then $\frac{1}{a} < \frac{1}{b}$.

6 Is it always true that $\frac{1}{(x-1)(x-2)} + \frac{1}{(x-2)(x-3)} = \frac{2x-4}{(x-1)(x-2)(x-3)}$ for $x = 4, 5, 6, \ldots$?

7 Is it always true that $\frac{1}{n} < \frac{1}{n-1}$?

8 Are all squares rhombuses? Are all rhombuses squares?

9 A circle can always be drawn through the four vertices of a rectangle. Is this true for all quadrilaterals?

10 Is the angle sum of all polygons with n sides $S_n = 180°(n - 2)$?

11 Do the diagonals of a kite always intersect inside the kite?

12 Is it always true that $|x - y| = |y - x|$?

13 Is it always true that if $n > m$ then $nk > mk$?

DID YOU KNOW?

Bertrand Russell

Bertrand Russell was a philosopher and mathematician who studied logic. He is famous for inventing a paradox in 1901 that is understandably called Russell's Paradox. It was a proof by contradiction for an idea in set theory.

AT THE END OF EACH CHAPTER

- **Test yourself** contains chapter revision exercises.

- **Practice sets** (after several chapters) provide a comprehensive variety of mixed exam-style questions from various chapters, including short-answer, free-response and multiple-choice questions.

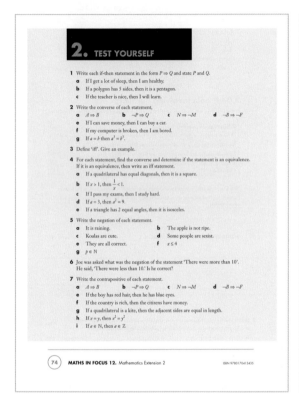

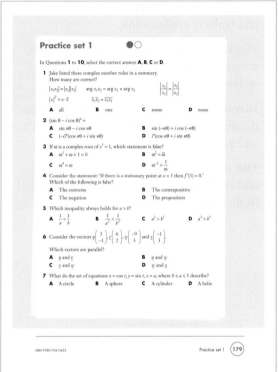

AT THE END OF THE BOOK

Answers and **Index** (worked solutions on the teacher website).

MATHS IN FOCUS 12. Mathematics Extension 2

ISBN 9780170413435

NELSONNET STUDENT WEBSITE

Margin icons link to worksheets and chapter quizzes found on the *NelsonNet* student website, **www.nelsonnet.com.au.** These include:

- **Worksheets** that are write-in enabled PDFs
- *ExamView* **quizzes:** interactive and self-marking

NELSONNET TEACHER WEBSITE

The *NelsonNet* teacher website, also at **www.nelsonnet.com.au**, contains:

- A **teaching program**, in Microsoft Word and PDF formats
- **Topic tests**, in Microsoft Word and PDF formats
- **Worked solutions** to each exercise set
- **Chapter PDFs** of the textbook
- **ExamView** exam-writing software and questionbanks
- **Resource Finder:** search engine for *NelsonNet* resources

Note: Complimentary access to these resources is only available to teachers who use this book as a core educational resource in their classroom. Contact your Cengage Education Consultant for information about access codes and conditions.

NELSONNETBOOK

NelsonNetBook is the web-based interactive version of this book found on *NelsonNet*.

- To each page of NelsonNetBook you can add notes, voice and sound bites, highlighting, weblinks and bookmarks

- **Zoom** and **Search** functions

- Chapters can be customised for different groups of students

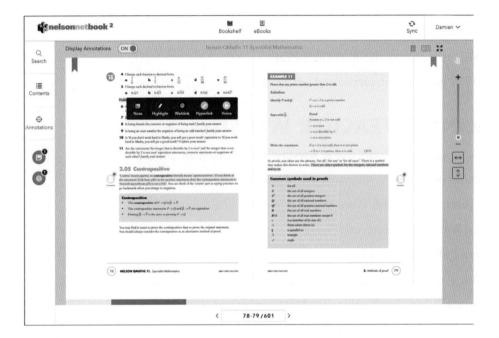

MATHS IN FOCUS 12. Mathematics Extension 2

ISBN 9780170413435

MATHEMATICAL VERBS

A glossary of 'doing words' commonly found in mathematics problems

analyse: study in detail the parts of a situation

apply: use knowledge or a procedure in a given situation

classify, identify: state the type, name or feature of an item or situation

comment: express an observation or opinion about a result

compare: show how two or more things are similar or different

construct: draw an accurate diagram

describe: state the features of a situation

estimate: make an educated guess for a number, measurement or solution, to find roughly or approximately

evaluate, calculate: find the value of a numerical expression, for example, 3×8^2 or $4x + 1$ when $x = 5$

expand: remove brackets in an algebraic expression, for example, expanding $3(2y + 1)$ gives $6y + 3$

explain: describe why or how

factorise: opposite to **expand**, to insert brackets by taking out a common factor, for example, factorising $6y + 3$ gives $3(2y + 1)$

give reasons: show the rules or thinking used when solving a problem. See also **justify**

increase: make larger

interpret: find meaning in a mathematical result

justify: give reasons or evidence to support your argument or conclusion. See also **give reasons**

rationalise: make rational, remove surds

show that, prove: (in questions where the answer is given) use calculation, procedure or reasoning to prove that an answer or result is true

simplify: give a result in its most basic, shortest, neatest form, for example, simplifying a ratio or algebraic expression

sketch: draw a rough diagram that shows the general shape or ideas, less accurate than **construct**

solve: find the value(s) of an unknown pronumeral in an equation or inequality

substitute: replace a variable by a number and evaluate

verify: check that a solution or result is correct, usually by substituting back into the equation or referring back to the problem

write, state: give the answer, formula or result without showing any working or explanation (This usually means that the answer can be found mentally, or in one step)

1.

COMPLEX NUMBERS

Imaginary numbers were first noticed by Hero of Alexandria in the 1st century CE. In 1545, the Italian mathematician Girolamo Cardano wrote about them, but believed negative numbers did not have a square root. Imaginary numbers were largely ignored until the 18th century when they were studied by Leonhard Euler and Carl Friedrich Gauss.

Imaginary numbers are useful for solving physics and engineering problems involving heat conduction, elasticity, hydrodynamics and the flow of electric current.

CHAPTER OUTLINE

IN THIS CHAPTER YOU WILL:

- learn about complex numbers of the form $z = a + ib$ where a, b are real numbers and $i = \sqrt{-1}$
- determine the real and imaginary parts of a complex number
- perform arithmetic operations with complex numbers
- realise the denominator of a complex number
- find the square root of a complex number
- find the complex conjugate of a complex number
- represent complex numbers in the complex plane
- add and subtract complex numbers in the complex plane
- find the modulus and argument of a complex number
- convert a complex number in Cartesian form to polar form
- multiply and divide complex numbers in polar form
- prove properties of complex numbers involving modulus and argument
- prove properties of complex numbers using Euler's formula

TERMINOLOGY

Argand diagram: A diagram used to represent geometrically the complex number $z = a + ib$ as the point $P(a, b)$ or the vector $\underset{\sim}{z}$ or $\overrightarrow{OP}$ on the complex plane.

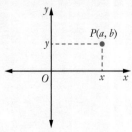

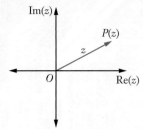

argument: If a complex number, z, is represented by a vector in the complex plane, then the argument of z, denoted **arg z** or θ, is the angle that the vector makes with the positive x-axis. For $z = a + ib$, $\tan \theta = \dfrac{b}{a}$.

$\mathbb{C}$: The set of complex numbers.

Cartesian (rectangular) form: The notation $z = a + ib$ for a complex number.

complex conjugate: The conjugate of $z = a + ib$ is $\overline{z} = a - ib$.

complex number: A number that can be written in the form $a + ib$, where a and b are real numbers. It is a member of $\mathbb{C}$, the set of complex numbers.

complex plane: A number plane for graphing complex numbers. Also called **Argand diagram**.

exponential form: The notation $z = re^{i\theta}$ for a complex number, which involves Euler's formula.

Euler's formula: For any real θ, $e^{i\theta} = \cos \theta + i \sin \theta$

imaginary number i: The number such that $i = \sqrt{-1}$, which implies $i^2 = -1$.

imaginary part: The imaginary part of $z = a + ib$ is $\text{Im}(z) = b$.

modulus: If a complex number, z, is represented by a vector in the complex plane, then the modulus of z, denoted **mod z**, $|z|$ or r, is the length of the vector. For $z = a + ib$, $r = |z| = \sqrt{a^2 + b^2}$.

polar form: The notation $z = r(\cos \theta + i \sin \theta)$ for a complex number. Also known as modulus–argument form.

real part: The real part of $z = a + ib$ is $\text{Re}(z) = a$.

vector: A quantity with both magnitude and direction, represented graphically by an arrow with specific length and direction.

1.01 Complex numbers

Imaginary numbers

Imaginary numbers arose because mathematicians wanted to solve equations such as $x^2 + 1 = 0$, which have no solutions in the set of real numbers. If we define $\sqrt{-1} = i$ then all quadratic equations will have solutions.

> ### The imaginary number i
> The **imaginary number** i is the number such that $i = \sqrt{-1}$.

An **imaginary number** is a number that can be written in the form bi, where b is a constant and real number.

For example, $3i$, $-i$ and $\sqrt{2}i$ are imaginary numbers.

Complex
numbers

Complex
conjugates

Complex
number
operations

Complex
conjugates
and inverses

MATHS IN FOCUS 12. Mathematics Extension 2

ISBN 9780170413435

EXAMPLE 1

a   Express each imaginary number in terms of i.

    i $\sqrt{-9}$ **ii** $\sqrt{-75}$ **iii** $\sqrt{-\dfrac{10}{49}}$

b   Simplify:

    i i^2 **ii** i^3 **iii** i^4 **iv** i^{10}

c   Solve the equation $x^2 + 1 = 0$.

Solution

a   **i** $\sqrt{-9} = \sqrt{-1} \times \sqrt{9}$ **ii** $\sqrt{-75} = \sqrt{-1} \times \sqrt{25} \times \sqrt{3}$

                $= i \times 3$ $= i \times 5 \times \sqrt{3}$

                $= 3i$ $= 5i\sqrt{3}$

    iii $\sqrt{-\dfrac{10}{49}} = \dfrac{\sqrt{-1} \times \sqrt{10}}{\sqrt{49}}$

                        $= \dfrac{i\sqrt{10}}{7}$

b   **i** $i^2 = (\sqrt{-1})^2$ **ii** $i^3 = i^2 \times i$ **iii** $i^4 = i^2 \times i^2$

                $= -1$ $= (-1) \times i$ $= (-1) \times (-1)$

                                $= -i$ $= 1$

    iv $i^{10} = i^4 \times i^4 \times i^2$ or $i^{10} = (i^2)^5$

                $= 1 \times 1 \times (-1)$ $= (-1)^5$

                $= -1$ $= -1$

c $x^2 + 1 = 0$

          $x^2 = -1$

           $x = \pm\sqrt{-1}$

           $x = \pm i$

Complex numbers

We have seen that a quadratic equation can have solutions with rational and irrational parts, for example, $x^2 - 6x + 7 = 0$ has solutions $x = 3 \pm \sqrt{2}$. Similarly, some quadratic equations have solutions with real and imaginary parts, for example, $x = 5 \pm 3i$. In this case, we say that the solutions are **complex**.

A **complex number** is a number that can be written in the form $a + ib$, where a and b are real numbers.

A complex number is often denoted by the letter z, so $z = a + ib$.

ISBN 9780170413435   **1. Complex numbers**

Real and imaginary parts of a complex number

The real part of $z = a + ib$ is denoted by $\text{Re}(z)$, where $\text{Re}(z) = a$

The imaginary part of $z = a + ib$ is denoted by $\text{Im}(z)$, where $\text{Im}(z) = b$

If $\text{Re}(z) = 0$ then we say that z is **purely imaginary**.
For example, $2i$, $-5i$ and $-\sqrt{10}i$ are purely imaginary.

If $\text{Im}(z) = 0$ then we say that z is **purely real** or just **real**.
For example, 12, $\sqrt{7}$ and $2 - \sqrt{3}$ are purely real.

The set of complex numbers is shown as $\mathbb{C}$, and includes $\mathbb{R}$, the set of real numbers.
All real numbers are also complex numbers.

EXAMPLE 2

a Solve the equation $x^2 + 2x + 10 = 0$ using the quadratic formula.

b Solve the equation $x^2 - 4x + 7 = 0$ by completing the square.

c Factorise $x^2 - 2x + 5$ as a difference of 2 squares, then solve the equation $x^2 - 2x + 5 = 0$.

Solution

a $x^2 + 2x + 10 = 0$

$$x = \frac{-b \pm \sqrt{b^2 - 4ac}}{2a}$$

$$= \frac{-2 \pm \sqrt{2^2 - 4(1)(10)}}{2(1)}$$

$$= \frac{-2 \pm \sqrt{-36}}{2}$$

$$= \frac{-2 \pm 6i}{2}$$

$$= -1 \pm 3i$$

b
$$x^2 - 4x + 7 = 0$$
$$x^2 - 4x + 4 + 7 = 0 + 4$$
$$(x - 2)^2 + 7 = 4$$
$$(x - 2)^2 = -3$$
$$x - 2 = \pm\sqrt{-3}$$
$$x - 2 = \pm i\sqrt{3}$$
$$x = 2 \pm i\sqrt{3}$$

ISBN 9780170413435

c First complete the square, then note that $+4 = -4i^2$.

$$x^2 - 2x + 5 = x^2 - 2x + 1 - 1 + 5$$
$$= (x - 1)^2 + 4$$
$$= (x - 1)^2 - 4i^2$$
$$= (x - 1 - 2i)(x - 1 + 2i)$$

Hence solving:

$$x^2 - 2x + 5 = 0$$

$$(x - 1 - 2i)(x - 1 + 2i) = 0$$

$$\therefore x = 1 + 2i \text{ or } x = 1 - 2i$$

EXAMPLE 3

State the real and the imaginary parts of each complex number.

i $z = \dfrac{3 - i\sqrt{5}}{2}$

ii $z = x - iy + 5 + 3i$ where $x, y \in \mathbb{R}$

Solution

i $z = \dfrac{3 - i\sqrt{5}}{2}$ can be expressed as $z = \dfrac{3}{2} - \dfrac{i\sqrt{5}}{2}$.

Therefore $\operatorname{Re}(z) = \dfrac{3}{2}$, $\operatorname{Im}(z) = -\dfrac{\sqrt{5}}{2}$.

ii $z = x - iy + 5 + 3i$ can be expressed as $z = (x + 5) + i(3 - y)$.

Therefore $\operatorname{Re}(z) = x + 5$, $\operatorname{Im}(z) = 3 - y$.

Complex conjugates

Earlier, we saw that the quadratic equation $x^2 - 6x + 7 = 0$ has solutions $x = 3 \pm \sqrt{2}$. Recall that $3 + \sqrt{2}$ and $3 - \sqrt{2}$ form a **conjugate pair** and the product $(3 + \sqrt{2})(3 - \sqrt{2}) = 7$ is **rational**. Similarly, the solutions to the quadratic equation $x^2 - 4x + 7 = 0$ are $x = 2 \pm i\sqrt{3}$. Note that $2 + i\sqrt{3}$ and $2 - i\sqrt{3}$ form a **complex conjugate** pair and their product is **real**:

$$(2 + i\sqrt{3})(2 - i\sqrt{3}) = 2^2 - (i\sqrt{3})^2$$

$$= 4 - 3i^2$$

$$= 4 - 3(-1)$$

$$= 4 + 3$$

$$= 7$$

Complex conjugate

For a complex number $z = a + ib$ (where a and b are real numbers), its **complex conjugate** is denoted by $\bar{z}$ and $\bar{z} = a - ib$.

The product $z\bar{z}$ is real.

EXAMPLE 4

a State the conjugate of each complex number.

  i $w = \dfrac{-1 + i\sqrt{3}}{2}$ **ii** $z = \dfrac{2x - 5i - ix + 3y}{x^2 + y^2}$ where $x, y \in \mathbb{R}$

b If $z = a + ib$ where $a, b \in \mathbb{R}$, prove that the product $z\bar{z}$ is always real.

> $a, b \in \mathbb{R}$ means 'a and b are real numbers'

Solution

a **i** $w = \dfrac{-1 + i\sqrt{3}}{2}$ so the conjugate $\bar{w}$ is $\bar{w} = \dfrac{-1 - i\sqrt{3}}{2}$.

  ii Regrouping $z = \dfrac{2x - 5i - ix + 3y}{x^2 + y^2}$ we have $z = \dfrac{2x + 3y - i(5 + x)}{x^2 + y^2}$ so the

  conjugate is $\bar{z} = \dfrac{2x + 3y + i(5 + x)}{x^2 + y^2}$.

b Proof:

$$z\bar{z} = (a + ib)(a - ib)$$
$$= a^2 - i^2 b^2$$
$$= a^2 - (-1)b^2$$
$$= a^2 + b^2$$

Since both $a, b \in \mathbb{R}$, then $a^2 + b^2$ is also real, therefore $z\bar{z}$ is always real.

  ISBN 9780170413435

Operations with complex numbers

Just like real numbers, complex numbers can be added, subtracted, multiplied and divided. When operating with surds, we group or equate the rational and irrational parts. When operating with complex numbers, we group or equate the real and imaginary parts.

Equivalence of complex numbers

For complex numbers $a + ib$ and $c + id$ (where a, b, c and d are real numbers), $a + ib = c + id$ if and only if $a = c$ AND $b = d$.

EXAMPLE 5

a If x and y are real, solve the equation $2x - 6i - 2yi + 10 = 0$.

b If $W = \dfrac{x + yi - 3 + ix - 2y + 5i}{x^2 + y^2}$ where x, y are real, find a relationship between x and y if $\mathrm{Re}(W) = 0$.

Solution

a $2x - 6i - 2yi + 10 = 0$ means $2x - 6i - 2yi + 10 = 0 + 0i$

$(2x + 10) + i(-6 - 2y) = 0 + 0i$

Equating real and imaginary parts:

$$2x + 10 = 0 \qquad -6 - 2y = 0$$
$$2x = -10 \qquad -2y = 6$$
$$x = -5 \qquad y = -3$$

b $W = \dfrac{x + yi - 3 + ix - 2y + 5i}{x^2 + y^2} = \dfrac{(x - 3 - 2y) + i(y + x + 5)}{x^2 + y^2}$

If $\mathrm{Re}(W) = 0$ then

$$\frac{x - 3 - 2y}{x^2 + y^2} = 0$$

$$\therefore x - 2y - 3 = 0$$

EXAMPLE 6

Simplify each expression.

a $3 - 7i + 4 + 9i$

b $6i(1 + 2i) - 4(5 - 4i)$

c $(2 + 3i)(2 - 3i) - (4 - i)^2$

Solution

a $3 - 7i + 4 + 9i = 7 + 2i$

b $6i(1 + 2i) - 4(5 - 4i) = 6i + 12i^2 - 20 + 16i$

$$= 22i + 12(-1) - 20$$

$$= 22i - 12 - 20$$

$$= -32 + 22i$$

> Note that we can replace i^2 by -1 when it arises.

c $(2 + 3i)(2 - 3i) - (4 - i)^2 = 4 - 9i^2 - (16 - 8i + i^2)$

$$= 4 - 9(-1) - (16 - 8i - 1)$$

$$= 4 + 9 - (15 - 8i)$$

$$= 13 - 15 + 8i$$

$$= -2 + 8i$$

EXAMPLE 7

Find the quadratic equation with roots $\alpha = 2 - i\sqrt{5}$ and $\beta = 2 + i\sqrt{5}$.
Express your answer in the form $az^2 + bz + c = 0$ where a, b, c are real.

Solution

A quadratic equation with roots α and β can take the form $(x - \alpha)(x - \beta) = 0$.

Expanding, we have $z^2 - (\alpha + \beta)z + \alpha\beta = 0$.

Now substitute $\alpha = 2 - i\sqrt{5}$ and $\beta = 2 + i\sqrt{5}$:

$$z^2 - [(2 - i\sqrt{5}) + (2 + i\sqrt{5})]z + (2 - i\sqrt{5})(2 + i\sqrt{5}) = 0$$

$$z^2 - 4z + (4 - 5i^2) = 0$$

$$z^2 - 4z + (4 + 5) = 0$$

$$z^2 - 4z + 9 = 0$$

Realising the denominator

When simplifying surds with a rational denominator, we multiply the numerator and denominator by the conjugate surd to **rationalise the denominator**. Similarly, when simplifying numbers with a complex denominator, we multiply the numerator and denominator by the complex conjugate to **realise the denominator**.

Realising the denominator

To **realise a denominator** z of a complex number, we multiply the number by $\dfrac{\bar{z}}{\bar{z}}$.

EXAMPLE 8

Simplify each complex number by realising the denominator.

a $\dfrac{1}{1+i}$

b $\dfrac{2+2i\sqrt{3}}{\sqrt{3}-i}$

c $\dfrac{1}{\sqrt{2}+i\sqrt{2}}+\dfrac{1}{1-i}$

Solution

a
$$\frac{1}{1+i}=\frac{1}{1+i}\times\frac{1-i}{1-i}$$
$$=\frac{1-i}{1-i^2}$$
$$=\frac{1-i}{1+1}$$
$$=\frac{1-i}{2}$$
$$=\frac{1}{2}-\frac{1}{2}i$$

b
$$\frac{2+2i\sqrt{3}}{\sqrt{3}-i}=\frac{2+2i\sqrt{3}}{\sqrt{3}-i}\times\frac{\sqrt{3}+i}{\sqrt{3}+i}$$
$$=\frac{2\sqrt{3}+2i+6i+2i^2\sqrt{3}}{3-i^2}$$
$$=\frac{2\sqrt{3}+8i-2\sqrt{3}}{3+1}$$
$$=\frac{8i}{4}$$
$$=2i$$

c
$$\frac{1}{\sqrt{2}+i\sqrt{2}}+\frac{1}{1-i}=\left(\frac{1}{\sqrt{2}+i\sqrt{2}}\times\frac{\sqrt{2}-i\sqrt{2}}{\sqrt{2}-i\sqrt{2}}\right)+\left(\frac{1}{1-i}\times\frac{1+i}{1+i}\right)$$
$$=\frac{\sqrt{2}-i\sqrt{2}}{2+2}+\frac{1+i}{1+1}$$
$$=\frac{\sqrt{2}-i\sqrt{2}}{4}+\frac{1+i}{2}$$
$$=\frac{\sqrt{2}-i\sqrt{2}+2+2i}{4}$$
$$=\frac{\sqrt{2}+2}{4}+\frac{-i\sqrt{2}+2i}{4}$$
$$=\frac{2+\sqrt{2}}{4}+\left(\frac{2-\sqrt{2}}{4}\right)i$$

POWERS OF i

We have seen the powers of i: i, i^2, i^3, i^4. Evaluate i^5, i^6, ..., i^{16}.

Can you see a pattern?

Evaluate i^{100} and i^0.

Can you see a way to evaluate i^{-1}?

Exercise 1.01 Complex numbers

1 Simplify each expression, writing the answer in terms of i.

a $\sqrt{-4}$ **b** $\sqrt{-7}$ **c** $\sqrt{-\dfrac{1}{9}}$

d $\sqrt{-12}$ **e** $\sqrt{-\dfrac{6}{25}}$ **f** $\sqrt{(-2)^2 - 4(3)(3)}$

g i^7 **h** i^{13} **i** i^{99}

j $i + i^2 + i^3 + \ldots + i^{149} + i^{150}$ **k** $\dfrac{i^4}{i}$ **l** $\dfrac{1}{i^3}$

2 Find the roots of each quadratic equation.

a $x^2 = -4$ **b** $x^2 + 9 = 0$

c $z^2 = -\dfrac{1}{36}$ **d** $5z^2 + 100 = 0$

3 Use the quadratic formula to solve each quadratic equation.

a $x^2 + 2x + 3 = 0$ **b** $x^2 - x + 6 = 0$

c $z^2 + 3z + 3 = 0$ **d** $3z^2 - 5z + 9 = 0$

4 Complete the square to solve each quadratic equation.

a $x^2 - 2x + 3 = 0$ **b** $x^2 - 4x + 11 = 0$

c $z^2 + 8z + 20 = 0$ **d** $z^2 - 2z + 4 = 0$

5 For each quadratic equation, factorise the expression as a difference of two squares, then solve the equation.

a $x^2 - 2x + 2 = 0$ **b** $v^2 - 6v + 12 = 0$

c $w^2 + 4w + 10 = 0$ **d** $z^2 + 2z + 7 = 0$

e $z^2 + z + 1 = 0$ **f** $z^2 - 3z + 4 = 0$

6 State the real and imaginary parts for each complex number, where $a, b, x, y \in \mathbb{R}$.

a $\sqrt{3} + i$ **b** $\dfrac{5 - i\sqrt{2}}{2}$ **c** $6i - 3$

d $x - iy + 3 + 2i$ **e** $\dfrac{a + 2ib}{a^2 + 4b^2}$ **f** $\dfrac{x - i - 4 + ix - 6y + yi}{x^2 + y^2}$

7 Determine the conjugate for each complex number in Question **6**.

8 Evaluate each product.

a $(2 + 3i)(2 - 3i)$

b $(1 - i\sqrt{2})(1 + i\sqrt{2})$

c $(5i + 4)(5i - 4)$

d $\left(\dfrac{1}{2} - \dfrac{i\sqrt{3}}{2}\right)\left(\dfrac{1}{2} + \dfrac{i\sqrt{3}}{2}\right)$

e $\left(\dfrac{4+i}{3}\right)\left(\dfrac{4-i}{3}\right)$

f $\left(\dfrac{\sqrt{2} - 2i\sqrt{2}}{8}\right)\left(\dfrac{\sqrt{2} + 2i\sqrt{2}}{8}\right)$

9 For each complex number z, find $z\bar{z}$. Assume $a, b, x, y \in \mathbb{R}$.

a $z = 5 + 6i$

b $z = \sqrt{3} - i$

c $z = 4i - 3$

d $z = \dfrac{1 + i\sqrt{3}}{2}$

e $z = \dfrac{1}{17} - \dfrac{4i}{17}$

f $z = \sqrt{5} + i\sqrt{3}$

g $z = 2a - 3ib$

h $z = x + y + i(x - y)$

10 If $z = w - iv$ where $w, v \in \mathbb{R}$, prove that $z\bar{z}$ is real.

11 If $z = a + ib$ and $w = c + id$ where $a, b, c, d \in \mathbb{R}$, prove that $\overline{z + w} = \bar{z} + \bar{w}$.

12 By equating real and imaginary parts, solve each equation for x and y.

a $2x + 8i - 4 + iy = 0$

b $3x + 2iy = 9 - 8i$

c $x + y + 2xi - yi = 7 + 8i$

d $3x - 2y - 8 + xi + 3yi - 10i = 0$

13 Given $z = 3y - 6i + xi - 8 + yi$ and $\text{Im}(z) = 0$, find a relationship between x and y.

14 Simplify each expression.

a $4 - 3i + 7i - 8$

b $2(3 + i) - i(7 - 2i)$

c $(2 - 9i)^2$

d $(4 + i)(5 - 3i)$

e $(\sqrt{5} - 4i)(\sqrt{5} + 4i)$

f $3(8i - 1)(2 + i)$

g $(\sqrt{2} + i)(\sqrt{2} - i\sqrt{3}) - i(\sqrt{6} + 2i)$

h $(x - iy)^2 - (x + iy)^2$

15 Show that:

a $(1 + i\sqrt{3})(\sqrt{3} + i)$ is purely imaginary

b $(\sqrt{2} + i\sqrt{2})(-\sqrt{2} + i\sqrt{2})$ is real

16 a Find a quadratic equation $az^2 + bz + c = 0$ that has the complex conjugate roots:

i $2 \pm i$

ii $\sqrt{3} \pm 5i$

iii $\dfrac{1}{2} \pm \dfrac{\sqrt{3}}{2}i$

iv $-4 \pm i\sqrt{5}$

In each case, verify that the coefficients a, b and c are real.

b Copy and complete the statement: A quadratic equation with complex conjugate roots will have _____ coefficients.

17 Realise the denominator of each complex number z, then state $\text{Im}(z)$.

a $z = \dfrac{1}{2 - i}$

b $z = \dfrac{1 + i}{1 - 2i}$

c $z = \dfrac{5 - 7i}{3 + 4i}$

d $z = \dfrac{\sqrt{3} - i\sqrt{2}}{\sqrt{3} + i\sqrt{2}}$

18 Show that:

a $\dfrac{-2+2i}{1+i}$ is purely imaginary

b $\dfrac{1+i\sqrt{3}}{(1-i\sqrt{3})^2}$ is real

19 Simplify each expression.

a $\dfrac{1}{(1+2i)^2}$

b $\dfrac{1}{(\sqrt{3}+i\sqrt{3})^2} - \dfrac{1}{(\sqrt{3}-i\sqrt{3})^2}$

20 a Find a quadratic equation in the form $az^2 + bz + c = 0$ with non-conjugate roots $3-i$ and $2 + 9i$. Identify the values of a, b and c. Are they real?

b Copy and complete the statement: A quadratic equation with complex non-conjugate roots will have some coefficients that are not _____.

21 Find the values of x and y, where x and y are real, if:

a $x + iy = \dfrac{2-3i}{3+4i}$

b $(x + iy)(1 - 5i) = 2 + i$

22 If $z = 5 - 2i$ and $w = -3 + i$, evaluate each expression.

a zw **b** $z - w$ **c** $\dfrac{z}{w}$ **d** $z^2 - w^2$

1.02 Square root of a complex number

Square root of a complex number

To find $\sqrt{a+ib}$, let $a + ib = (x + iy)^2$, $x, y \in \mathbb{R}$, then equate real and imaginary parts.

EXAMPLE 9

Find $\sqrt{5+12i}$.

Solution

Let $\sqrt{5+12i} = x + iy$ where x and y are real.

Then, squaring both sides,

$$5 + 12i = (x + iy)^2$$
$$= x^2 + 2xyi + i^2y^2$$
$$= x^2 - y^2 + 2xyi$$

Equating real and imaginary parts,

$$5 = x^2 - y^2$$

$$12 = 2xy$$

$$xy = 6$$

Method 1: Solving by inspection

By inspection, we can see that $x = 3, y = 2$ OR $x = -3, y = -2$.

Therefore $\sqrt{5 + 12i} = 3 + 2i$ or $-3 - 2i$.

It is conventional to take the solution with the positive real part ($x > 0$) for the square root of a complex number unless both roots are required.

So $\sqrt{5 + 12i} = 3 + 2i$.

Method 2: Solving algebraically

Solving simultaneously:

$$x^2 - y^2 = 5 \qquad [1]$$

$$xy = 6 \qquad [2]$$

From [2]:

$$y = \frac{6}{x} \qquad [3]$$

Substitute into [1]:

$$x^2 - \left(\frac{6}{x}\right)^2 = 5$$

$$x^2 - \frac{36}{x^2} = 5$$

$$x^4 - 36 = 5x^2$$

$$x^4 - 5x^2 - 36 = 0$$

$$(x^2 - 9)(x^2 + 4) = 0$$

$$x^2 - 9 = 0 \qquad x \in \mathbb{R}$$

$$x = \pm 3$$

Substitute into [3] to find y:

When $x = 3, y = \dfrac{6}{3} = 2$

It is conventional to take the solution with the positive real part ($x > 0$) for the square root of a complex number unless both roots are required.

So $\sqrt{5 + 12i} = 3 + 2i$

ISBN 9780170413435

Exercise 1.02 Square root of a complex number

1 Evaluate each square root.

    a $\sqrt{3+4i}$ **b** $\sqrt{5-12i}$ **c** $\sqrt{8+6i}$ **d** $\sqrt{4i}$

2 Find 2 square roots of each complex number.

    a $15-8i$ **b** $-3-4i$ **c** $21-20i$

    d $-24+10i$ **e** $-9i$ **f** i

3 Find the possible values of z such that:

    a $z^2 = 9 + 40i$ **b** $z^2 = -7 + 24i$ **c** $z^2 = 12 - 16i$

4 a Find $\sqrt{-3+4i}$.

    b Hence, use the quadratic formula to solve the complex quadratic equation $z^2 - 3z + (3 - i) = 0$.

5 Solve each quadratic equation over the set of complex numbers.

    a $x^2 - (2 + 3i)x + (-5 + i) = 0$ **b** $v^2 - 2iv - 3v + 6i = 0$

    c $iz^2 - z + 2i = 0$

1.03 The Argand diagram

It is possible to represent the complex number $z = a + ib$ geometrically on a number plane, with a horizontal axis denoted by x or Re(z) and a vertical axis denoted by y or Im(z). This plane is called an **Argand diagram** or **complex plane**, after the French mathematician Jean-Robert Argand (1768–1822).

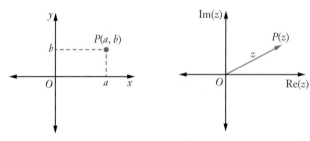

The complex number $z = a + ib$ can be represented by the point $P(a, b)$ or the vector $\overrightarrow{OP}$.

A **vector** is a quantity with both magnitude and direction, and is represented graphically by an arrow with a specific length and direction.

Real numbers are plotted along the x-axis so the axis is labelled Re(z), Re or sometimes x.

Purely imaginary numbers are plotted along the y-axis so the axis is labelled Im(z), Im or sometimes iy.

By convention, the complex number $z = 0 + 0i$ is just written as O on the Argand diagram. It is the origin.

All complex numbers $z = a + ib$ correspond to a unique point (a, b) or vector $\underset{\sim}{z}$ with head (a, b) and tail O on the complex plane.

Plotting complex numbers as points

Represent each complex number as a point on the complex plane.

a $u = 2 + 3i$

b $w = -1 - 4i$

c $v = (1 + 2i)(1 - 2i)$

d $z = \dfrac{3}{i}$

Solution

a, b u and w can be plotted as points $U(2, 3)$ and $W(-1, -4)$ respectively.

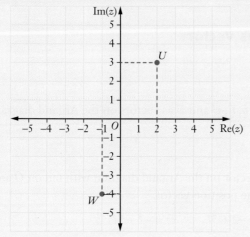

c, d v and z need to be expressed in the form $a + ib$:

$$v = (1 + 2i)(1 - 2i)$$
$$= 1 + 4$$
$$= 5$$
$$= 5 + 0i$$

$$z = \frac{3}{i}$$
$$= \frac{3}{i} \times \frac{i}{i}$$
$$= \frac{3i}{-1}$$
$$= -3i$$
$$= 0 - 3i$$

We can now plot v and z as $V(5, 0)$ and $Z(0, -3)$ respectively.

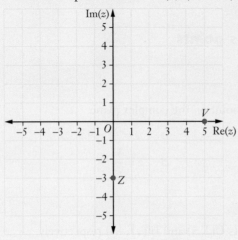

Plotting complex numbers as vectors

EXAMPLE 11

Given $u = 3 - 2i$ and $w = -2 + 5i$, plot $\underset{\sim}{u}, \overline{\underset{\sim}{u}}, \underset{\sim}{w}$ and $\overline{\underset{\sim}{w}}$ as vectors on the Argand diagram.

Solution

We can plot a complex number as a vector on the Argand diagram with tail on O and head on the corresponding point. This is the position vector.

$$u = 3 - 2i \qquad \overline{u} = 3 + 2i \qquad w = -2 + 5i \qquad \overline{w} = -2 - 5i$$

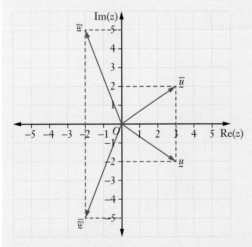

Note that the vectors of complex conjugates $\underset{\sim}{u}$ and $\overline{\underset{\sim}{u}}$, and $\underset{\sim}{w}$ and $\overline{\underset{\sim}{w}}$, are reflections of each other in the real axis.

Complex conjugates on the Argand diagram

On the Argand diagram, the vectors of a complex number z and its conjugate $\bar{z}$ are reflections in the real axis.

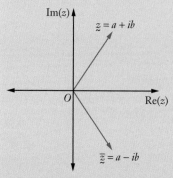

Adding and subtracting complex numbers on the complex plane

EXAMPLE 12

If $z = 1 + 3i$ and $w = 4 + i$, find $z + w$ and plot z, w, $-z$ and $z + w$ on the Argand diagram.

Solution

$$-z = -(1 + 3i) \qquad z + w = (1 + 3i) + (4 + i)$$
$$= -1 - 3i \qquad\qquad = 5 + 4i$$

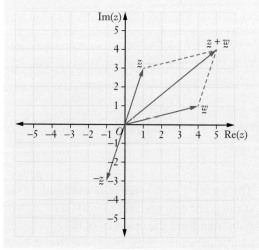

1. Complex numbers

Note that the vector $-z$ is the same length as z but in the opposite direction. Note also that the points corresponding with O, z, w and $z + w$ form a parallelogram. The vector addition $z + w$ is the vector from O along the diagonal of the parallelogram to $z + w$.

A similar result holds for subtraction.

The vector $-z$

The vectors z and $-z$ are the same length but in opposite directions.

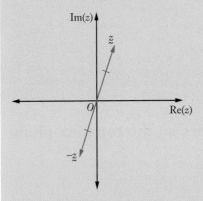

The parallelogram rule for adding and subtracting vectors

Given vectors z and w, the vector sum $z + w$ and difference $z - w$ correspond with the diagonals of a parallelogram.

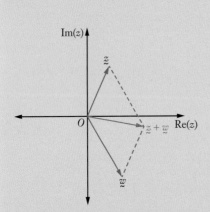

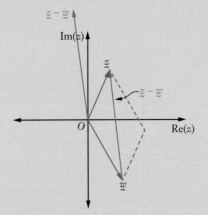

The vector $z + w$ has tail at O and head at the point representing $z + w$. It is the diagonal of the parallelogram where z and w are adjacent sides, from the tails of z and w.

The vector $z - w$ has tail at O and head at the point representing $z - w$. It is translated from the diagonal of the parallelogram from the head of w to the head of z.

Multiplying complex numbers by a constant

We can also multiply a complex number by a constant real number (scalar), which will alter the length of its vector. For instance, the vector $2\underset{\sim}{v}$ is twice the length of $\underset{\sim}{v}$.

EXAMPLE 13

The complex numbers u, z and v are plotted as vectors on the Argand diagram.

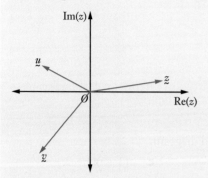

Plot the points A, B, C and D corresponding to the vectors $-\underset{\sim}{u}, \dfrac{1}{2}\underset{\sim}{v}, \underset{\sim}{u}+\underset{\sim}{z}$ and $\underset{\sim}{u}-\underset{\sim}{z}$ respectively on the same diagram.

Solution

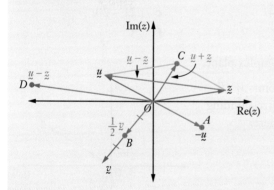

Exercise 1.03 The Argand diagram

1 Plot each complex number as a point on the complex plane.

   a $2 - 4i$ **b** $1 + 5i$ **c** -4 **d** $3i$

2 Express each point shown as a complex number.

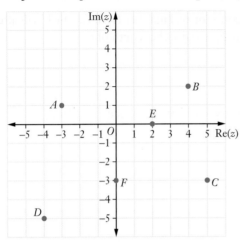

3 Plot each complex number as a vector on an Argand diagram.

a $(4 - 3i) + (-2 + 2i)$ **b** $(5 - i) - (3 - 5i)$ **c** $(3 + i)(1 + 2i)$

d $(1 + i\sqrt{3})(1 - i\sqrt{3})$ **e** $4(2 - 5i) - 2i(-3 - i)$ **f** $\dfrac{1}{2 - i}$

g $\dfrac{4 + 8i}{1 + i}$

4 Plot each complex number as a vector in the complex plane.

a $15 - 8i$ **b** $-3 - 4i$ **c** $21 - 20i$

d $-24 + 10i$ **e** $-9i$ **f** 3

5 The vectors $\underset{\sim}{v}$ and $\underset{\sim}{u}$ are shown on the complex plane.

Use the parallelogram rule to plot the points W, P and Z representing complex numbers corresponding to the vectors $\underset{\sim}{w}, \underset{\sim}{p}$ and $\underset{\sim}{z}$ respectively, where:

a $\underset{\sim}{w} = \underset{\sim}{v} + \underset{\sim}{u}$

b $\underset{\sim}{p} = -\underset{\sim}{v}$

c $\underset{\sim}{z} = \underset{\sim}{u} - \underset{\sim}{v}$

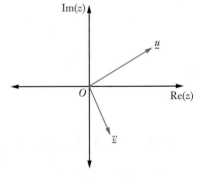

6 On an Argand diagram, plot the vectors corresponding to each complex number z with its conjugate $\bar{z}$.

a $z = 1 - i\sqrt{3}$ **b** $z = -\dfrac{1}{\sqrt{2}} + \dfrac{i}{\sqrt{2}}$ **c** $z = -4 - 3i$

7 The vectors corresponding to 4 complex numbers are shown on the Argand diagram.

   a State the complex number representing each vector z, v, k and w.

   b State the complex conjugate of each vector in part **a**.

   c Plot $\bar{z}, \bar{v}, \bar{k}$ and $\bar{w}$ on an Argand diagram.

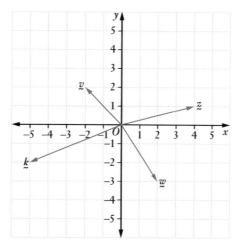

8 Represent each number as a vector on an Argand diagram.

   a $z = (2 + i)(1 - i)$ **b** $w = \dfrac{1}{-2 + i}$ **c** $v = (\sqrt{3} - i)^2$

9 For $z = 2 + 3i$:

   a Evaluate the complex numbers $w = iz$, $v = i^2 z$ and $u = i^3 z$.

   b Plot the corresponding vectors z, w, v and u on an Argand diagram.

   c What effect does multiplying a complex number by i have on the position of its vector?

1.04 Modulus and argument

The **modulus** of a complex number is the length or magnitude of the corresponding vector on the complex plane.

The modulus, r, of $z = a + ib$ is written as **mod** z or $|z|$.

$$r = |z| = \sqrt{a^2 + b^2}$$

The **argument** of a complex number is the angle the vector makes with the positive x-axis in an anti-clockwise direction.

The argument, θ, of $z = a + ib$, where $z \neq 0 + 0i$, is written as **arg** z.

$\tan \theta = \dfrac{b}{a}$ where $\theta = \arg z$

Note that if $z = 0 + 0i$, then arg z is undefined.

As with solutions to trigonometric equations, the angle θ can take multiple values in an unrestricted domain. For this reason, by convention, mathematicians define the **principal argument** of z, **Arg** z, as the unique angle in the interval $(-\pi, \pi]$.

Modulus and argument

For $z = a + ib$,

- $r = |z| = \sqrt{a^2 + b^2}$

- $\tan\theta = \dfrac{b}{a}$ where $\theta = \arg z$

Arg z is the **principal argument** of z in the interval $(-\pi, \pi]$.

EXAMPLE 14

If $z = 1 - i\sqrt{3}$, find r and Arg z.

Solution

For $z = 1 - i\sqrt{3}$, $a = 1$, $b = -\sqrt{3}$

$$r = \sqrt{a^2 + b^2}$$
$$= \sqrt{1^2 + \left(-\sqrt{3}\right)^2}$$
$$= \sqrt{4}$$
$$= 2$$

$$\tan\theta = \frac{b}{a} = \frac{-\sqrt{3}}{1} = -\sqrt{3}$$

If we graphed $z = 1 - i\sqrt{3}$ or $(1, -\sqrt{3})$ on the complex plane, it would be in the 4th quadrant, so $\theta = -\dfrac{\pi}{3}$ (in the interval $(-\pi, \pi]$).

Therefore Arg $z = -\dfrac{\pi}{3}$.

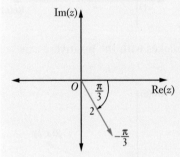

Polar or modulus–argument form of a complex number

We saw that a complex number $z = a + ib$ can be expressed in terms of its modulus r and its argument θ.

We know from trigonometry and the parametric equations of a circle that

$$\cos\theta = \frac{a}{r} \text{ and } \sin\theta = \frac{b}{r}$$

or $a = r\cos\theta$ and $b = r\sin\theta$. So we can write:

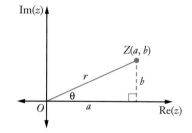

$$z = a + ib$$
$$= r\cos\theta + ir\sin\theta$$
$$= r(\cos\theta + i\sin\theta)$$

So z is now expressed in terms of r and θ.

This is called the **polar form** or **modulus–argument form** of a complex number because it is based on the distance and angle from the 'pole' or 'centre' that is the origin.

Polar form of a complex number

$z = r(\cos\theta + i\sin\theta)$

where r is the modulus and θ is the principal argument.

When z is expressed in terms of x and y, that is $x + iy$, this is called the **Cartesian form** or **rectangular form** because it is based on the number plane.

Note: Sometimes $z = r(\cos\theta + i\sin\theta)$ is shortened to $z = r\operatorname{cis}\theta$, where $\operatorname{cis}\theta$ stands for $\cos\theta + i\sin\theta$.

EXAMPLE 15

Express each complex number in polar form.

a $z = -\sqrt{2} + i\sqrt{2}$

b $z = 2\left(\cos\dfrac{\pi}{6} - i\sin\dfrac{\pi}{6}\right)$

Solution

a $a = -\sqrt{2},\ b = \sqrt{2}$

$$r = \sqrt{\left(-\sqrt{2}\right)^2 + \left(\sqrt{2}\right)^2}$$
$$= \sqrt{4}$$
$$= 2$$

$\tan\theta = \dfrac{\sqrt{2}}{-\sqrt{2}} = -1$ and $z = -\sqrt{2} + i\sqrt{2}$ lies in the 2nd quadrant, so $\theta = \dfrac{3\pi}{4}$

(in the interval $(-\pi, \pi]$).

ISBN 9780170413435

In polar form, $z = 2\left(\cos\dfrac{3\pi}{4} + i\sin\dfrac{3\pi}{4}\right)$.

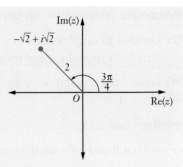

b $z = 2\left(\cos\dfrac{\pi}{6} - i\sin\dfrac{\pi}{6}\right)$ is not in polar form since it is not in the form

$z = r(\cos\theta + i\sin\theta)$ where the cos and sin terms are separated by
a + sign. However, we can use the relations $\cos(-\theta) = \cos\theta$
and $\sin(-\theta) = -\sin\theta$ to express z in the correct form.

$$z = 2\left(\cos\dfrac{\pi}{6} - i\sin\dfrac{\pi}{6}\right)$$

$$= 2\left[\cos\left(-\dfrac{\pi}{6}\right) + i\sin\left(-\dfrac{\pi}{6}\right)\right]$$

which is now in polar form.

The conjugate in polar form

The conjugate of $z = r(\cos\theta + i\sin\theta)$ is $\bar{z} = r(\cos\theta - i\sin\theta)$ which can be written in
polar form as

$$\bar{z} = r[\cos(-\theta) + i\sin(-\theta)].$$

EXAMPLE 16

Express $z = 3\left(\cos\dfrac{5\pi}{3} + i\sin\dfrac{5\pi}{3}\right)$ in Cartesian form.

Solution

To convert from polar form to Cartesian form, simply evaluate and expand.

$$z = 3\left(\cos\dfrac{5\pi}{3} + i\sin\dfrac{5\pi}{3}\right)$$

$$= 3\left[\dfrac{1}{2} + i\left(-\dfrac{\sqrt{3}}{2}\right)\right]$$

$$= \dfrac{3}{2} - \dfrac{3\sqrt{3}\,i}{2}$$

   ISBN 9780170413435

Exercise 1.04 Modulus and argument

1 Find the modulus of each complex number.

   a $1+i$ **b** $2+4i$ **c** $7-2i$ **d** $\sqrt{3}+i\sqrt{2}$

   e $-\dfrac{1}{7}+\dfrac{6}{7}i$ **f** $-5-i\sqrt{2}$ **g** i

2 Find the argument of each complex number.

   a $1+i$ **b** $\sqrt{3}+i$ **c** $\sqrt{2}-i\sqrt{2}$ **d** $-\dfrac{\sqrt{3}}{2}+\dfrac{i}{2}$

   e $-1-i$ **f** 4 **g** i

3 Express each complex number in polar form.

   a $1-i$ **b** $-1+i\sqrt{3}$ **c** $\dfrac{-2-2i}{3}$ **d** $\dfrac{1}{2}+\dfrac{i\sqrt{3}}{2}$

   e $\dfrac{\sqrt{2}+i\sqrt{2}}{7}$ **f** $-2\sqrt{3}-2i$ **g** $-\sqrt{6}$

4 Convert each complex number to Cartesian form.

   a $z=2\left(\cos\dfrac{\pi}{3}+i\sin\dfrac{\pi}{3}\right)$ **b** $z=\dfrac{1}{2}\left(\cos\dfrac{\pi}{6}+i\sin\dfrac{\pi}{6}\right)$

   c $z=3\left(\cos\dfrac{\pi}{4}+i\sin\dfrac{\pi}{4}\right)$ **d** $z=\cos\left(-\dfrac{\pi}{3}\right)+i\sin\left(-\dfrac{\pi}{3}\right)$

   e $z=\sqrt{2}\left[\cos\left(-\dfrac{2\pi}{3}\right)+i\sin\left(-\dfrac{2\pi}{3}\right)\right]$ **f** $z=2\left[\cos\left(-\dfrac{5\pi}{6}\right)+i\sin\left(-\dfrac{5\pi}{6}\right)\right]$

5 Convert each complex number to rectangular form.

   a $z=\dfrac{1}{\sqrt{3}}\left(\cos\dfrac{\pi}{4}-i\sin\dfrac{\pi}{4}\right)$ **b** $z=\sqrt{3}\left(\cos\dfrac{5\pi}{6}-i\sin\dfrac{5\pi}{6}\right)$

   c $z=-2\left(\cos\dfrac{\pi}{2}+i\sin\dfrac{\pi}{2}\right)$ **d** $z=-\left[\cos\left(-\dfrac{\pi}{4}\right)+i\sin\left(-\dfrac{\pi}{4}\right)\right]$

   e $z=3\sqrt{2}\left[\cos\left(-\pi\right)+i\sin\left(-\pi\right)\right]$ **f** $z-i\left[\cos\left(\dfrac{\pi}{3}\right)-i\sin\left(-\dfrac{\pi}{3}\right)\right]$

   g $z=\sqrt{3}\left[\cos\left(\dfrac{\pi}{6}\right)+i\sin\left(\dfrac{\pi}{3}\right)\right]$ **h** $z=2\left[\cos\left(-\dfrac{5\pi}{4}\right)-i\sin\left(\dfrac{\pi}{4}\right)\right]$

6 Write each complex number graphed in modulus–argument form.

a
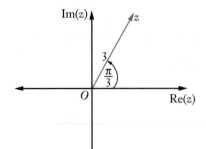

b
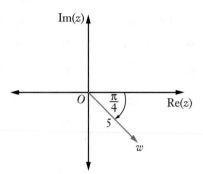

c
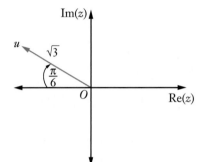

d
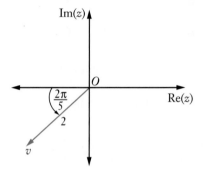

e
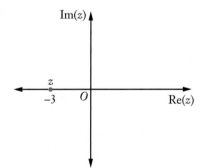

f
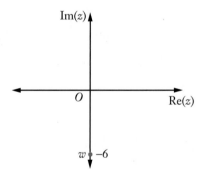

7 Plot each number on an Argand diagram and express it in modulus–argument form.

a $-2i$

b $-\sqrt{3}$

c $\dfrac{i}{3}$

d $\dfrac{1}{2}\left(\cos\dfrac{\pi}{6} - i\sin\dfrac{\pi}{6}\right)$

e $-\left(\cos\dfrac{3\pi}{4} + i\sin\dfrac{3\pi}{4}\right)$

f $-2 - 2\sqrt{3}\,i$

g 2

8 Express each number in polar form. Where necessary, give an approximation to the argument in radians.

a $2 + i$

b $-5 + 7i$

c $\dfrac{4 - i}{5}$

d $4\left(\cos\dfrac{7\pi}{6} + i\sin\dfrac{7\pi}{6}\right)$

e $2\left(\cos\dfrac{9\pi}{4} + i\sin\dfrac{9\pi}{4}\right)$

f $-\sqrt{5} - i\sqrt{10}$

g $-\left(\cos\dfrac{4\pi}{3} - i\sin\dfrac{4\pi}{3}\right)$

h $\sqrt{2}\left(\sin\dfrac{\pi}{3} + i\cos\dfrac{\pi}{3}\right)$

i $-\dfrac{i}{4} + \dfrac{\sqrt{3}}{4}$

j $-\left(\sin\dfrac{\pi}{4} - i\cos\dfrac{3\pi}{4}\right)$

1.05 Properties of moduli and arguments

There are many advantages in using polar form on operations with complex numbers, especially when multiplying or dividing them. They rely on the properties outlined below.

Polar complex number operations

Properties of moduli and arguments

Let $z_1 = r_1(\cos\theta_1 + i\sin\theta_1)$ and $z_2 = r_2(\cos\theta_2 + i\sin\theta_2)$.

Property 1: Product of 2 complex numbers

$z_1 z_2 = r_1 r_2\left[\cos(\theta_1 + \theta_2) + i\sin(\theta_1 + \theta_2)\right]$

Property 2: Quotient of 2 complex numbers

$\dfrac{z_1}{z_2} = \dfrac{r_1}{r_2}\left[\cos(\theta_1 - \theta_2) + i\sin(\theta_1 - \theta_2)\right]$, $z_2 \neq 0$

Property 3: Power of a complex number

Let $z = r(\cos\theta + i\sin\theta)$.

$z^n = r^n(\cos n\theta + i\sin n\theta)$ where n is an integer.

This is an extension of de Moivre's theorem, which you will meet in Chapter 4.

Property 4: Reciprocal of a complex number

$z^{-1} = r^{-1}(\cos\theta - i\sin\theta)$, $z \neq 0$

Property 5: Negative power of a complex number

$z^{-n} = r^{-n}[\cos(-n\theta) + i\sin(-n\theta)]$, $z \neq 0$ where n is an integer.

Let $z_1 = r_1(\cos\theta_1 + i\sin\theta_1)$, $z_2 = r_2(\cos\theta_2 + i\sin\theta_2)$, ..., $z_n = r_n(\cos\theta_n + i\sin\theta_n)$.

Property 6: Product of complex numbers

$z_1 z_2 z_3 \dots z_n = r_1 r_2 r_3 \dots r_n[\cos(\theta_1 + \theta_2 + \theta_3 + \dots + \theta_n) + i\sin(\theta_1 + \theta_2 + \theta_3 + \dots + \theta_n)]$ where n is an integer.

We will prove properties 1 and 4 below. The others can be done as an exercise.

Proof of Property 1: Product of 2 complex numbers

Using the identities $\cos (A + B) = \cos A \cos B - \sin A \sin B$
and $\sin (A + B) = \sin A \cos B + \cos A \sin B$ we have:

$$z_1 z_2 = r_1(\cos \theta_1 + i \sin \theta_1) \times r_2(\cos \theta_2 + i \sin \theta_2)$$
$$= r_1 r_2(\cos \theta_1 \cos \theta_2 + \cos \theta_1 \, i \sin \theta_2 + i \sin \theta_1 \cos \theta_2 + i \sin \theta_1 \, i \sin \theta_2)$$
$$= r_1 r_2[\cos \theta_1 \cos \theta_2 - \sin \theta_1 \sin \theta_2 + i(\cos \theta_1 \sin \theta_2 + \sin \theta_1 \cos \theta_2)]$$
$$= r_1 r_2[\cos (\theta_1 + \theta_2) + i \sin (\theta_1 + \theta_2)]$$

Proof of Property 4: Reciprocal of a complex number

Realising the denominator we have:

$$z^{-1} = \frac{1}{z}$$
$$= \frac{1}{r(\cos \theta + i \sin \theta)} \times \frac{\cos \theta - i \sin \theta}{\cos \theta - i \sin \theta}$$
$$= \frac{\cos \theta - i \sin \theta}{r(\cos^2 \theta + \sin^2 \theta)}$$
$$= \frac{\cos \theta - i \sin \theta}{r(1)}$$
$$= r^{-1} (\cos \theta - i \sin \theta)$$

These properties can also be written separately in terms of their moduli and arguments.

Properties of moduli and arguments

Property 1: Product of 2 complex numbers

$|z_1||z_2| = |z_1 z_2|$ and $\arg z_1 z_2 = \arg z_1 + \arg z_2$

Property 2: Quotient of 2 complex numbers

$\dfrac{|z_1|}{|z_2|} = \left|\dfrac{z_1}{z_2}\right|$ and $\arg \dfrac{z_1}{z_2} = \arg z_1 - \arg z_2$

Property 3: Power of a complex number

$|z|^n = |z^n|$ and $\arg z^n = n \times \arg z$ for $n \in \mathbb{Z}$

Property 4: Reciprocal of a complex number

$|z^{-1}| = \dfrac{1}{|z|}$ and $\arg z^{-1} = -\arg z,\ z \neq 0$

MATHS IN FOCUS 12. Mathematics Extension 2

ISBN 9780170413435

Property 5: Negative power of a complex number

$\left| z^{-n} \right| = \dfrac{1}{\left| z \right|^n}$ and $\arg z^{-n} = -n \arg z$, $z \neq 0$ for $n \in \mathbb{Z}$

Property 6: Product of complex numbers

$\left| z_1 \right|\left| z_2 \right|\left| z_3 \right| \dots \left| z_n \right| = \left| z_1 z_2 z_3 \dots z_n \right|$ and $\arg z_1 z_2 z_3 \dots z_n = \arg z_1 + \arg z_2 + \dots + \arg z_n$
where $n \in \mathbb{Z}$

EXAMPLE 17

For $z_1 = 2\left(\cos \dfrac{\pi}{5} + i \sin \dfrac{\pi}{5} \right)$ and $z_2 = 5\left(\cos \dfrac{\pi}{7} + i \sin \dfrac{\pi}{7} \right)$, evaluate:

a $z_1 z_2$ **b** $\dfrac{z_1}{z_2}$ **c** $\dfrac{1}{z_1}$ **d** $(z_2)^7$

Solution

a Using $z_1 z_2 = r_1 r_2 [\cos (\theta_1 + \theta_2) + i \sin (\theta_1 + \theta_2)]$

$z_1 z_2 = 2\left(\cos \dfrac{\pi}{5} + i \sin \dfrac{\pi}{5} \right) \times 5\left(\cos \dfrac{\pi}{7} + i \sin \dfrac{\pi}{7} \right)$

$= (2 \times 5)\left[\cos \left(\dfrac{\pi}{5} + \dfrac{\pi}{7} \right) + i \sin \left(\dfrac{\pi}{5} + \dfrac{\pi}{7} \right) \right]$

$= 10 \left[\cos \dfrac{12\pi}{35} + i \sin \dfrac{12\pi}{35} \right]$

b Using $\dfrac{z_1}{z_2} = \dfrac{r_1}{r_2}\left[\cos (\theta_1 - \theta_2) + i \sin (\theta_1 - \theta_2) \right]$

$\dfrac{z_1}{z_2} = \dfrac{2\left(\cos \dfrac{\pi}{5} + i \sin \dfrac{\pi}{5} \right)}{5\left(\cos \dfrac{\pi}{7} + i \sin \dfrac{\pi}{7} \right)}$

$= \dfrac{2}{5}\left[\cos \left(\dfrac{\pi}{5} - \dfrac{\pi}{7} \right) + i \sin \left(\dfrac{\pi}{5} - \dfrac{\pi}{7} \right) \right]$

$= \dfrac{2}{5}\left[\cos \dfrac{2\pi}{35} + i \sin \dfrac{2\pi}{35} \right]$

c   Using $z^{-1} = r^{-1}(\cos \theta - i \sin \theta)$

$$\frac{1}{z_1} = \frac{1}{2\left(\cos \dfrac{\pi}{5} + i \sin \dfrac{\pi}{5}\right)}$$

$$= \frac{1}{2}\left(\cos \frac{\pi}{5} - i \sin \frac{\pi}{5}\right)$$

d   Using $z^n = r^n(\cos n\theta + i \sin n\theta)$

$$(z_2)^7 = 5^7\left(\cos \frac{7\pi}{7} + i \sin \frac{7\pi}{7}\right)$$

$$= 5^7(\cos \pi + i \sin \pi)$$

$$= 5^7[-1 + i(0)]$$

$$= -5^7$$

EXAMPLE 18

Find arg z if $z = \dfrac{(\cos 2\beta + i \sin 2\beta)(\cos \alpha - i \sin \alpha)}{\cos 3\theta + i \sin 3\theta}$.

Solution

We can use the properties above if the complex numbers are in polar form.

Note $\cos \alpha - i \sin \alpha = \cos(-\alpha) + i \sin(-\alpha)$.

$$z = \frac{(\cos 2\beta + i \sin 2\beta)(\cos \alpha - i \sin \alpha)}{\cos 3\theta + i \sin 3\theta}$$

$$= \frac{(\cos 2\beta + i \sin 2\beta)[\cos(-\alpha) + i \sin(-\alpha)]}{\cos 3\theta + i \sin 3\theta}$$

$$= \cos[2\beta + (-\alpha) - 3\theta] + i \sin[2\beta + (-\alpha) - 3\theta]$$

$$= \cos[2\beta - \alpha - 3\theta] + i \sin[2\beta - \alpha - 3\theta]$$

So arg $z = 2\beta - \alpha - 3\theta$.

Properties of conjugates

Property 7: Reciprocal of a complex number with modulus 1

If $z = \cos \theta + i \sin \theta$, then $z^{-1} = \overline{z}$.

Property 8: Product of complex conjugate pairs

For any complex number z,

$z\overline{z} = |z|^2$ and arg $z\overline{z} = 0$.

Property 9: Modulus and argument of a complex conjugate

$|\overline{z}| = |z|$ and arg $\overline{z} = -\arg z$

Property 10: Sum of conjugates of 2 complex numbers

$\overline{z_1} + \overline{z_2} = \overline{z_1 + z_2}$

MATHS IN FOCUS 12. Mathematics Extension 2   ISBN 9780170413435

Property 11: Product of conjugates of 2 complex numbers

$\overline{z_1} \, \overline{z_2} = \overline{z_1 z_2}$

Property 12: Sum of conjugate pairs

$z + \overline{z} = 2\,\mathrm{Re}(z)$

Property 13: Difference of conjugate pairs

$z - \overline{z} = 2i\,\mathrm{Im}(z)$

The triangle inequality

For 2 complex numbers z and w,

$\left| z + w \right| \le \left| z \right| + \left| w \right|$

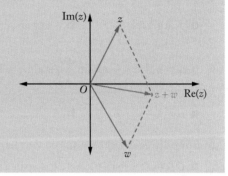

Proof of the triangle inequality

Since $\left| z \right|, \left| w \right|$ and $\left| z + w \right|$ are all non-negative quantities, we can use the argument that if 2 numbers $a, b \ge 0$ and $a^2 \ge b^2$ then $a \ge b$. Recall also the property of complex conjugates that $\overline{u + v} = \overline{u} + \overline{v}$ from page 13, Question 11.

Consider:

$$\left| z + w \right|^2 = (z + w)(\overline{z} + \overline{w}) \qquad \text{from Property 8}$$

$$= z\overline{z} + w\overline{w} + z\overline{w} + w\overline{z}$$

$$= \left| z \right|^2 + \left| w \right|^2 + z\overline{w} + w\overline{z}$$

$$\le \left| z \right|^2 + \left| w \right|^2 + \left| z\overline{w} \right| + \left| w\overline{z} \right|$$

$$= \left| z \right|^2 + \left| w \right|^2 + \left| z \right|\left| \overline{w} \right| + \left| w \right|\left| \overline{z} \right| \qquad \text{using Property 1}$$

$$= \left| z \right|^2 + \left| w \right|^2 + \left| z \right|\left| w \right| + \left| w \right|\left| z \right| \qquad \text{since } \left| z \right| = \left| \overline{z} \right|$$

$$= \left| z \right|^2 + \left| w \right|^2 + 2\left| z \right|\left| w \right|$$

$$= \left(\left| z \right| + \left| w \right| \right)^2$$

Therefore $\left| z + w \right|^2 \le \left(\left| z \right| + \left| w \right| \right)^2$ and we can deduce that $\left| z + w \right| \le \left| z \right| + \left| w \right|$.

Exercise 1.05 Properties of moduli and arguments

1 Verify that each equation is true for $z = -1 + 2i$.

   a $|z|^2 = z\bar{z}$ **b** $|z|^2 = |z^2|$

2 Let $z_1 = 3 - i$ and $z_2 = 2 + 5i$. Verify that each property is true.

   a $|z_1 z_2| = |z_1||z_2|$ **b** $\left|\dfrac{z_1}{z_2}\right| = \dfrac{|z_1|}{|z_2|}$ **c** $\overline{z_1 + z_2} = \bar{z_1} + \bar{z_2}$ **d** $\overline{z_1 z_2} = \bar{z_1}\,\bar{z_2}$

3 Let $z_1 = 1 - i$ and $z_2 = \sqrt{3} + i$. By first expressing in polar form, evaluate exactly:

   a $\arg z_1 z_2$ **b** $\arg \dfrac{z_1}{z_2}$ **c** $\arg (z_2)^5$ **d** $\arg (z_1)^{-2}$

4 If $z_1 = \sqrt{3}\left(\cos \dfrac{\pi}{3} + i \sin \dfrac{\pi}{3}\right)$ and $z_2 = 2\left(\cos \dfrac{\pi}{5} + i \sin \dfrac{\pi}{5}\right)$, find:

   a $z_1 z_2$ **b** $\dfrac{z_1}{z_2}$ **c** $\dfrac{1}{z_2}$ **d** $(z_1)^5$

5 Simplify each expression.

   a $(\cos \alpha + i \sin \alpha)(\cos 2\alpha + i \sin 2\alpha)$ **b** $\dfrac{\cos 3\beta - i \sin 3\beta}{\cos 2\lambda + i \sin 2\lambda}$

   c $\dfrac{(\cos 5 + i \sin 5)(\cos 3 - i \sin 3)}{\cos 2 - i \sin 2}$

6 Let $z = a + ib$ where $a, b \in \mathbb{R}$. Prove that:

   a $|z|^2 = z\bar{z}$ **b** $|z|^2 = |z^2|$

7 Let $z_1 = a + ib$ and $z_2 = c + id$ where $a, b, c, d \in \mathbb{R}$. Prove that:

   a $|z_1 z_2| = |z_1||z_2|$ **b** $\left|\dfrac{z_1}{z_2}\right| = \dfrac{|z_1|}{|z_2|}$ **c** $\overline{z_1 + z_2} = \bar{z_1} + \bar{z_2}$ **d** $\overline{z_1 z_2} = \bar{z_1}\,\bar{z_2}$

8 Let $z = r(\cos \theta + i \sin \theta)$. Show that:

   a $z + \bar{z}$ is real   **b** $z\bar{z}$ is real

9 Is it true that $\dfrac{1}{|z|^2} = \dfrac{1}{|z^2|}$ for all complex numbers z? Give reasons for your answer.

10 Simplify for $z = \cos \theta + i \sin \theta$:

   a $z + \dfrac{1}{z}$ **b** $z^2 + \dfrac{1}{z^2}$ **c** $z^3 + \dfrac{1}{z^3}$

   d $z^2 - \dfrac{1}{z^2}$ **e** $z^3 - \dfrac{1}{z^3}$

11 Simplify each expression.

   a $|(7 - 4i)(3 + i)|$ **b** $\left|(1 + i\sqrt{3})^2 (\sqrt{2} + i\sqrt{2})\right|$ **c** $\left|\dfrac{3 - 2i}{5 + i}\right|$

   d $\left|\dfrac{(2 + i)^2}{(4 - 3i)^2}\right|$ **e** $\left|\dfrac{1}{(1 - i)^6}\right|$

MATHS IN FOCUS 12. Mathematics Extension 2   ISBN 9780170413435

12 Use the properties of arguments to evaluate exactly:

a $\arg\left[(1+i)(1-i\sqrt{3})\right]$

b $\arg\left[(\sqrt{3}-i)(-\sqrt{2}-i\sqrt{2})\right]$

c $\arg[i(3+3i)]$

d $\arg\left(\dfrac{2-2i}{\sqrt{3}+i}\right)$

e $\arg\left(\dfrac{-1+i}{\sqrt{2}+i\sqrt{2}}\right)$

13 Let $z = 3\left(\cos\dfrac{\pi}{6} + i\sin\dfrac{\pi}{6}\right)$ and $w = 2\left(\cos\dfrac{\pi}{4} + i\sin\dfrac{\pi}{4}\right)$.

a Find z and w in exact Cartesian form.

b Hence find the exact values of $\dfrac{w}{z}$ in both polar and Cartesian form.

c Hence, by equating real and imaginary parts, find the exact value of:

i $\cos\dfrac{\pi}{12}$

ii $\sin\dfrac{\pi}{12}$

14 Using a similar proof to the one for the triangle inequality on page 33, prove that
$$|z - w| \geq |z| - |w|.$$

ISBN 9780170413435

1. Complex numbers

35

1.06 Euler's formula

Leonhard Euler is considered to be one of the greatest mathematicians in history and his innovative thoughts were centuries before his time. One of his most famous results was a formula linking the real world with the imaginary world.

> **Euler's formula**
>
> For any real θ,
>
> $e^{i\theta} = \cos \theta + i \sin \theta$
>
> This is called the **exponential form** of a complex number.

If $\theta = \pi$, then we have the extraordinary result:

$$e^{i\pi} = -1$$

This links an imaginary number with a real number.

Writing a complex number in terms of e enables us to use the index laws to prove and simplify many results and expressions like the ones seen in the previous section, only much more efficiently. **Euler's formula** can be derived using polynomial expansions of $\sin \theta$ and $\cos \theta$. This appears in the next exercise.

EXAMPLE 19

Use Euler's formula to write each expression in exponential form.

a $\cos \dfrac{\pi}{4} + i \sin \dfrac{\pi}{4}$

b $3(\cos 2 - i \sin 2)$

Solution

a   Using $e^{i\theta} = \cos \theta + i \sin \theta$:

For $\cos \dfrac{\pi}{4} + i \sin \dfrac{\pi}{4}$, $\theta = \dfrac{\pi}{4}$

$\therefore \cos \dfrac{\pi}{4} + i \sin \dfrac{\pi}{4} = e^{\frac{i\pi}{4}}$

b   First express $3(\cos 2 - i \sin 2)$ in the form $r(\cos \theta + i \sin \theta)$.

$3(\cos 2 - i \sin 2) = 3[\cos (-2) + i \sin (-2)]$

$\qquad\qquad\qquad\quad = 3e^{i(-2)}$

$\qquad\qquad\qquad\quad = 3e^{-2i}$

EXAMPLE 20

Convert each complex number to polar form.

a $e^{\frac{5i\pi}{6}}$

b $4e^{-\frac{2i\pi}{9}}$

Solution

a Using $e^{i\theta} = \cos\theta + i\sin\theta$

$$e^{\frac{5i\pi}{6}} = \cos\frac{5\pi}{6} + i\sin\frac{5\pi}{6}$$

b $4e^{-\frac{2i\pi}{9}} = 4\left[\cos\left(-\frac{2\pi}{9}\right) + i\sin\left(-\frac{2\pi}{9}\right)\right]$

This can also be written as $4\left(\cos\frac{2\pi}{9} - i\sin\frac{2\pi}{9}\right)$.

EXAMPLE 21

If $z = re^{i\theta}$ prove that $z \times \bar{z} = |z|^2$.

Solution

Since $z = r(\cos\theta + i\sin\theta)$ then $\bar{z} = r(\cos\theta - i\sin\theta) = r[\cos(-\theta) + i\sin(-\theta)] = re^{-i\theta}$

$$\therefore \ z \times \bar{z} = re^{i\theta} \times re^{-i\theta}$$
$$= r^2 e^{i\theta + (-i\theta)}$$
$$= r^2 e^0$$
$$= r^2$$
$$= |z|^2$$

Exercise 1.06 Euler's formula

1 Convert each complex number to:

    i polar form   **ii** Cartesian form

a $e^{2i\pi}$ **b** $\sqrt{2}\,e^{\frac{-i\pi}{3}}$ **c** $5e^{3i}$ **d** $-e^{\frac{i\pi}{2}}$

2 Express each complex number in exponential form $re^{i\theta}$.

a $\sqrt{3}\left(\cos\dfrac{3\pi}{4}+i\sin\dfrac{3\pi}{4}\right)$

b $2\left[\cos\left(-\dfrac{\pi}{3}\right)+i\sin\left(-\dfrac{\pi}{3}\right)\right]$

c $\dfrac{1}{2}\left(\cos\dfrac{\pi}{5}-i\sin\dfrac{\pi}{5}\right)$

d $\sqrt{3}\left(\cos\dfrac{7\pi}{6}+i\sin\dfrac{7\pi}{6}\right)$

e $6(\cos 1 + i\sin 1)$

f $4-4i$

g $-\sqrt{3}+i$

h $\dfrac{-1+i\sqrt{3}}{2}$

i i

j $\dfrac{1}{2}$

38   **MATHS IN FOCUS 12.** Mathematics Extension 2   ISBN 9780170413435

1.07 Applying Euler's formula

In this section, we will see the elegance of the formula $e^{i\theta} = \cos\theta + i\sin\theta$ in simplifying complex numbers and executing proofs. Since $e^{i\theta}$ has an index, the usual index laws apply.

EXAMPLE 22

Simplify each expression.

a $\quad e^{5i} \times e^{-3i}$

b $\quad \dfrac{e^{-\frac{5i\pi}{6}}}{e^{i\pi}}$

Solution

a $\quad e^{5i} \times e^{-3i} = e^{5i + (-3i)}$

$\qquad\qquad\quad = e^{2i}$

b $\quad \dfrac{e^{-\frac{5i\pi}{6}}}{e^{i\pi}} = e^{-\frac{5i\pi}{6} - i\pi}$

$\qquad\quad = e^{-\frac{11i\pi}{6}}$

$\qquad\quad = e^{\frac{i\pi}{6}}$

Note: $e^{i\pi} = -1$ so we could write

$$\dfrac{e^{-\frac{5i\pi}{6}}}{e^{i\pi}} = \dfrac{e^{-\frac{5i\pi}{6}}}{-1}$$

$$= -e^{-\frac{5i\pi}{6}}$$

But this is not strictly in the form $e^{i\theta}$.

EXAMPLE 23

Use Euler's formula to simplify $\left(\cos\dfrac{\pi}{4}+i\sin\dfrac{\pi}{4}\right)\left(\cos\dfrac{2\pi}{3}+i\sin\dfrac{2\pi}{3}\right)$.

Solution

$$\left(\cos\frac{\pi}{4}+i\sin\frac{\pi}{4}\right)\left(\cos\frac{2\pi}{3}+i\sin\frac{2\pi}{3}\right)=e^{\frac{i\pi}{4}}\times e^{\frac{2i\pi}{3}}$$

$$=e^{\frac{i\pi}{4}+\frac{2i\pi}{3}}$$

$$=e^{\frac{11i\pi}{12}}$$

$$=\cos\frac{11\pi}{12}+i\sin\frac{11\pi}{12}$$

EXAMPLE 24

If $z=re^{i\theta}$, prove that $\arg z^n = n\arg z$.

Solution

$$z=re^{i\theta}$$

$$z^n=(re^{i\theta})^n=r^n e^{in\theta}$$

$$\arg z^n = \arg(r^n e^{in\theta})$$

$$=n\theta$$

$$=n\arg z \qquad \text{since } \theta = \arg z$$

DID YOU KNOW?

Euler's formula

We can use Euler's formula to evaluate $(i)^i$.

Did you know it is real?

ISBN 9780170413435

Exercise 1.07 Applying Euler's formula

1 Use Euler's formula to simplify each expression.

a $\left(\cos\dfrac{\pi}{6}+i\sin\dfrac{\pi}{6}\right)\left(\cos\dfrac{5\pi}{6}+i\sin\dfrac{5\pi}{6}\right)$

b $3\left(\cos\dfrac{\pi}{4}+i\sin\dfrac{\pi}{4}\right)\times\sqrt{2}\left(\cos\dfrac{\pi}{3}+i\sin\dfrac{\pi}{3}\right)$

c $\dfrac{\cos\left(-\dfrac{7\pi}{8}\right)+i\sin\left(-\dfrac{7\pi}{8}\right)}{\cos\dfrac{3\pi}{4}+i\sin\dfrac{3\pi}{4}}$

d $\dfrac{-5\left(\cos\dfrac{\pi}{2}+i\sin\dfrac{\pi}{2}\right)}{\sqrt{5}\left(\cos\dfrac{2\pi}{3}+i\sin\dfrac{2\pi}{3}\right)}$

2 Simplify each expression, writing the answer in exponential form.

a $(\cos\theta_1+i\sin\theta_1)(\cos\theta_2+i\sin\theta_2)$

b $\dfrac{\cos\theta_1+i\sin\theta_1}{\cos\theta_2+i\sin\theta_2}$

c $\left[\dfrac{1}{2}\left(\cos\dfrac{3\pi}{4}-i\sin\dfrac{3\pi}{4}\right)\right]^2$

d $\left[\sqrt{2}\left(\cos\dfrac{2\pi}{5}+i\sin\dfrac{2\pi}{5}\right)\right]^{-1}$

e $\dfrac{\sqrt{10}\left(\cos2\alpha-i\sin2\alpha\right)}{\sqrt{2}\left(\cos5\lambda+i\sin5\lambda\right)}$

f $(-2+2i)(1-i\sqrt{3})$

g $-i\left(\dfrac{-\sqrt{3}+i}{2}\right)$

h $\dfrac{2\sqrt{2}+i2\sqrt{2}}{5}$

3 If $z_1=e^{i\theta_1}$ and $z_2=e^{i\theta_2}$, prove that:

a $\arg z_1z_2=\arg z_1+\arg z_2$

b $\arg\dfrac{z_1}{z_2}=\arg z_1-\arg z_2$

4 a Using the fact that $e^{i\theta}=\cos\theta+i\sin\theta$, write down a similar result for $e^{-i\theta}$.

b Hence prove that:

i $\sin\theta=\dfrac{e^{i\theta}-e^{-i\theta}}{2i}$

ii $\cos\theta=\dfrac{e^{i\theta}+e^{-i\theta}}{2}$

1. TEST YOURSELF

1 Express each complex number in terms of i.

   a $\sqrt{-25}$ **b** $\sqrt{-18}$ **c** $\sqrt{-\dfrac{8}{9}}$

2 Simplify each expression.

   a i^5 **b** $\sqrt{(-4)^2 - 4(2)(7)}$ **c** $\dfrac{1}{i^6}$

   d i^{91} **e** $\dfrac{-2 \pm \sqrt{(2)^2 - 4(1)(5)}}{2}$

3 Solve each equation.

   a $x^2 + 49 = 0$ **b** $(x + 3)^2 + 4 = 0$

4 Solve each equation using the quadratic formula.

   a $x^2 - 4x + 9 = 0$ **b** $x^2 + 6x + 15 = 0$ **c** $2x^2 + 3x + 9 = 0$

5 Solve each equation by completing the square.

   a $x^2 - 2x + 2 = 0$ **b** $x^2 + 8x + 20 = 0$ **c** $x^2 - x + 3 = 0$

6 Factorise each expression in the equation as a difference of 2 squares, then solve the equation.

   a $x^2 + 4x + 8 = 0$ **b** $x^2 - 8x + 25 = 0$ **c** $x^2 + 10x + 41 = 0$

7 State $\mathrm{Re}(z)$ and $\mathrm{Im}(z)$ for each complex number.

   a $z = \dfrac{\sqrt{3} - 2i}{4}$

   b $z = (-3 - 7i) + (5 - 2i)$

   c $z = \dfrac{(4x - 3iy) + (x + yi)}{x^2 + y^2}, \; x, y \in \mathbb{R}$

8 State the complex conjugate of each complex number.

   a $z = -6 + 11i$

   b $w = \dfrac{3 - i\sqrt{2}}{2}$

   c $u = \dfrac{-a + 2i + ai - 7b}{a^2 + b^2}, \; a, b \in \mathbb{R}$

9 If $z = p - 3iq$ where $p, q \in \mathbb{R}$, prove that:

   a $z\bar{z}$ is always real

   b $z + \bar{z}$ is always real

   c   the expression $\sqrt{(\mathrm{Re}(z))^2 + (\mathrm{Im}(z))^2}$ is always real.

10 If x and y are real, solve each equation for x and y.

   a $4x - 6i - 3yi - 12 = 0$

   b $5x + 2xi + i - 3y + yi + 20 - 4i = 0$

11 If $V = \dfrac{2x + 2yi - 5 + 3ix - 2y + 7i}{x^2 + y^2}$ is always real, where x, y are real, find a relationship between x and y.

12 Simplify each expression, giving your answer in the form $u + iv$ where $u, v \in \mathbb{R}$.

   a $8i + 5 - 4i + 10$ **b** $-3(2 - 7i) + 2i(6 - i)$

   c $(1 - 3i)(4 + 9i)$ **d** $(2 - 5i)^2 - (-3 + 4i)(-3 - 4i)$

13 Find the quadratic equation with the given roots α and β.
Use the formula $(x - \alpha)(x - \beta) = x^2 - (\alpha + \beta)x + \alpha\beta$.

   a $\alpha = 1 - i\sqrt{2}$ and $\beta = 1 + i\sqrt{2}$

   b $\alpha = -3 - 5i$ and $\beta = -3 + 5i$

   c $\alpha = \sqrt{7} + 3i$ and $\beta = \sqrt{7} - 3i$

   d $\alpha = \dfrac{-1}{2} + \dfrac{i\sqrt{3}}{2}$ and $\beta = \dfrac{-1}{2} - \dfrac{i\sqrt{3}}{2}$

14 Simplify each expression by realising the denominator.

   a $\dfrac{1}{1 - 2i}$ **b** $\dfrac{\sqrt{2} - i\sqrt{2}}{1 + i}$ **c** $\dfrac{1}{2\sqrt{3} + i} - \dfrac{1}{5 + i}$ **d** $\dfrac{\sqrt{3} + 2i}{\sqrt{3} - 2i}$

15 Find the 2 square roots of each complex number.

   a $3 - 4i$ **b** $8 + 6i$ **c** $15 - 8i$ **d** $12 + 5i$

16 Represent each of the following complex numbers as a point on the complex plane.

   a $u = -1 - i\sqrt{3}$ **b** $w = \sqrt{2} + i\sqrt{2}$

   c $v = (2 - 3i)(2 + 3i)$ **d** $z = \dfrac{1 + i}{1 - i}$

17 Given $z = -2 + 5i$ and $v = 3 - 4i$, plot $z, \bar{z}, v$ and $\bar{v}$ as vectors on an Argand diagram.

18 If $u = 4 - i$ and $z = -1 + 2i$, find $u + z$ and plot $u, z, -u$ and $\underset{\sim}{u} + \underset{\sim}{z}$ on an Argand diagram.

19 The complex numbers z, v and w are plotted on the complex plane. Copy the plane and plot the points corresponding with $-z$, $\dfrac{1}{2}w$, $z + v$ and $z - v$ on the plane.

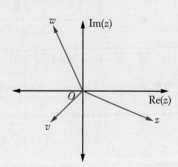

20 Find $|z|$ and Arg(z) in exact form for each complex number.

a $z = \sqrt{2} - i\sqrt{2}$ **b** $z = \sqrt{3} + i$

c $z = -2 + 2i\sqrt{3}$ **d** $z = \dfrac{-1 - i}{2}$

21 Express each complex number in polar form.

a $z = \dfrac{-\sqrt{3} + i}{2}$

b $z = \sqrt{2}\left(\cos\dfrac{2\pi}{3} - i\sin\dfrac{2\pi}{3}\right)$

c $z = -\dfrac{1}{2}\left(\cos\dfrac{7\pi}{6} + i\sin\dfrac{7\pi}{6}\right)$

22 Express each complex number in Cartesian form.

a $z = 2\left(\cos\dfrac{\pi}{4} + i\sin\dfrac{\pi}{4}\right)$

b $z = \dfrac{1}{2}\left(\cos\dfrac{\pi}{3} + i\sin\dfrac{\pi}{6}\right)$

c $z = \sqrt{2}\left(i\cos\dfrac{5\pi}{6} + \sin\dfrac{5\pi}{6}\right)$

23 If $z_1 = \dfrac{1}{\sqrt{3}}\left(\cos\dfrac{3\pi}{4} + i\sin\dfrac{3\pi}{4}\right)$ and $z_2 = 3\left(\cos\dfrac{\pi}{6} + i\sin\dfrac{\pi}{6}\right)$ find, in exact modulus–argument form:

a $z_1 z_2$ **b** $\dfrac{z_1}{z_2}$ **c** $\dfrac{1}{z_2}$ **d** $(z_1)^{11}$

24 Find arg z for each complex number.

a $z = \dfrac{\cos\theta + i\sin\theta}{\cos 2\beta + i\sin 2\beta}$

b $z = (\cos 3\alpha - i\sin 3\alpha)(\cos 2\lambda + i\sin 2\lambda)$

c $z = \dfrac{\left(\cos\dfrac{\delta}{2} + i\sin\dfrac{\delta}{2}\right)\left(\cos\dfrac{\alpha}{4} - i\sin\dfrac{\alpha}{4}\right)}{\cos 2\phi - i\sin 2\phi}$

d $z = (\cos 3\varepsilon - i\sin 3\varepsilon)^{-1}$

25 Find mod z for each complex number.

a $z = (1 + 7i)(2 - 3i)$ **b** $z = \dfrac{2 + i}{\sqrt{3} - 4i}$ **c** $z = \dfrac{1}{(2 - i)^2}$

ISBN 9780170413435

26 $z = 2\left(\cos\dfrac{\pi}{4} + i\sin\dfrac{\pi}{4}\right)$ and $w = 4\left(\cos\dfrac{\pi}{3} + i\sin\dfrac{\pi}{3}\right)$.

    a   Find z and w in exact Cartesian form.

    b   Hence find the value of zw in both polar and Cartesian form.

    c   Hence find the value of:

        i $\cos\dfrac{7\pi}{12}$ **ii** $\sin\dfrac{7\pi}{12}$

27 Use Euler's formula to write each expression in exponential form.

    a $z = \cos\dfrac{3\pi}{5} + i\sin\dfrac{3\pi}{5}$ **b** $\sqrt{2}\left(\cos 3 - i\sin 3\right)$

28 Convert each complex number to polar form.

    a $e^{\frac{i\pi}{4}}$ **b** $\dfrac{e^{-2i}}{2}$

29 If $z_1 = r_1 e^{i\theta_1}$ and $z_2 = r_2 e^{i\theta_2}$, prove that $\arg\left(\dfrac{z_1}{z_2}\right) = \arg(z_1) - \arg(z_2)$.

30 Simplify each expression.

    a $e^{\frac{5i\pi}{6}} \times e^{-\frac{7i\pi}{6}}$ **b** $\left(e^{-\frac{2i\pi}{3}}\right)^{12}$

31 Use the fact that $\cos\theta + i\sin\theta = e^{i\theta}$ to simplify $\dfrac{\cos\dfrac{2\pi}{7} + i\sin\dfrac{2\pi}{7}}{\cos\left(-\dfrac{3\pi}{14}\right) + i\sin\left(-\dfrac{3\pi}{14}\right)}$.

32 If $z = re^{i\theta}$, prove that $\arg z^{-n} = -n \arg z$.

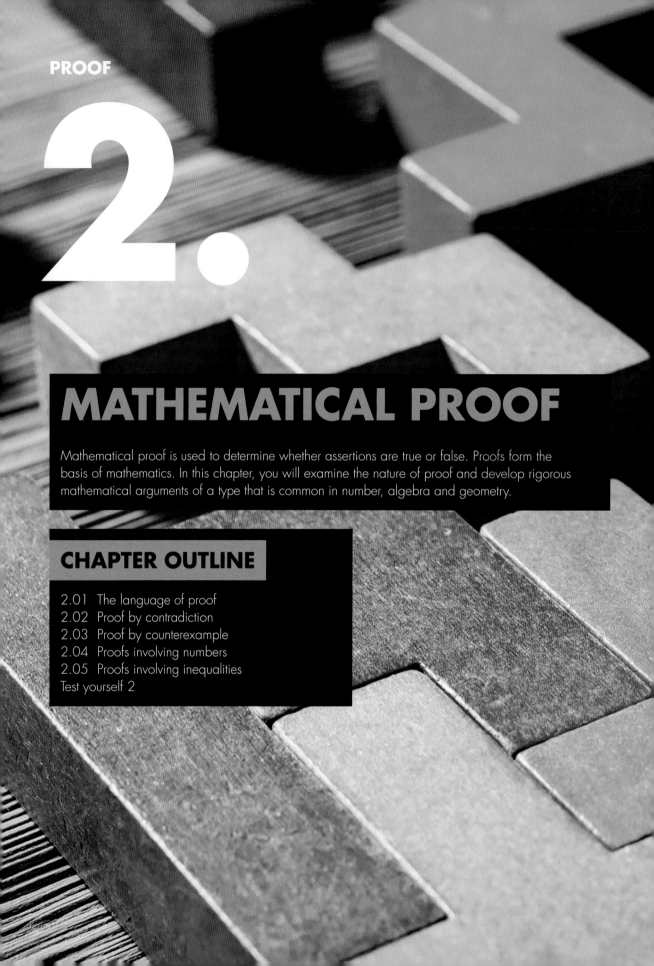

PROOF

2.

MATHEMATICAL PROOF

Mathematical proof is used to determine whether assertions are true or false. Proofs form the basis of mathematics. In this chapter, you will examine the nature of proof and develop rigorous mathematical arguments of a type that is common in number, algebra and geometry.

CHAPTER OUTLINE

IN THIS CHAPTER YOU WILL:

- learn the formal language and symbols of proof, including ∀, ∃, ⇒, ⇔, ¬P, iff and ∈
- use the proof concepts of implication, negation, equivalence and equality
- state the contrapositive and converse of a statement
- use proof by contradiction and counterexample
- prove results involving numbers and inequalities
- prove further results involving inequalities based on previous results

TERMINOLOGY

contrapositive: The contrapositive statement to $P \Rightarrow Q$ (if P then Q) is $\neg Q \Rightarrow \neg P$ (if not Q then not P). If $P \Rightarrow Q$ is true, then the contrapositive statement $\neg Q \Rightarrow \neg P$ is also true.

converse: The converse to $P \Rightarrow Q$ (if P then Q) is $Q \Rightarrow P$ (if Q then P).

counterexample: An example that shows that a statement is not true.

equivalence: $P \Leftrightarrow Q$ (P if and only if Q, also written P iff Q). If $P \Rightarrow Q$ and $Q \Rightarrow P$ then we can write $P \Leftrightarrow Q$.

implication statement: $P \Rightarrow Q$ (if P then Q, or P implies Q)

negation: The negation of a statement P is *not P*, written $\neg P$, P' or $\bar{P}$.

proof by contradiction: A proof that assumes that the direct opposite of a statement is true and then a logical argument is used to show the assumption is false.

QED: *quod erat demonstrandum*, Latin for 'demonstrated as required'.

statement, proposition or **premise:** A sentence that is either true or false (but not both).

triangle inequality: $|x| + |y| \geq |x + y|, \forall\, x, y \in \mathbb{R}$.

2.01 The language of proof

Converse

Contrapositive

Quantifiers

A proof in mathematics is most often an argument that shows a statement or **premise** is true for all cases. This is different from an example where the statement may be true only for a particular case. There are different types of proof, such as deductive, inductive and proof by contradiction or counterexample. They are based on **reason**, a way of thinking attributed most often to the Classical Greek mathematicians and philosophers such as Euclid, Socrates, Plato and Aristotle.

In mathematics, deductive proof and proof by contradiction are often used, whereas inductive proof is common in science. Proof by counterexample is used to show a statement is *not* true.

Statement, proposition or premise

A statement, proposition or premise is a sentence that is either true or false (but not both).

'It is raining' is a statement. 'Is it raining?' is not a statement but a question. We denote a statement by a capital letter; for example, P: It is raining.

A statement could be mathematical; for example, Q: For all real x, $\dfrac{d}{dx}(x^2 + 2x) = 2x + 2$. Q is a true statement.

Sometimes the truth of a statement depends on the context. For instance, the statement that $\dfrac{1}{a} < \dfrac{1}{b}$ when $a > b$ is true only if a and b are both positive but it is not true for all numbers.

We will now define some terminology used in logical proof.

Implication or 'if-then' statement

One of the most common arguments is the **if-then statement**, also known as the **implication statement** or **conditional statement**. This can be written using symbols.

Implication

If P and Q stand for the 2 parts of an **if-then statement**, we can write the statement 'if P then Q' as $P \Rightarrow Q$.

The notation $P \Rightarrow Q$ can also be read as 'P implies Q'.

EXAMPLE 1

Write the following statement in mathematical notation:

> If I don't study, then I will fail.

Solution

Define P and Q as follows:

> P: I don't study.

> Q: I will fail.

Then we can write:

> P implies Q, or $P \Rightarrow Q$.

Converse

To find the **converse** of a statement, we reverse the implication.

Converse

For the statement 'If P then Q', the **converse** is 'If Q then P', or the converse of 'P implies Q' is 'Q implies P'.

In symbols, the converse of $P \Rightarrow Q$ is $Q \Rightarrow P$.

The converse of a true statement may or may not be true. Equally, the converse may be true but the original statement may or may not be true.

For example, both Pythagoras' theorem and its converse are true.

Pythagoras' theorem: In a right-angled triangle, the square of the hypotenuse is equal to the sum of the squares of the other 2 sides.

Converse: If the square of the longest side of a triangle is equal to the sum of the squares of the other 2 sides, then the triangle is right angled.

However, consider the statement 'If you are a cow, then you eat grass'. This is a true statement.

The converse is 'If you eat grass, then you are a cow'. The converse is not true because there are other animals that eat grass.

EXAMPLE 2

Express each sentence as a mathematical statement and write the converse of each statement.

a If you drink alcohol, then your brain is damaged.

b If you do your homework, then you have more knowledge.

Solution

a Let P and Q be as follows.

P: You drink alcohol.

Q: Your brain is damaged.

Then we can write:

P implies Q, or $P \Rightarrow Q$.

Reverse the implication so we have the converse.

Q implies P, or $Q \Rightarrow P$.

In words, the converse is: If your brain is damaged, then you drink alcohol.

b Let P and Q be as follows.

P: You do your homework

Q: You have more knowledge.

P implies Q, or $P \Rightarrow Q$.

Reverse the implication so we have the converse.

Q implies P, or $Q \Rightarrow P$.

In words, the converse is: If you have more knowledge, then you do your homework.

It is worth pointing out here that we are operating with statements logically without claiming that such statements are actually true. For example, we are examining the converse of 'If you drink alcohol, then your brain is damaged' without saying whether this statement is correct.

MATHS IN FOCUS 12. Mathematics Extension 2

ISBN 9780170413435

Equivalence

If a statement is true and its converse is also true (such as Pythagoras' theorem) then we call this an **equivalence**.

The term 'if and only if' is used to describe an equivalence. This can be abbreviated to 'iff'.

> **Equivalence**
>
> If $P \Rightarrow Q$ and $Q \Rightarrow P$ then we can write $P \Leftrightarrow Q$.
>
> In words, $P \Leftrightarrow Q$ means 'P if and only if Q' or 'P iff Q'.

EXAMPLE 3

This pair of statements A and B relate to a quadrilateral.

    A: It is a parallelogram.

    B: Its diagonals bisect each other.

Determine whether they form an equivalence and $A \Leftrightarrow B$.

Solution

Test $A \Rightarrow B$ and $B \Rightarrow A$.

$A \Rightarrow B$: If it is a parallelogram, then the diagonals bisect each other. This is true.

$B \Rightarrow A$: If the diagonals bisect each other, then it is a parallelogram. This is true.

So $A \Rightarrow B$ and $B \Rightarrow A$. We can say $A \Leftrightarrow B$. It is an equivalence.

In words the statement becomes: It is a parallelogram *if and only if* the diagonals bisect each other.

Negation

The **negation** of a statement P is to say *not P*.

> **Negation**
>
> The negation of P, or *not P*, is written $\neg P$, P', $\sim P$ or $\overline{P}$.
>
> If P is true, then $\neg P$ is false. If P is false, then $\neg P$ is true.

When statements are in words, we need to be careful with the negation. For example, some people might think that the negation of the statement 'All of the students passed' is 'All of the students failed', but this is incorrect. It is 'Not all of the students passed', or 'Some of the students did not pass'.

EXAMPLE 4

Write the negation of each statement.

a A: It is wet.

b B: The number is even.

c C: All birds are black.

d D: The parcel weighs less than 250 g.

e E: Trees have leaves.

f F: No boys can sing.

g G: Some plums are sour.

Solution

a A: It is wet.

The negation is $\neg A$: it is *not* wet.

b B: The number is even.

The negation is $\neg B$: The number is *not* even.

c C: All birds are black.

The negation is $\neg C$: *Not* all birds are black, OR There is at least one bird that is not black.

> The word 'some' means 'not none' or 'at least one', so we could also say 'Some birds are not black'.

d D: The parcel weighs less than 250 g.

The negation is $\neg D$: The parcel doesn't weigh less than 250 g; that is, The parcel weighs 250 g or more, OR The parcel weighs at least 250 g.

e E: Trees have leaves.

This one is tricky. We will interpret the statement 'Trees have leaves' to mean *All* trees have leaves. The negation is $\neg E$: Not all trees have leaves, or There is at least one tree that does not have leaves.

f F: No boys can sing.

The negation is $\neg F$: It is not true that no boys can sing. That is, there is at least one boy who can sing.

g G: Some plums are sour.

The negation is $\neg G$: It is not true to say that some plums are sour. That is, no plums are sour.

 ISBN 9780170413435

Contrapositive

Contrapositive

For the statement $P \Rightarrow Q$, the **contrapositive** statement is $\neg Q \Rightarrow \neg P$.

In words, the contrapositive statement of 'If P then Q' is 'If not Q then not P'.

EXAMPLE 5

Write the contrapositive of each statement.

a   If you are tired when driving, then you have road accidents.

b $X \Rightarrow Y$

Solution

a P: You are tired when driving

   Q: You have road accidents.

   We can write $P \Rightarrow Q$.

   $\neg P$: You are not tired when driving

   $\neg Q$: You do not have road accidents

   So the contrapositive is $\neg Q \Rightarrow \neg P$

   or: If you do not have road accidents, then you are not tired when driving.

> Again, we are examining the logic of these statements without claiming that such statements are actually true.

b   The contrapositive of $X \Rightarrow Y$ is $\neg Y \Rightarrow \neg X$.

The contrapositive can be a useful way to check whether or not a statement is true.

Contrapositive and equivalence

If the original statement $P \Rightarrow Q$ is true, then the contrapositive statement $\neg Q \Rightarrow \neg P$ is also true.

If the original statement $P \Rightarrow Q$ is not true, then the contrapositive statement $\neg Q \Rightarrow \neg P$ is also not true.

$P \Rightarrow Q$ and $\neg Q \Rightarrow \neg P$ are an equivalence.

Consider the statement: If a polygon is a triangle, then its interior angle sum is 180°.

Explain why the contrapositive of the statement is true. State the contrapositive.

Solution

P: A polygon is a triangle.

Q: Its interior angle sum is 180°.

Since $P \Rightarrow Q$ is true, then the contrapositive $\neg Q \Rightarrow \neg P$ is also true.

$\neg P$: A polygon is not a triangle.

$\neg Q$: Its interior angle sum is not 180°.

So the contrapositive is $\neg Q \Rightarrow \neg P$.

or: If a polygon's interior angle sum is not 180°, then it is not a triangle.

Quantifiers

Mathematical notation is useful because it is brief and precise. Mathematical symbols called **quantifiers** are commonly used in mathematics proofs.

Quantifiers

$\forall$ for all

$\exists$ there exists

We use $\forall$ to specify a whole set of numbers or items we are talking about. We use $\exists$ to refer to a particular set of numbers or items that have a specific property.

Sets of numbers

We have already met the set $\mathbb{R}$ of real numbers and the set $\mathbb{C}$ of complex numbers, which includes all the real numbers. The **natural numbers** $\{1, 2, 3, ...\}$ form the set $\mathbb{N}$.
The **integers** $\{..., -2, -1, 0, 1, 2, ...\}$ form the set $\mathbb{Z}$, which includes all of $\mathbb{N}$.
The **rational numbers** form the set $\mathbb{Q}$, which includes all of $\mathbb{Z}$.
The **real numbers** form the set $\mathbb{R}$, which includes all of $\mathbb{Q}$.

Some definitions of natural numbers include the number 0 in the set $\mathbb{N}$.

Sets of numbers

$\mathbb{N}$ the set of natural numbers

$\mathbb{Z}$ the set of integers

$\mathbb{Q}$ the set of rational numbers

$\mathbb{R}$ the set of real numbers

$\mathbb{C}$ the set of complex numbers

Other symbols

$\in$ is an element of, belongs to

:   such that

EXAMPLE 7

Use quantifiers to express each sentence as a mathematical statement.

a   For all integers n, there exists an integer M such that $M = 2n$.

b   For all rational numbers r, there exist 2 integers p and q such that $r = \dfrac{p}{q}$.

Solution

a   Since n belongs to the set of integers, we can write $n \in \mathbb{Z}$.

Also, M is an integer so we can write $M \in \mathbb{Z}$.

Using the quantifiers we can write the sentence mathematically as:

$\forall\, n \in \mathbb{Z}, \exists\, M \in \mathbb{Z} : M = 2n$

('for all n belonging to the set of integers, there exists M belonging to the set of integers such that $M = 2n$')

b $\forall\, r \in \mathbb{Q}, \exists\, p, q \in \mathbb{Z} : r = \dfrac{p}{q}$

('for all r belonging to the set of rational numbers, there exists p and q belonging to the set of integers such that $r = \dfrac{p}{q}$')

It is important to use mathematical notation correctly. For example, '=' means 'is equal to' and should only be used in mathematical equations such as $2x + 1 = x^2 - 2$, not in word statements. For instance, if P and Q are statements and P iff Q, then we would write $P \Leftrightarrow Q$, not $P = Q$.

Exercise 2.01 The language of proof

1 Write each $P \Rightarrow Q$ implication statement in mathematical notation, identifying P and Q.

    a If there are crumbs, then ants will come.

    b If a quadrilateral has equal diagonals, then it is a square.

    c If people are unemployed, then they are bored.

2 Write the converse of each statement.

    a If you live in Cooma, then you go skiing.

    b If you like maths, then you have friends.

    c If you are a politician, then you can debate.

    d If an animal can fly, then it is a bird.

3 For each true statement, find the converse and decide whether it is true.

    a If you are a carnivore, then you eat meat.

    b If you are seasick, then you are on a boat.

    c If a shape is a square, then it has equal sides.

    d If an animal is a honeybee, then it can sting.

4 For each true statement, find the converse and decide whether or not it is an equivalence. If it is, write it in the form 'P iff Q'.

    a If $x - 5 = 4$ then $x = 9$.

    b If a quadrilateral is a rhombus, then it has diagonals that are perpendicular.

    c If $a > b > 0$, then $\dfrac{1}{a} < \dfrac{1}{b}$.

    d If you have a driver's licence, then you passed a driving test.

5 Write the negation of each statement.

    a It is white.   **b** I know everything.

    c Fish swim in the ocean.   **d** All babies are cute.

    e There are more than 5.   **f** There is none.

    g No one passed the test.   **h** Some teachers are mean.

    i The potatoes weigh less than 3 kg.   **j** Cassie is small.

6 Write the contrapositive statement for each statement.

    a If you live in a mansion, then you are rich.

    b If you are in the army, then you have boots.

    c If you are old, then you are wise.

    d If $x = 3$ then $x^2 = 9$.

    e If an animal is a horse, then it has four legs.

    f If you are a woman, then you are superior.

7 Given each contrapositive statement, state the original statement.

   a If the water is not rising, then it is not global warming.

   b If you do not speed, then you do not have accidents.

   c If there is not a drought, then animals do not die.

   d If a number is not rational, then it is not a fraction.

   e If Sam does not pass his exams, then he is lazy.

   f If a number is negative, then it does not have a square root.

8 Write the contrapositive of each true statement, then verify that its contrapositive is also true.

   a If $n \geq 1$, then $\dfrac{1}{n} \geq \dfrac{1}{n+1}$.

   b If a line is horizontal, then its gradient is zero.

   c In the outback, the bulldust is red. ← Bulldust is the fine powdery dust that often covers the ground in the Australian bush.

   d All blue whales are mammals.

9 Write the contrapositive of each statement, then state whether the statement and its contrapositive are both true or both false.

   a If you exercise, then your heart rate increases.

   b If a plant does not get sufficient water, then it dies.

   c If a triangle is isosceles, then it has 2 equal angles.

   d If a number is an integer, then it is real.

   e If $x > 2$, then $x^2 > 4$.

10 Write the contrapositive of each statement, then determine whether the statement is true or false.

   a If an animal is a bird, then it has a beak.

   b A quadrilateral with 2 pairs of opposite angles equal is a rhombus.

   c All fish have fins.

   d If $x \leq 5$, then $x^2 \leq 25$.

   e If a number is odd, then it is prime.

11 Isobel was trying to work out whether the following statement was true by stating the contrapositive. She decided it was true. Can you explain where she went wrong?

   Statement: If a quadrilateral has 4 equal angles, then it is a square.
   Contrapositive: If a quadrilateral does not have 4 equal angles, then it is not a square.

12 If $\neg A \Rightarrow \neg B$ is a true statement, which of the following statements is always true?

   A $A \Rightarrow B$ **B** $\neg B \Rightarrow \neg A$ **C** $A \Leftrightarrow B$ **D** $B \Rightarrow A$

13 Use symbols and quantifiers to write each sentence in mathematical notation.

a For all natural numbers x, there exists a natural number y such that y is greater than x.

b If x is rational, then there exist 2 integers p and q, where q is not zero, such that $x = \dfrac{p}{q}$.

c For all non-zero integers a, there exists a rational number b such that $b = \dfrac{1}{a}$.

d For any 2 ordered pairs (x, y) and (w, v) where x, y, w and v are real, and $x < w, y < v$, there exists an ordered pair (c, d) such that c is between x and w and d is between y and v.

e For all real non-negative numbers x, there exists a real non-negative number y such that $y = \sqrt{x}$.

14 Write each mathematical statement in words.

a $\forall\, m \in \mathbb{N}, \exists\, n \in \mathbb{Z}: n + m = 0$

b $\forall\, a, b \in \mathbb{Z}, b \neq 0, \exists\, p, q \in \mathbb{Q}$ such that $\dfrac{1}{a + b\sqrt{2}} = p + q\sqrt{2}$

15 The statement M iff N can be written as:

A $M \Rightarrow N$ **B** $M \Leftarrow N$ **C** $N \Leftrightarrow M$ **D** $\neg M \Rightarrow \neg N$

iStock.com/duncan1890

2.02 Proof by contradiction

Sometimes it is easier to prove a statement is true by assuming the opposite is true, then showing that this leads to a contradiction.

Proof by contradiction

A **proof by contradiction** works by taking an assumption that the direct opposite of a statement, or **negation**, is true, and then showing by a logical argument that the assumption is false.

EXAMPLE 8

Use proof by contradiction to show that $\sqrt{3}$ is irrational.

Solution

Proof

Assume that $\sqrt{3}$ is rational.

That is, $\exists\, p, q \in \mathbb{Z}, q \neq 0$ (p and q are integers) such that:

$$\sqrt{3} = \frac{p}{q} \quad \text{where } p, q \text{ have no common factors}$$

Squaring both sides: $\quad 3 = \dfrac{p^2}{q^2}$

Rearranging: $\quad p^2 = 3q^2$

So p^2 is a multiple of 3, which implies that p is a multiple of 3.

Therefore we can write $p = 3k$ for some $k \in \mathbb{Z}$.

So $p^2 = 9k^2$ but $p^2 = 3q^2$ also.

$\therefore 3q^2 = 9k^2$

$\quad q^2 = 3k^2$

So q^2 is a multiple of 3, which implies that q is a multiple of 3.

$\therefore p$ and q are both multiples of 3.

Contradiction of the assumption that p and q have no common factors.

Therefore the original assumption was wrong.

Therefore $\sqrt{3}$ is irrational. QED.

The abbreviation **QED** stands for *quod erat demonstrandum*, which is Latin for 'demonstrated as required.' It is commonly put at the end of a proof to show that the proof is complete.

Use proof by contradiction to show that $\log_2 3$ is irrational.

Solution

Proof

Assume that $\log_2 3$ is rational.

That is, $\exists\, p, q \in \mathbb{Z}, q \neq 0$ such that:

$\log_2 3 = \dfrac{p}{q}$ where p, q have no common factors.

Expressing as an index: $2^{\frac{p}{q}} = 3$

Raising both sides to the power of q: $2^p = 3^q$

Now, since 2 is even then 2^p is also even.

Also since 3 is odd then 3^q is also odd.

Therefore $2^p \neq 3^q$.

Contradiction, since $2^p = 3^q$ but an even number cannot be equal to an odd number.

Therefore the original assumption was wrong.

Therefore $\log_2 3$ is irrational. QED.

Exercise 2.02 Proof by contradiction

1 Prove by contradiction that each of the following numbers is irrational.

    a $\sqrt{2}$ **b** $\sqrt{5}$ **c** $\sqrt{7}$ **d** $\sqrt[3]{2}$ **e** $\sqrt[3]{5}$

2 Prove by contradiction that each of the following numbers is irrational.

    a $\log_2 5$ **b** $\log_2 7$ **c** $\log_3 8$

3 Complete this proof by contradiction that the angle in a semicircle is 90°.

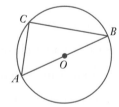

    Proof

    Assume that A, B, C lie on a circle of centre O.

    Assume that $\angle ACB \neq 90°$.

4 Prove each statement by contradiction.

    a The bisector of the angle between the equal sides of an isosceles triangle bisects the third side.

    b The opposite angles of a parallelogram are equal.

    c The diagonals of a kite intersect at right angles.

5 Prove each statement by contradiction.

a A triangle with sides 8, 15, 17 is a right-angled triangle.

b A triangle with sides 4, 5, 6 is not a right-angled triangle.

c The number $I = 5 + \sqrt{2}$ is irrational. [Hint: isolate the $\sqrt{2}$.]

d The number $I = \sqrt{2} + \sqrt{3}$ is irrational. [Hint: square both sides.]

e A number in the form $a + b\sqrt{2}$ where $a, b \in \mathbb{N}$ is irrational.

6 Prove by contradiction that each statement is true.

a Given a number in the form $\dfrac{1}{a + b\sqrt{2}}$ where $a, b, \in \mathbb{N}, b \neq 0, \exists\, p, q \in \mathbb{Q}$:

$$\frac{1}{a + b\sqrt{2}} = p + q\sqrt{2}.$$

b $\forall\, a \in \mathbb{N}, \exists\, b \in \mathbb{Z}: a + b = 0$

2.03 Proof by counterexample

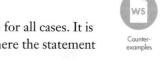

WS

Counter-
examples

It can be difficult to construct a rigorous proof to show a statement is true for all cases. It is much easier to show a statement is false. You only need to find *one* case where the statement is false to disprove the statement. This is called a **counterexample**.

Counterexample

A **counterexample** is an example that shows that a statement is not true for all cases.

EXAMPLE 10

Provide a counterexample to show that the following statement is not always true:

$\forall\, n \in \mathbb{Z}, n^2 > n$

Solution

The statement says that the square of an integer is always greater than the integer. It appears to be true for almost all cases. However, if we choose $n = 0$, it is not true.

$0^2 \not> 0$. In fact $0^2 = 0$.

Therefore the statement is false.

The statement is also not true for $n = 1$.

EXAMPLE 11

Determine whether this statement is true:

All camels have one hump.

Solution

Most camels have one hump, but the Bactrian camel has two humps.

Therefore the statement is false.

EXAMPLE 12

Explain what is wrong with the following argument.

Teacher: Smoking is bad for you.

Student: I have a counterexample: social media is bad for you.

Solution

The student has found another habit that is bad for you rather than a counterexample to show that smoking is not bad for you.

Exercise 2.03 Proof by counterexample

1 Find a counterexample to prove that each statement is not true.

a $\forall n \in \mathbb{R}, n^2 \geq n$

b $\forall n \in \mathbb{R}, n^2 + n \geq 0$

c All prime numbers are odd.

d $\forall x, y \in \mathbb{Z}$ and $n \in \mathbb{N}, x^n + y^n = (x + y)(x^{n-1} - x^{n-2}y + x^{n-3}y^2 - \ldots x^2 y^{n-2} + y^{n-1})$

e $\forall x \in \mathbb{Z}, x + \dfrac{1}{x} \geq 2$

2 Find a counterexample for each statement to demonstrate it is false.

a If $n^2 = 100$, then $n = 10$.

b The statement $x^3 - 6x^2 + 11x - 6 = 0$ is true for $x = 1, 2, 3, \ldots$

c All lines that never meet are parallel.

d If an animal lays eggs, then it is a bird.

3 Decide whether each statement is true or false. If it is false, provide a counterexample.

   a If a quadrilateral has diagonals that are perpendicular, then it is a square.

   b If $p \leq 3$, then $\dfrac{1}{p} \geq \dfrac{1}{3}$

   c $(x + y)^2 \geq x^2 + y^2$ for all $x, y \in \mathbb{R}$

   d If $pq = rq$ then $r = p$

   e All rectangles are similar.

4 In each case, determine whether the counterexample shows that the statement is false.

   a *Statement:* $x^2 = 3x - 2$ for $x \in \mathbb{N}$

   Counterexample: For $x = 3$, LHS $= 3^2 = 9$

   RHS $= 3(3) - 2 = 7$ ∴ LHS $\neq$ RHS

   The statement is not true.

   b *Statement:* All dogs are domesticated.

   Counterexample: Cats are domesticated.

   The statement is not true.

5 Decide whether the statement below is true or false. If it is false, find a counterexample.

   If $a > b$ then $\dfrac{1}{a} < \dfrac{1}{b}$.

6 Is it always true that $\dfrac{1}{(x-1)(x-2)} + \dfrac{1}{(x-2)(x-3)} = \dfrac{2x-4}{(x-1)(x-2)(x-3)}$ for $x = 4, 5, 6, \ldots$?

7 Is it always true that $\dfrac{1}{n} < \dfrac{1}{n-1}$?

8 Are all squares rhombuses? Are all rhombuses squares?

9 A circle can always be drawn through the four vertices of a rectangle. Is this true for all quadrilaterals?

10 Is the angle sum of all polygons with n sides $S_n = 180°(n - 2)$?

11 Do the diagonals of a kite always intersect inside the kite?

12 Is it always true that $|x - y| = |y - x|$?

13 Is it always true that if $n > m$ then $nk > mk$?

DID YOU KNOW?

Bertrand Russell

Bertrand Russell was a philosopher and mathematician who studied logic. He is famous for inventing a paradox in 1901 that is understandably called Russell's Paradox. It was a proof by contradiction for an idea in set theory.

---

1 = −1?

Mathematics is based on definitions and proof. It is very important to be accurate with the definitions when we set up a proof and to use rigour when applying the mathematical algorithms so that the conclusions are valid.

Study the proof that $1 = -1$ below and see if you can find the error. There must be a flaw somewhere because $1 \neq -1$!

Proof that 1 = −1

Consider the complex number $i = \sqrt{-1}$.

Then we know that:

$$i = i$$

$$\sqrt{-1} = \sqrt{-1}$$

$$\sqrt{\frac{1}{-1}} = \sqrt{\frac{-1}{1}}$$

$$\frac{\sqrt{1}}{\sqrt{-1}} = \frac{\sqrt{-1}}{\sqrt{1}}$$

Then cross-multiplying we have:

$$\sqrt{1} \times \sqrt{1} = \sqrt{-1} \times \sqrt{-1}$$

$$\therefore \quad 1 = -1 \text{ QED}$$

2.04 Proofs involving numbers

In this section we will develop some techniques for proving common properties of numbers.

Properties of positive integers

An **even number** can be described by the formula $2n$ where $n \in \mathbb{N}$.

An **odd number** can be described by the formula $2n - 1$ where $n \in \mathbb{N}$.

A **square number** can be described by the formula n^2 where $n \in \mathbb{N}$.

A number, $X (X \in \mathbb{N})$, is **divisible** by another number, p $(p \in \mathbb{N})$, if $\exists\, Y\, (Y \in \mathbb{N})$ such that $X = pY$.

Note that in general, properties such as even, odd, multiples, factors, refer to the **positive integers** only. For instance, we would not usually say -7 is an odd number or that $\frac{4}{9}$ is a square number.

EXAMPLE 13

Consider 2 odd numbers M and N. Prove that their sum is even.

Solution

Proof

Let $M = 2m - 1$ and $N = 2n - 1$ for some $m, n \in \mathbb{N}$.

Then $M + N = (2m - 1) + (2n - 1)$

$$= 2m + 2n - 2$$

$$= 2(m + n - 1)$$

$$= 2k \quad \text{where } k \in \mathbb{N}$$

$\therefore M + N$ is divisible by 2.

$\therefore$ The sum is even. QED.

EXAMPLE 14

If $S_n = \dfrac{n^2 + n}{2}, n \in \mathbb{N}$, prove that $S_n - S_{n-1} = n$.

Solution

Proof

Let $S_n = \dfrac{n^2 + n}{2}, n \in \mathbb{N}$.

Then $\quad S_{n-1} = \dfrac{(n-1)^2 + (n-1)}{2}$

Then $S_n - S_{n-1} = \dfrac{n^2 + n}{2} - \dfrac{(n-1)^2 + (n-1)}{2}$

$$= \dfrac{n^2 + n - [n^2 - 2n + 1 + (n-1)]}{2}$$

$$= \dfrac{n^2 + n - [n^2 - n]}{2}$$

$$= \dfrac{n^2 + n - n^2 + n}{2}$$

$$= \dfrac{2n}{2}$$

$$= n \quad \text{QED}$$

Exercise 2.04 Proofs involving numbers

1 Copy and complete this proof that the product of 2 odd numbers is an odd number.

Proof

Let the 2 odd numbers be M and N.

Let $M = 2m - 1$ and $N = $ _____ for some $m, n \in$ _____

Then $M \times N = ($ _____ $) \times (2n - 1)$

$\qquad = 4mn + $ _____ $+ 1$

$\qquad = 2($ _____ $) + $ _____

$\qquad = 2\square + $ _____ where $\square \in \mathbb{N}$

Since $2\square$ is _____ then $2\square + $ _____ is _____ .

Therefore the _____ is _____ . QED.

2 Use a similar structure as the proof in Question **1** to prove that:

a the sum of 2 even numbers is even

b the product of 2 even numbers is even

c the difference between 2 even numbers is even

d the difference between 2 odd numbers is even

e the sum of an even and an odd number is odd

f the product of an even number and an odd number is even

g the difference between an even number and an odd number is odd

h the square of an odd number is odd

i the square of an even number is even

3 Prove each property.

a $\forall\, a, b \in \mathbb{N}, \exists\, p, q \in \mathbb{Q}: \dfrac{1}{a + b\sqrt{3}} = p + q\sqrt{3}$

b $\forall\, a, b, c, d \in \mathbb{N}, \exists\, p, q \in \mathbb{Q}: \dfrac{a + b\sqrt{2}}{c + d\sqrt{2}} = p + q\sqrt{2}$

4 Prove each formula.

a If $S_n = n^2 + n, n \in \mathbb{N}$, then $S_n - S_{n-1} = 2n$.

b If $S_n = n^2, n \in \mathbb{N}$, then $S_n - S_{n-1} = 2n - 1$.

c If $S_n = 3n - n^2, n \in \mathbb{N}$, then $S_n - S_{n-1} = 4 - 2n$.

d If $S_n = 2n^2 + n, n \in \mathbb{N}$, then $S_n - S_{n-1} = 4n - 1$.

e If $S_n = \dfrac{(n+1)(n+3)}{2}, n \in \mathbb{N}$, then $S_n - S_{n-1} = \dfrac{2n+3}{2}$.

5 For each function, prove the given property.

   a If $f(x) = x^2$, then $f(x) = f(-x)$

   b If $f(x) = x^3$, then $f(-x) = -f(x)$

   c If $f(x) = \dfrac{x^3}{x^2 - 1}$, then $f(-x) = -f(x)$

   d If $f(x) = x \sin x$, then $f(x) = f(-x)$

   e If $f(x) = x^2 \cos x$, then $f(x) = f(-x)$

   f If $f(x) = xe^{-x^2}$, then $f(-x) = -f(x)$

6 Prove each equation for all positive integers m, n, k.

   a $\dfrac{n(n+1)(2n+1)}{6} + (n+1)^2 = \dfrac{(n+1)(n+2)(2n+3)}{6}$

   b $\dfrac{n(n+1)(n+2)}{3} + (n+1)(n+2) = \dfrac{(n+1)(n+2)(n+3)}{3}$

   c $2^{k+1} - 1 + 2^{k+1} = 2^{k+2} - 1$

   d $k \times 2^k + (k+2)2^k = (k+1)2^{k+1}$

   e $\dfrac{m}{m+1} + \dfrac{1}{(m+1)(m+2)} = \dfrac{m+1}{m+2}$

7 Prove the following for all positive integers n.

   a If $3^n - 1 = 2X$ for some $X \in \mathbb{N}$ (that is, $3^n - 1$ is divisible by 2), then $3^{n+1} - 1 = 2Y$ for some $Y \in \mathbb{N}$ (that is, $3^{n+1} - 1$ is also divisible by 2).

   b If $4^n - 1 = 3X$ for some $X \in \mathbb{N}$, then $4^{n+1} - 1 = 3Y$ for some $Y \in \mathbb{N}$.

   c If $n^3 + 2n = 3X$ for some $X \in \mathbb{N}$, then $(n+1)^3 + 2(n+1) = 3Y$ for some $Y \in \mathbb{N}$.

   d If $n^2 + 2n$ is a multiple of 8 for $n \in \mathbb{N}$ where n **is even**, then $(n+2)^2 + 2(n+2)$ is a multiple of 8.

8 Prove each equation.

   a $(x+5)^3(2x-1)^2 + (x+5)^4(2x-1) = (2x-1)(x+5)^3(3x+4)$

   b $\dfrac{4x^3(2-3x)^5 + 15x^4(2-3x)^4}{(2-3x)^{10}} = \dfrac{x^3(8+3x)}{(2-3x)^6}$

9 Prove that $\forall\, x \in \mathbb{R},\, x \neq 0,\, \dfrac{|x|}{x}$ can take 2 values and find for what values of x they occur.

2.05 Proofs involving inequalities

There are many famous results in mathematics that use inequalities. An inequality begins with a very simple yet powerful definition.

Inequality definition

For any two real numbers a and b, $a > b$ if $a - b > 0$.

We can use this definition to prove an inequality, $a > b$, by **considering the difference**, $a - b$, and showing it is positive for all cases.

EXAMPLE 15

For $n \in \mathbb{N}$, prove $\dfrac{1}{n} > \dfrac{1}{n+1}$.

Solution

Proof

Consider the difference, $\dfrac{1}{n} - \dfrac{1}{n+1}$

Then $\dfrac{1}{n} - \dfrac{1}{n+1} = \dfrac{(n+1) - n}{n(n+1)}$

$$= \dfrac{1}{n(n+1)}$$

Now since $n > 0$ and $n + 1 > 0$ then the product $n(n + 1) > 0$.

Therefore $\dfrac{1}{n(n+1)} > 0$.

Since the difference is positive then $\dfrac{1}{n} > \dfrac{1}{n+1} \; \forall \, n \in \mathbb{N}$.

QED.

ISBN 9780170413435

EXAMPLE 16

a Prove that $\dfrac{p^2 + q^2}{2} \geq pq \ \forall \ p, q \in \mathbb{R}$.

b Hence prove $x^2 + y^2 + z^2 \geq xy + yz + zx \ \forall \ x, y, z \in \mathbb{R}$.

Solution

a **Proof**

Consider the difference,

$$\frac{p^2 + q^2}{2} - pq = \frac{p^2 + q^2 - 2pq}{2}$$

$$= \frac{p^2 - 2pq + q^2}{2}$$

$$= \frac{(p - q)^2}{2}$$

Now since $(p - q)^2 \geq 0 \ \forall \ p, q \in \mathbb{R}$, [equality when $p = q$] then $\dfrac{(p - q)^2}{2} \geq 0$.

Therefore $\dfrac{p^2 + q^2}{2} \geq pq \ \forall \ p, q \in \mathbb{R}$.

b We have just proved the result $\dfrac{p^2 + q^2}{2} \geq pq$, where the p and q represent any real variables. Rearranging, we can also write $p^2 + q^2 \geq 2pq$.

This means that, using the result from part **a**, we can write:

$x^2 + y^2 \geq 2xy$

$y^2 + z^2 \geq 2yz$

$z^2 + x^2 \geq 2zx$

Then adding all 3 LHS and adding all 3 RHS we have:

$2x^2 + 2y^2 + 2z^2 \geq 2(xy + yz + zx)$

Now dividing both sides by 2 we have

$x^2 + y^2 + z^2 \geq xy + yz + zx \ \forall \ x, y, z \in \mathbb{R}$. QED.

EXAMPLE 17

a Prove $\dfrac{x}{y}+\dfrac{y}{x}\geq 2 \;\forall\; x,y \in \mathbb{R}, x,y > 0.$

b Hence prove $\dfrac{1}{x}+\dfrac{1}{y}\geq \dfrac{4}{x+y} \;\forall\; x,y \in \mathbb{R}, x,y > 0.$

Solution

a **Proof**

Consider the difference

$$\dfrac{x}{y}+\dfrac{y}{x}-2 = \dfrac{x^2+y^2-2xy}{xy}$$

$$= \dfrac{(x-y)^2}{xy}$$

Now, since $(x-y)^2 \geq 0$ [equality when $x=y$] and $xy > 0$, $\forall\; x,y \in \mathbb{R}, x,y > 0$,

then $\dfrac{(x-y)^2}{xy} \geq 0.$

Therefore $\dfrac{x}{y}+\dfrac{y}{x}\geq 2 \;\forall\; x,y \in \mathbb{R}, x,y > 0.$

b **Proof**

Consider the product

$$(x+y)\left(\dfrac{1}{x}+\dfrac{1}{y}\right) = 1+\dfrac{x}{y}+\dfrac{y}{x}+1$$

$$= 2+\dfrac{x}{y}+\dfrac{y}{x}$$

$$\geq 2+2 \text{ since } \dfrac{x}{y}+\dfrac{y}{x} \geq 2 \qquad \text{from } \textbf{a}$$

$$\geq 4$$

Now, since $x+y > 0 \;\forall\; x,y \in \mathbb{R}, x,y > 0$, then dividing by $(x+y)$ we have

$\dfrac{1}{x}+\dfrac{1}{y}\geq \dfrac{4}{x+y}.$ \qquad QED

Recall the definition of absolute value.

Definition of absolute value

$$|x| = \begin{cases} x, & \text{for } x \geq 0 \\ -x, & \text{for } x < 0 \end{cases}$$

We can deduce the following results by considering positive and negative cases of a and b:

Properties of absolute value

$|a| \geq a, \forall \, a \in \mathbb{R}$

$|a||b| = |ab|, \forall \, a, b \in \mathbb{R}$

$|a|^2 = a^2, \forall \, a \in \mathbb{R}$

$|ab| \geq ab, \forall \, a, b \in \mathbb{R}$

$a^2 \geq b^2 \Leftrightarrow |a| \geq |b|, \forall \, a, b \in \mathbb{R}$

This leads us to the important **triangle inequality**.

The triangle inequality

$$|x| + |y| \geq |x + y|, \forall \, x, y \in \mathbb{R}$$

Proof

Consider the expression

$$\begin{aligned}
\left(|x| + |y|\right)^2 &= |x|^2 + 2|x||y| + |y|^2 \\
&\geq |x|^2 + 2xy + |y|^2 \\
&\geq x^2 + 2xy + y^2 \\
&\geq (x + y)^2 \\
&\geq \left(|x + y|\right)^2
\end{aligned}$$

Then, since both sides are positive, it follows that

$$|x| + |y| \geq |x + y|, \forall \, x, y \in \mathbb{R} \qquad \text{QED}$$

The triangle inequality can also be written in the form

$$|x| - |y| \leq |x - y|, \forall \, x, y \in \mathbb{R}$$

Chapter 1, *Complex numbers*, and Chapter 3, *3D vectors*, show how the triangle inequality describes the 2 diagonals of the parallelogram rule, or that the sum of the lengths of 2 sides of a triangle is always greater than the length of the 3rd side.

ISBN 9780170413435

Exercise 2.05 Proofs involving inequalities

1 By considering the difference, prove that $a^2 + b^2 \geq 2ab$, $\forall\, a, b \in \mathbb{R}$.

2 By choosing a suitable substitution for p and q in the result $p^2 + q^2 \geq 2pq$, prove each statement below.

   a $x^4 + y^4 \geq 2x^2y^2$, $\forall\, x, y \in \mathbb{R}$

   b $x^2y^2 + w^2v^2 \geq 2xywv$, $\forall\, x, y, w, v \in \mathbb{R}$

   c $\dfrac{1}{x^2} + \dfrac{1}{y^2} \geq \dfrac{2}{xy}$, $\forall\, x, y \in \mathbb{R}, x, y \neq 0$

   d $\dfrac{1}{x^4} + \dfrac{1}{y^4} \geq \dfrac{2}{x^2y^2}$, $\forall\, x, y \in \mathbb{R}, x, y \neq 0$

3 The AM–GM inequality, where AM = arithmetic mean, GM = geometric mean, states that $\dfrac{x + y}{2} \geq \sqrt{xy}$, $\forall\, x, y \in \mathbb{R}$, such that $x, y > 0$. Prove this result.

4 Prove each property.

   a If $a > b$ and $b > c$ then $a > c$

   b If $a > b$ and $c > 0$ then $ac > bc$

   c If $a > b$ and $c < 0$ then $ac < bc$

   d If $a > b$ and $b > 0$ then $ab > b^2$

   e If $a > b > 0$ and $c > d > 0$ then $ac > bd$

   f If $a > b$ and $b > c$ then $ac + bd > ad + bc$

   g If $a > b$ and $b > 0$ then $\dfrac{1}{a} < \dfrac{1}{b}$

5 By considering the difference, prove each statement is true $\forall\, n \in \mathbb{N}$.

   a If $T_n = \dfrac{1}{n}$, then $T_n > T_{n+1}$ **b** If $T_n = \dfrac{n}{n+1}$, then $T_n < T_{n+1}$

   c If $T_n = \dfrac{n^2 - 1}{n^2 + 1}$, then $T_n < T_{n+1}$

6 Prove each result, assuming that $a^2 + b^2 \geq 2ab$.

   a $(a + b)^2 \geq 4ab$, $\forall\, a, b \in \mathbb{R}$

   b $ab + cd \geq 2\sqrt{abcd}$, $\forall\, a, b, c, d \in \mathbb{R}$

   c $(a+b)\left(\dfrac{1}{a} + \dfrac{1}{b}\right) \geq 4$, $\forall\, a, b \in \mathbb{R}, a, b > 0$

   d $\dfrac{\sqrt{ab} + \sqrt{cd}}{2} \geq \sqrt[4]{abcd}$, $\forall\, a, b, c, d \in \mathbb{R}, a, b, c, d > 0$

7 Prove that, $\forall\, a, b \in \mathbb{R}, a, b > 0$:

   a $a + \dfrac{1}{a} \geq 2$ **b** $\dfrac{a}{b} + \dfrac{b}{a} \geq 2$ **c** $\dfrac{1}{a^2} + \dfrac{1}{b^2} \geq \dfrac{4}{a^2 + b^2}$

   d $\dfrac{a + b + c + d}{4} > \sqrt[4]{abcd}$, $\forall\, a, b, c, d \in \mathbb{R}, a, b, c, d > 0$

8 Prove each inequality.

a $x\sqrt{x} + 1 \geq x + \sqrt{x}, \forall x \in \mathbb{R}, x \geq 0$

b **i** For $1 \leq p \leq q$, where $p, q \in \mathbb{N}$, that $p(q - p + 1) \geq q$

ii Hence for $1 \leq r \leq s$ where $r, s \in \mathbb{N}$, that $\sqrt{s} \leq \sqrt{r(s - r + 1)} \leq \dfrac{s + 1}{2}$.

9 Prove that $(kn - mp)^2 \geq (k^2 - m^2)(n^2 - p^2), \forall k, m, n, p \in \mathbb{R}$.

Hence prove that $(k^3 - m^3)^2 \geq (k^2 - m^2)(k^4 - m^4)$.

10 Use similar techniques to those shown in Examples 16 and 17 to prove that:

a **i** $x^4 + y^4 + w^4 + z^4 \geq 4xywz$ for real x, y, z, w, and hence:

ii if $x^4 + y^4 + w^4 + z^4 < 4$ and $x, y, z, w > 0$ then $\dfrac{1}{x^4} + \dfrac{1}{y^4} + \dfrac{1}{w^4} + \dfrac{1}{z^4} > 4$

iii $(x + y + z + w)^4 \geq 256xywz$

b Assume $a, b, c \in \mathbb{R}$ such that $a, b, c > 0$.

Prove that:

$$(a + b)(b + c)(c + a) = c(a - b)^2 + b(c - a)^2 + a(b - c)^2 + 8abc$$

Hence prove that:

$$(a + b)(b + c)(c + a) > 8abc.$$

c Prove for $a, b, c \in \mathbb{R}$ such that $a, b, c > 0$:

$$(a + b + c)\left(\dfrac{1}{a} + \dfrac{1}{b} + \dfrac{1}{c}\right) \geq 9$$

d Prove for $a, b, c \in \mathbb{R}$ such that $a, b, c > 0$:

i $a^3 + b^3 + c^3 - 3abc = (a + b + c)(a^2 + b^2 + c^2 - ab - bc - ca)$

ii $(a - b)^2 + (c - a)^2 + (b - c)^2 = 2(a^2 + b^2 + c^2 - ab - bc - ca)$

Hence prove that:

iii $a^3 + b^3 + c^3 \geq 3abc$　　　**iv** $\dfrac{a + b + c}{3} \geq \sqrt[3]{abc}$　　　**v** $\dfrac{a}{b} + \dfrac{b}{c} + \dfrac{c}{a} \geq 3$

e Prove for $a, b, c \in \mathbb{R}$ such that $a, b, c > 0$:

$$(a + b + c)^3 \geq 27abc.$$

f Prove for $a, b, c, d \in \mathbb{R}$ such that $a, b, c, d > 0$:

$$\dfrac{a}{b} + \dfrac{b}{c} + \dfrac{c}{d} + \dfrac{d}{a} \geq 4.$$

When does equality hold?

11 Use the fact that $|x| + |y| \geq |x + y|, \forall x, y \in \mathbb{R}$, to prove that:

a $|z + w| + |z - w| \geq |2z|, \forall z, w \in \mathbb{R}$

b $|x + z| + |y - z| \geq |x + y|, \forall x, y, z \in \mathbb{R}$

c $|x - z| \geq |x - y| - |z - y|, \forall x, y, z \in \mathbb{R}$

1 Write each if-then statement in the form $P \Rightarrow Q$ and state P and Q.

    a If I get a lot of sleep, then I am healthy.

    b If a polygon has 5 sides, then it is a pentagon.

    c If the teacher is nice, then I will learn.

2 Write the converse of each statement,

    a $A \Rightarrow B$ **b** $\neg P \Rightarrow Q$ **c** $N \Rightarrow \neg M$ **d** $\neg B \Rightarrow \neg F$

    e If I can save money, then I can buy a car.

    f If my computer is broken, then I am bored.

    g If $a = b$ then $a^3 = b^3$.

3 Define 'iff'. Give an example.

4 For each statement, find the converse and determine if the statement is an equivalence. If it is an equivalence, then write an iff statement.

    a If a quadrilateral has equal diagonals, then it is a square.

    b If $x > 1$, then $\dfrac{1}{x} < 1$.

    c If I pass my exams, then I study hard.

    d If $a = 3$, then $a^2 = 9$.

    e If a triangle has 2 equal angles, then it is isosceles.

5 Write the negation of each statement.

    a It is raining.   **b** The apple is not ripe.

    c Koalas are cute.   **d** Some people are sexist.

    e They are all correct.   **f** $x \leq 4$

    g $p \in \mathbb{N}$

6 Joe was asked what was the negation of the statement 'There were more than 10'. He said, 'There were less than 10.' Is he correct?

7 Write the contrapositive of each statement.

    a $A \Rightarrow B$ **b** $\neg P \Rightarrow Q$ **c** $N \Rightarrow \neg M$ **d** $\neg B \Rightarrow \neg F$

    e If the boy has red hair, then he has blue eyes.

    f If the country is rich, then the citizens have money.

    g If a quadrilateral is a kite, then the adjacent sides are equal in length.

    h If $x = y$, then $x^2 = y^2$

    i If $a \in \mathbb{N}$, then $a \in \mathbb{Z}$

8 Explain why a statement and its contrapositive are equivalences. Give an example.

9 Determine whether each statement is true by considering its contrapositive.

a If $a^2 \neq b^2$ then $a \neq b$.

b If the car's battery is flat, then the car does not start.

c If a number is an integer, then it is rational.

d If a quadrilateral has diagonals that bisect each other at right angles, then it is a rhombus.

e If $a > b$, then $ab > b^2$.

f If an animal lives in the water, then it is a fish.

10 Write each sentence in mathematical notation.

a For all positive real numbers x and y, if x is greater than y, then the square of x is greater than the square of y.

b There exists a rational number c between integers a and b where $a < b$, such that c is the average of a and b.

c Let n be a positive integer such that for all n, $1 + 2 + 3 + \ldots + n = \dfrac{n(n+1)}{2}$.

11 Write each mathematical statements in words.

a $\forall\, n, m \in \mathbb{Z}, n, m > 0$, if $n < m$, then $\dfrac{1}{n} > \dfrac{1}{m}$

b $\forall\, a, b \in \mathbb{R}, a^2 + b^2 \geq 2ab$

c $\forall\, p, q \in \mathbb{Q}, p < q, \exists\, r \in \mathbb{R}: p < r < q$

12 Prove by contradiction that:

a $\sqrt{11}$ is irrational

b $\log_3 4$ is irrational

c $2 + \sqrt{5}$ is irrational

d The diagonals of a square are perpendicular.

e A triangle with sides $(t^2 - 1)$, $2t$, $(t^2 + 1)$ where $t > 0$ forms a right-angled triangle.

13 Find a counterexample to prove that each statement is false.

a $\forall\, x, y \in \mathbb{Q}, x, y \neq 0$, if $x^2 = y^2$ then $x = y$

b $\forall\, n \in \mathbb{N}, n > \dfrac{1}{n}$.

c If an animal sheds its skin, then it is a snake.

d If $a^2 + b^2 = c^2$ then a, b, c are the lengths of the sides of a right-angled triangle.

e $\forall\, k \in \mathbb{N}, k(k - 1) + 17$ is prime.

f $\forall\, c \in \mathbb{R}, (c \leq 1) \Rightarrow (c^2 \leq 1)$

2. Mathematical proof

14 It is asserted that if $a \geq b$ and $c \geq d$, then $ac \geq bd$, $\forall\, a, b, c, d \in \mathbb{R}$. Either prove that it is true or find a counterexample to prove it false.

15 If $x > y\ \forall\, x, y \in \mathbb{R}$, $x, y \neq 0$, then $\dfrac{1}{x^2} < \dfrac{1}{y^2}$. Prove this result or find a counterexample to disprove it.

16 Prove each property.

   a   If $m \in \mathbb{N}$, then $m(m + 1)$ is always even.

   b   If $n \in \mathbb{N}$, then $n(n + 1)(n + 2)$ is always divisible by 6.

   c   If $n \in \mathbb{N}$ and n is odd, then $n(n + 2) + (n + 2)(n + 4)$ is always even.

17 Factorise $k^6 - m^6$ in two ways. Hence prove $k^4 + k^2m^2 + m^4 = (k^2 + km + m^2)(k^2 - km + m^2)$.

   Note: $a^2 - b^2 = (a + b)(a - b)$, $a^3 - b^3 = (a - b)(a^2 + ab + b^2)$

18 Prove each statement.

   a   If $S_n = \dfrac{4^n - 1}{3}$ and $n \in \mathbb{N}$, $n \geq 1$, then $S_n - S_{n-1} = 4^{n-1}$

   b   If $S_n = 2^{n+1} - n - 2$ and $n \in \mathbb{N}$, $n \geq 1$, then $S_{n+1} - S_n = 2^{n+1} - 1$

   c $\forall\, k \in \mathbb{N}$, $\dfrac{k}{3k+1} + \dfrac{1}{(3k+1)(3k+4)} = \dfrac{k+1}{3k+4}$

   d   If $f(x) = x^2 \sin x$, then $f(-x) = -f(x)$

19 Prove each property.

   a   If $a > b$ and $a, b \neq 0$ then $ab^2 > b^3$

   b   If $a - b > b - c$ then $b < \dfrac{a+c}{2}$

   c $\left| x \right| \geq x$ for $x \in \mathbb{R}$

20 Prove each inequality by considering the difference.

   a   If $a, b \in \mathbb{R}$, then $\dfrac{a^2 + b^2}{2} \geq ab$

   b   If $k \in \mathbb{N}$ and $T_k = \dfrac{k}{2k+1}$, then $T_k < T_{k+1}$

21 Prove each inequality for $a, b, c, d \in \mathbb{R}$ such that $a, b, c, d > 0$.

   a $\sqrt{\dfrac{a^2 + b^2}{2}} \geq \dfrac{a+b}{2}$

   b $\sqrt{\dfrac{a^2 + b^2 + c^2 + d^2}{4}} \geq \sqrt[4]{abcd}$

   c   If $a + b = 1$ then:

      i $\dfrac{1}{a} + \dfrac{1}{b} \geq 4$

      ii $\dfrac{1}{a^2} + \dfrac{1}{b^2} \geq 8$ $\left[\text{Hint: consider the product } (x + y)\left(\dfrac{1}{x} + \dfrac{1}{y}\right)\right]$

22 Prove $\dfrac{9}{a+b+c} \leq \left(\dfrac{1}{a}+\dfrac{1}{b}+\dfrac{1}{c}\right)$ for $a, b, c \in \mathbb{R}$ such that $a, b, c > 0$.

23 Prove for $a, b, c \in \mathbb{R}$ such that $a, b, c > 0$ and $a + b + c = 1$:

a $\left(\dfrac{1}{a}+\dfrac{1}{b}+\dfrac{1}{c}\right) \geq 9$ **b** $\left(\dfrac{1}{a^2}+\dfrac{1}{b^2}+\dfrac{1}{c^2}\right) \geq 27$

24 Prove each inequality.

a $|x| - |y| \leq |x - y|, \forall\, x, y \in \mathbb{R}$

b $|x| + |y| + |z| \geq |x + y + z|, \forall\, x, y, z \in \mathbb{R}$

c $|x - y| - |z - y| \leq |x - z|, \forall\, x, y, z \in \mathbb{R}$

3.

3D VECTORS

Aircraft vectoring is a service provided to individual aircraft by air traffic control. The controller determines the best pattern for the aircraft to fly on take-off, during a flight and at landing.

The aircraft follows specific headings (directions) at particular times and for set durations. Aircraft are vectored to apply adequate separation, improve traffic flow, to comply with noise regulations, to avoid hazardous weather, and to assist in arrival and departure schedules.

CHAPTER OUTLINE

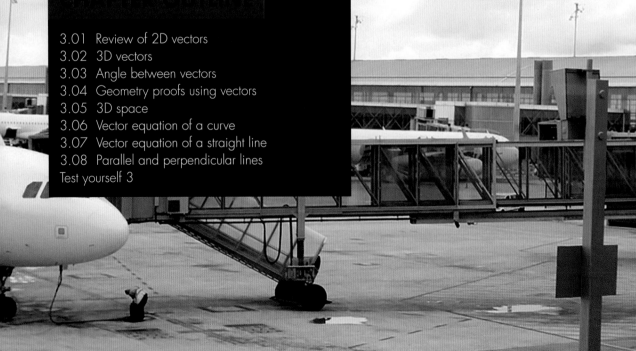

IN THIS CHAPTER YOU WILL:

- extend our knowledge and skills with vectors in 2 dimensions to vectors in 3 dimensions
- add and subtract 3D vectors, and multiply by a vector by a scalar
- calculate the magnitude of a 3D vector
- determine the scalar (dot) product of 2 vectors
- determine the angle between 2 vectors
- use vectors to prove geometrical relationships
- determine lines and curves in 3 dimensions, including spheres, using parametric equations
- determine parallel and perpendicular vectors
- determine whether a given point lies on a vector

TERMINOLOGY

Cartesian equation: An equation for a line or plane in terms of x, y and z.

component form: Representation of a vector $\begin{pmatrix} a \\ b \end{pmatrix}$ in the form $a\underset{\sim}{i} + b\underset{\sim}{j}$ where $\underset{\sim}{i}$ is a unit vector in the x direction and $\underset{\sim}{j}$ is a unit vector in the y direction.

parallel vectors: Vectors that have the same or opposite direction.

parametric equations: A set of equations that express a set of quantities as explicit functions of a number of independent variables, known as **parameters**.

perpendicular vectors: Lines that have a right angle between their directions.

scalar: A quantity that has magnitude but no direction.

scalar (or dot) product: The product of 2 vectors as a scalar or value (not a vector).

unit vector: A vector with magnitude 1. The standard unit vectors are $\underset{\sim}{i}$ in the x direction and $\underset{\sim}{j}$ in the y direction.

vector: A quantity with both magnitude and direction. A vector can be represented as $\underset{\sim}{a}, \boldsymbol{a}$ or $\overrightarrow{AB}$.

3.01 Review of 2D vectors

A **vector** is a quantity with both magnitude and direction.

All geometric situations can be transformed into algebraic language using vectors. Instead of combining points and lines using geometry, we perform algebraic operations on vectors. Vectors obey algebraic laws similar to those we use for numbers. For example, if $\underset{\sim}{a}$ and $\underset{\sim}{b}$ are vectors, then $\underset{\sim}{a} + \underset{\sim}{b} = \underset{\sim}{b} + \underset{\sim}{a}$.

When $\underset{\sim}{v}$ is represented by interval AB, we write $\underset{\sim}{v} = \overrightarrow{AB}$. Thus, $\overrightarrow{AB}$ denotes the vector determined by the directed line segment AB.

We often choose a fixed origin O and then represent each vector $\underset{\sim}{u}$ as $\overrightarrow{OP}$. This is a directed line segment from O to P, called a **position vector**.

There are many examples where vectors are used in mathematics, for example displacement, velocity, acceleration, force and momentum. For example, force has both magnitude and direction and so can be represented by a directed line segment.

The **magnitude** or length of a vector $\underset{\sim}{v}$ is $|\underset{\sim}{v}|$. If $|\underset{\sim}{v}| = 1$ then $\hat{\underset{\sim}{v}}$ is called a **unit vector**.

The zero vector $\underset{\sim}{0}$ is a directed line segment from a point to itself. That is, $\overrightarrow{PP} = \underset{\sim}{0}$.

This vector has magnitude 0 and can have any direction.

Numbers or quantities having magnitude but **not** direction are referred to as **scalars**.

Vector addition

Given 2 vectors $u = \overrightarrow{AB}$ and $v = \overrightarrow{BC}$, then $u + v$ is the resultant vector given by $w = \overrightarrow{AC}$.

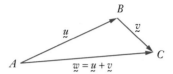

We can write $\overrightarrow{AB} + \overrightarrow{BC} = \overrightarrow{AC}$.

The parallelogram law gives the rule for the vector addition of u and v. The sum of the vectors u and v is the longer diagonal of the parallelogram.

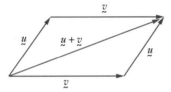

The laws of vector addition

$u + v = v + u$	Commutative law
$(u + v) + w = u + (v + w)$	Associative law
$u + 0 = u$	Additive identity
$\lvert u + v \rvert \leq \lvert u \rvert + \lvert v \rvert$	Triangle inequality

Vector subtraction

Given 2 vectors $u = \overrightarrow{AB}$ and $-v = \overrightarrow{BC}$, then $u + (-v)$ is given by $w = \overrightarrow{AC}$.

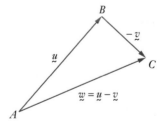

We can write $\overrightarrow{AB} + \overrightarrow{BC} = \overrightarrow{AC}$, so we get $u - v = w$.

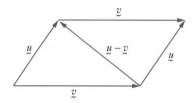

The parallelogram law gives the rule for the vector subtraction of u and v.

The difference between the vectors v and u is the shorter diagonal of the parallelogram, pointing from the second vector to the first.

Scalar multiplication of vectors

If a is real number, or scalar, and v is a vector, then av is a vector whose magnitude is $|a||v|$, having the same or opposite direction to v according to whether $a > 0$ or $a < 0$.

Graphically we can show scalar multiplication of vectors as below:

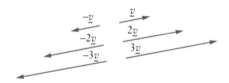

Components of vectors

A vector can be written as a **column vector**. For example, $\begin{pmatrix} 3 \\ 2 \end{pmatrix}$ is the vector 3 across and 2 up or $3i + 2j$, where i and j are the unit vectors in the x and y directions respectively. We say that $3i$ and $2j$ are the **components of the vector**. Vectors in column and component form can easily be added, subtracted and multiplied.

For example, $(3i + 2j) + (5i - 3j) = 8i - 1j$. Similarly, $\begin{pmatrix} 3 \\ 2 \end{pmatrix} + \begin{pmatrix} 5 \\ -3 \end{pmatrix} = \begin{pmatrix} 8 \\ -1 \end{pmatrix}$.

Magnitude of a position vector

Magnitude of the vector $u = \begin{pmatrix} x \\ y \end{pmatrix}$ is:

$$|u| = \sqrt{x^2 + y^2}$$

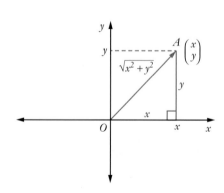

MATHS IN FOCUS 12. Mathematics Extension 2   ISBN 9780170413435

Direction of a position vector

The direction of vector $\begin{pmatrix} x \\ y \end{pmatrix}$ is θ, the angle of

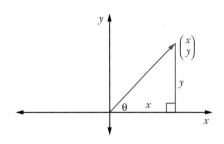

inclination of the vector with the positive x-axis, measured anticlockwise.

$$\tan \theta = \frac{y}{x}$$

$$0° \leq \theta < 360°$$

Scalar (dot) product of vectors

Two vectors can be combined as a scalar product. When we calculate the scalar product of 2 vectors the result is a **scalar** rather than a vector. The scalar product applies the directional growth of one vector to another.

Scalar product

The scalar product of $\underset{\sim}{u}$ and $\underset{\sim}{v}$ is:

$$\underset{\sim}{u} \cdot \underset{\sim}{v} = x_1 x_2 + y_1 y_2$$

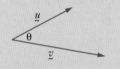

or

$$\underset{\sim}{u} \cdot \underset{\sim}{v} = |\underset{\sim}{u}||\underset{\sim}{v}| \cos \theta$$

where θ is the angle between $\underset{\sim}{u}$ and $\underset{\sim}{v}$.

As the symbol for the scalar product is the dot (·), we sometimes call it the **dot product**.

Proof of the cosine form

In component form,

$$\underset{\sim}{u} \cdot \underset{\sim}{v} = x_1 x_2 + y_1 y_2$$

Using the cosine rule,

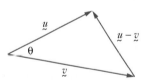

$$|\underset{\sim}{u} - \underset{\sim}{v}|^2 = |\underset{\sim}{u}|^2 + |\underset{\sim}{v}|^2 - 2|\underset{\sim}{u}||\underset{\sim}{v}| \cos \theta$$

Now substituting $\underset{\sim}{u} = x_1 \underset{\sim}{i} + y_1 \underset{\sim}{j}$ and $\underset{\sim}{v} = x_2 \underset{\sim}{i} + y_2 \underset{\sim}{j}$:

$$(x_1 - x_2)^2 + (y_1 - y_2)^2 = x_1^2 + y_1^2 + x_2^2 + y_2^2 - 2|\underset{\sim}{u}||\underset{\sim}{v}| \cos \theta$$

$$-2x_1 x_2 - 2y_1 y_2 = -2|\underset{\sim}{u}||\underset{\sim}{v}| \cos \theta$$

$$x_1 x_2 + y_1 y_2 = |\underset{\sim}{u}||\underset{\sim}{v}| \cos \theta$$

Hence, $\underset{\sim}{u} \cdot \underset{\sim}{v} = |\underset{\sim}{u}||\underset{\sim}{v}| \cos \theta$.

Parallel and perpendicular vectors

Two vectors $\underset{\sim}{u}$ and $\underset{\sim}{v}$ are **parallel vectors** if and only if they are scalar multiples of one another; that is, $\underset{\sim}{u} = k\underset{\sim}{v}$, k is a constant not equal to zero.

$\theta = 0°$ or $180°$ so $\underset{\sim}{u} \cdot \underset{\sim}{v} = |\underset{\sim}{u}||\underset{\sim}{v}|$ for vectors in like directions and $\underset{\sim}{u} \cdot \underset{\sim}{v} = -|\underset{\sim}{u}||\underset{\sim}{v}|$ for vectors in unlike directions.

In component form, $x_2\underset{\sim}{i} + y_2\underset{\sim}{j} = k(x_1\underset{\sim}{i} + y_1\underset{\sim}{j})$.

Two vectors $\underset{\sim}{u}$ and $\underset{\sim}{v}$ are **perpendicular vectors** if and only if their scalar product is equal to 0.

$\theta = 90°$ or $270°$ so $\underset{\sim}{u} \cdot \underset{\sim}{v} = |\underset{\sim}{u}||\underset{\sim}{v}|\cos\theta = 0$.

Exercise 3.01 Review of 2D vectors

1 How many different vectors are there below?

2 Copy and complete each statement.
- **a** Geometrically, the sum of 2 vectors is the longer diagonal …
- **b** Geometrically, the difference of 2 vectors is …
- **c** Two vectors are **parallel** if and only if …
- **d** Two vectors are **perpendicular** if and only if …

3 Determine the magnitude and direction of each vector in exact form.
- **a** $2\underset{\sim}{j}$
- **b** $5\underset{\sim}{i}$
- **c** $10\underset{\sim}{i} - 5\underset{\sim}{j}$
- **d** $-4\underset{\sim}{i} + 4\underset{\sim}{j}$
- **e** $2\underset{\sim}{i} + 2\sqrt{3}\underset{\sim}{j}$

4 Express the vector in terms of components if the magnitude and direction are, respectively:
- **a** 6 units, $45°$
- **b** 8 units, $30°$
- **c** 2 units, $135°$
- **d** 10 units, $-60°$
- **e** 6 units, $-150°$

5 Write each vector in Question 4 in column vector form.

6 Find the scalar (dot) product of each pair of vectors.
- **a** $\begin{pmatrix} 3 \\ 0 \end{pmatrix}$ and $\begin{pmatrix} 5 \\ 0 \end{pmatrix}$
- **b** $\begin{pmatrix} 2 \\ 0 \end{pmatrix}$ and $\begin{pmatrix} 0 \\ 7 \end{pmatrix}$
- **c** $\underset{\sim}{i} + 2\underset{\sim}{j}$ and $2\underset{\sim}{i} + \underset{\sim}{j}$
- **d** $\underset{\sim}{i} - 2\underset{\sim}{j}$ and $\underset{\sim}{i} + 2\underset{\sim}{j}$
- **e** $6\underset{\sim}{i} + 2\underset{\sim}{j}$ and $3\underset{\sim}{i} - 4\underset{\sim}{j}$

MATHS IN FOCUS 12. Mathematics Extension 2   ISBN 9780170413435

7 Find the scalar product of each pair of vectors, and hence find the angle between the vectors to the nearest degree.

a $\begin{pmatrix} 3 \\ 4 \end{pmatrix}$ and $\begin{pmatrix} 4 \\ -3 \end{pmatrix}$ **b** $2\underset{\sim}{i} + 8\underset{\sim}{j}$ and $\underset{\sim}{i} + 4\underset{\sim}{j}$

c $-\underset{\sim}{i} + \underset{\sim}{j}$ and $3\underset{\sim}{i} - 3\underset{\sim}{j}$ **d** $\begin{pmatrix} -5 \\ -4 \end{pmatrix}$ and $\begin{pmatrix} 4 \\ 0 \end{pmatrix}$

e $-2\underset{\sim}{i} + 3\underset{\sim}{j}$ and $\underset{\sim}{i} - 2\underset{\sim}{j}$

8 Which pair of vectors are perpendicular?

a $\underset{\sim}{u} = 5\underset{\sim}{i} + 2\underset{\sim}{j}, \underset{\sim}{v} = \underset{\sim}{j}, \underset{\sim}{w} = -2\underset{\sim}{i}$ **b** $\underset{\sim}{u} = \begin{pmatrix} 5 \\ 2 \end{pmatrix}, \underset{\sim}{v} = \begin{pmatrix} -2 \\ 1 \end{pmatrix}, \underset{\sim}{w} = \begin{pmatrix} -2 \\ 5 \end{pmatrix}$

9 Which pair of vectors are parallel?

a $\underset{\sim}{u} = \begin{pmatrix} 5 \\ 3 \end{pmatrix}, \underset{\sim}{v} = \begin{pmatrix} -4 \\ 6 \end{pmatrix}, \underset{\sim}{w} = \begin{pmatrix} 2 \\ -3 \end{pmatrix}$ **b** $\underset{\sim}{u} = 3\underset{\sim}{i} + \underset{\sim}{j}, \underset{\sim}{v} = -2\underset{\sim}{i} + 6\underset{\sim}{j}, \underset{\sim}{w} = 6\underset{\sim}{i} + 2\underset{\sim}{j}$

10 Consider the points $P(0, 2)$, $Q(2, 6)$ and $R(4, 10)$. Express $\overrightarrow{PQ}$ and $\overrightarrow{PR}$ as component vectors. Hence, show that P, Q and R are collinear.

DID YOU KNOW?

Vector images

A photograph taken with a digital camera is made up of many pixels, and each pixel may be a different colour. These images are called raster images. Enlarging raster images is a problem because they become pixellated when you zoom in. Raster images also require large files because data for each point must be saved.

A **vector image** is made up of points, lines and curves that are defined by mathematical equations. Vector images have the advantage of not pixellating with enlargement. They also require less storage space. Vector images have the same small file size no matter how much enlargement or reduction is used.

iStock.com/4kodiak

ISBN 9780170413435

3.02 3D vectors

As in the 2D plane, vectors are used to represent displacement, velocity and acceleration in 3D space. In 3 dimensions, the unit vectors are $\underset{\sim}{i}$, $\underset{\sim}{j}$ and $\underset{\sim}{k}$ in the directions of the x-, y- and z-axes respectively. In a 3D coordinate system, the x-y Cartesian plane 'sits on ground level' while the z-axis points upwards. Hence, the position vector of a typical point $P(x, y, z)$ from the origin is
$$\underset{\sim}{r} = \overrightarrow{OP} = x\underset{\sim}{i} + y\underset{\sim}{j} + z\underset{\sim}{k}.$$

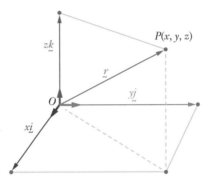

Using Pythagoras' theorem in 3D we can obtain:
$$r^2 = \left|\overrightarrow{OP}\right| = x^2 + y^2 + z^2$$

So, the unit vector is given by:
$$\hat{\underset{\sim}{r}} = \frac{\underset{\sim}{r}}{|\underset{\sim}{r}|} = \frac{1}{\sqrt{x^2 + y^2 + z^2}} (x\underset{\sim}{i} + y\underset{\sim}{j} + z\underset{\sim}{k})$$

Magnitude and unit vectors

The vector $\underset{\sim}{r} = x\underset{\sim}{i} + y\underset{\sim}{j} + z\underset{\sim}{k}$ has magnitude:
$$|r| = \sqrt{x^2 + y^2 + z^2}$$

The unit vector is:
$$\hat{\underset{\sim}{r}} = \frac{x\underset{\sim}{i} + y\underset{\sim}{j} + z\underset{\sim}{k}}{\sqrt{x^2 + y^2 + z^2}}$$

EXAMPLE 1

Find the position vector of $P(-1, 2, 3)$ in component form.

Solution

Components in the x, y and z directions are -1, 2 and 3 respectively.

Hence, $\overrightarrow{OP} = -\underset{\sim}{i} + 2\underset{\sim}{j} + 3\underset{\sim}{k}$

Addition, subtraction and scalar multiplication for vectors

Let $\underset{\sim}{u} = x_1\underset{\sim}{i} + y_1\underset{\sim}{j} + z_1\underset{\sim}{k}$ and $\underset{\sim}{v} = x_2\underset{\sim}{i} + y_2\underset{\sim}{j} + z_2\underset{\sim}{k}$

Adding, we get:

$$\underset{\sim}{u} + \underset{\sim}{v} = (x_1\underset{\sim}{i} + y_1\underset{\sim}{j} + z_1\underset{\sim}{k}) + (x_2\underset{\sim}{i} + y_2\underset{\sim}{j} + z_2\underset{\sim}{k})$$
$$= (x_1 + x_2)\underset{\sim}{i} + (y_1 + y_2)\underset{\sim}{j} + (z_1 + z_2)\underset{\sim}{k}$$

Subtracting, we get:

$$\underset{\sim}{u} - \underset{\sim}{v} = (x_1\underset{\sim}{i} + y_1\underset{\sim}{j} + z_1\underset{\sim}{k}) - (x_2\underset{\sim}{i} + y_2\underset{\sim}{j} + z_2\underset{\sim}{k})$$
$$= (x_1 - x_2)\underset{\sim}{i} + (y_1 - y_2)\underset{\sim}{j} + (z_1 - z_2)\underset{\sim}{k}$$

If we multiply by a scalar quantity c we get:

$$c\underset{\sim}{u} = c(x_1\underset{\sim}{i} + y_1\underset{\sim}{j} + z_1\underset{\sim}{k})$$
$$= (cx_1)\underset{\sim}{i} + (cy_1)\underset{\sim}{j} + (cz_1)\underset{\sim}{k}$$

EXAMPLE 2

If $\underset{\sim}{u} = 2\underset{\sim}{i} - 2\underset{\sim}{j} + \underset{\sim}{k}$, find:

a $|\underset{\sim}{u}|$

b $\hat{\underset{\sim}{u}}$

c a vector in the same direction as $\underset{\sim}{u}$ that is 4 units in length

Solution

a $|\underset{\sim}{u}| = \sqrt{2^2 + (-2)^2 + 1^2} = 3$

b $\hat{\underset{\sim}{u}} = \dfrac{1}{3}(2\underset{\sim}{i} - 2\underset{\sim}{j} + \underset{\sim}{k})$

c $4\hat{\underset{\sim}{u}} = \dfrac{4}{3}(2\underset{\sim}{i} - 2\underset{\sim}{j} + \underset{\sim}{k})$

The vector between 2 points

In expressing a vector in 3D from a point $A(x_1, y_1, z_1)$ to $B(x_2, y_2, z_2)$ we obtain the vector $\overrightarrow{AB}$.

$$\overrightarrow{AB} = -\overrightarrow{OA} + \overrightarrow{OB}$$
$$\overrightarrow{AB} = \overrightarrow{OB} - \overrightarrow{OA}$$
$$\overrightarrow{AB} = (x_2 - x_1)\underset{\sim}{i} + (y_2 - y_1)\underset{\sim}{j} + (z_2 - z_1)\underset{\sim}{k}$$

Find a unit vector $\hat{u}$ in the direction of the vector joining $A(0, 5, 2)$ to $B(1, 3, 0)$.

Solution

$$u = (1-0)\underset{\sim}{i} + (3-5)\underset{\sim}{j} + (0-2)\underset{\sim}{k}$$

$$|\underset{\sim}{u}| = \sqrt{(1-0)^2 + (3-5)^2 + (0-2)^2}$$

$$= 3$$

$$\hat{u} = \frac{1}{3}(\underset{\sim}{i} - 2\underset{\sim}{j} - 2\underset{\sim}{k})$$

Exercise 3.02 3D vectors

1 Find the magnitude of each 3D vector.

   a $4\underset{\sim}{i} + 3\underset{\sim}{j} - \underset{\sim}{k}$ **b** $8\underset{\sim}{i} - 6\underset{\sim}{j} + 5\underset{\sim}{k}$

   c $-4\underset{\sim}{i} + 5\underset{\sim}{j} - 2\sqrt{2}\underset{\sim}{k}$ **d** $-2\underset{\sim}{i} + 2\sqrt{3}\underset{\sim}{j} + 3\underset{\sim}{k}$

2 Find the unit vector for each vector in component form.

   a $\underset{\sim}{i} + \underset{\sim}{j} + \underset{\sim}{k}$ **b** $2\underset{\sim}{i} - \underset{\sim}{j} + 2\underset{\sim}{k}$

   c $3\underset{\sim}{i} + 4\underset{\sim}{j} - 12\underset{\sim}{k}$ **d** $\underset{\sim}{i} + 3\underset{\sim}{j} + 2\underset{\sim}{k}$

3 Write the vector $\overrightarrow{AB}$ joining each pair of points and determine its magnitude.

   a $A(0, 4, 0)$ and $B(3, 0, 0)$ **b** $A(1, 1, 1)$ and $B(-1, 1, 1)$

   c $A(2, 2, 3)$ and $B(1, 1, 2)$ **d** $A(-2, -2, -3)$ and $B(3, 4, 5)$

4 a The position vectors of points A and B are $\underset{\sim}{i} - \underset{\sim}{j} - 2\underset{\sim}{k}$ and $2\underset{\sim}{i} + \underset{\sim}{j} - 2\underset{\sim}{k}$ respectively. Find the magnitude of $\overrightarrow{AB}$.

   b The position vectors of points C and D are $2\underset{\sim}{i} - \underset{\sim}{j} - \underset{\sim}{k}$ and $2\underset{\sim}{i} - \underset{\sim}{j} - \underset{\sim}{k}$ respectively. Find the magnitude of $\overrightarrow{CD}$.

   c The position vectors of points X and Y are $\underset{\sim}{i} - 2\underset{\sim}{j} - 2\underset{\sim}{k}$ and $\underset{\sim}{i} + 2\underset{\sim}{j} - \underset{\sim}{k}$ respectively. Find the magnitude of $\overrightarrow{XY}$.

5 Find a vector that has, respectively, magnitude and the same direction as the vector:

   a 2 units, $\underset{\sim}{i}$ **b** 21 units, $\frac{3}{7}\underset{\sim}{i} - \frac{2}{7}\underset{\sim}{j} + \frac{6}{7}\underset{\sim}{k}$

   c 5 units, $\frac{3}{5}\underset{\sim}{i} - \frac{4}{5}\underset{\sim}{j}$ **d** $\sqrt{12}$ units, $\underset{\sim}{i} - \underset{\sim}{j} + \underset{\sim}{k}$

6 If $u = i - 2j + k$ and $v = i - j + 3k$, find:

   a $2u - v$ **b** $u + 2v$ **c** $u + v$ **d** $u - v$

7 Find the length of each vector from point A to point B.

   a $A(2, 1, 2), B(0, 4, 4)$ **b** $A(-1, 1, 2), B(2, 5, 0)$

   c $A(1, -2, 1), B(-1, 2, -1)$ **d** $A(3, 1, 4), B(-2, -2, 2)$

8 Two vectors u and v are given by $u = 2i + 2j + k$ and $v = 3i - 4k$. Find:

   a $|\hat{u}|$ **b** $|\hat{v}|$ **c** $\hat{u}$

   d $\hat{v}$ **e** $u + v$ **f** $u - v$

9 a   Find a vector of $\sqrt{7}$ units in the direction of $3i - 2j + k$.

   b   Find a vector of magnitude 3 in the direction of $4i - 4j + 4k$.

   c   Find a vector of 6 units in the direction of $2i - 2j + k$.

10 What is the angle between the vectors $-i + 2j + 2k$ and $i - 2j - 2k$?
What is the significance of this result?

3.03 Angle between vectors

The scalar product of u and v is:

$$u \cdot v = |u||v| \cos\theta$$

where θ is the angle between u and v.

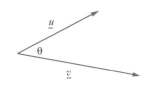

EXAMPLE 4

u has modulus 4 units, v has modulus 5 units, and the angle between them is 60°.

Find the scalar product of u and v.

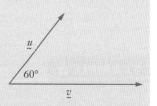

Solution

$u \cdot v = |u||v| \cos\theta$

    $= 4 \times 5 \times \cos 60°$

    $= 10$

EXAMPLE 5

Find the scalar product of the vectors $\underset{\sim}{u}$ and $\underset{\sim}{v}$.

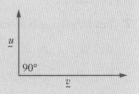

Solution

$$\underset{\sim}{u} \cdot \underset{\sim}{v} = \left|\underset{\sim}{u}\right|\left|\underset{\sim}{v}\right| \cos \theta$$

$$= \left|\underset{\sim}{u}\right|\left|\underset{\sim}{v}\right| \cos 90°$$

$$= 0$$

Note that the scalar product is 0 and the vectors are perpendicular.

Properties of the scalar product

The scalar product is commutative:

$$\underset{\sim}{u} \cdot \underset{\sim}{v} = \underset{\sim}{v} \cdot \underset{\sim}{u}$$

The scalar product is distributive over addition:

$$\underset{\sim}{u} \cdot (\underset{\sim}{v} + \underset{\sim}{w}) = \underset{\sim}{u} \cdot \underset{\sim}{v} + \underset{\sim}{u} \cdot \underset{\sim}{w}$$

We now consider how to find the scalar product of 2 vectors given in Cartesian form.

Given $\underset{\sim}{u} = a_1\underset{\sim}{i} + a_2\underset{\sim}{j} + a_3\underset{\sim}{k}$ and $\underset{\sim}{v} = b_1\underset{\sim}{i} + b_2\underset{\sim}{j} + b_3\underset{\sim}{k}$, the scalar (dot) product is:

$$\underset{\sim}{u} \cdot \underset{\sim}{v} = (a_1\underset{\sim}{i} + a_2\underset{\sim}{j} + a_3\underset{\sim}{k}) \cdot (b_1\underset{\sim}{i} + b_2\underset{\sim}{j} + b_3\underset{\sim}{k})$$

$$= a_1b_1 + a_2b_2 + a_3b_3$$

EXAMPLE 6

Given 2 vectors $\underset{\sim}{u} = 2\underset{\sim}{i} - 3\underset{\sim}{j} + 7\underset{\sim}{k}$ and $\underset{\sim}{v} = 4\underset{\sim}{i} - 5\underset{\sim}{j} - 3\underset{\sim}{k}$, find their scalar product.

Solution

$$\underset{\sim}{u} \cdot \underset{\sim}{v} = (2 \times 4) + [-3 \times (-5)] + [7 \times (-3)]$$

$$= 8 + 15 - 21$$

$$= 2$$

The angle, θ, between the 2 vectors $\underset{\sim}{u}$ and $\underset{\sim}{v}$ can be found using the scalar product $\underset{\sim}{u} \cdot \underset{\sim}{v} = |\underset{\sim}{u}||\underset{\sim}{v}| \cos \theta$.

On rearranging we get:

$$\cos \theta = \frac{\underset{\sim}{u} \cdot \underset{\sim}{v}}{|\underset{\sim}{u}||\underset{\sim}{v}|}$$

Angle between 2 vectors

The angle, θ, between the 2 vectors $\underset{\sim}{u}$ and $\underset{\sim}{v}$ can be found using:

$$\cos \theta = \frac{\underset{\sim}{u} \cdot \underset{\sim}{v}}{|\underset{\sim}{u}||\underset{\sim}{v}|}$$

EXAMPLE 7

Find the angle between the vectors $\underset{\sim}{u} = -5\underset{\sim}{i} + 3\underset{\sim}{j} - 10\underset{\sim}{k}$ and $\underset{\sim}{v} = 8\underset{\sim}{i} + 3\underset{\sim}{k}$, correct to the nearest degree.

Solution

$\underset{\sim}{u} \cdot \underset{\sim}{v} = (-5) \times 8 + 3 \times 0 + (-10) \times 3 = -70$

$|\underset{\sim}{u}| = \sqrt{(-5)^2 + 3^2 + (-10)^2} = \sqrt{134}$

$|\underset{\sim}{v}| = \sqrt{8^2 + 3^2} = \sqrt{73}$

$\cos \theta = \dfrac{\underset{\sim}{u} \cdot \underset{\sim}{v}}{|\underset{\sim}{u}||\underset{\sim}{v}|}$

$= \dfrac{-70}{\sqrt{134} \times \sqrt{73}}$

$= -0.707\,757\ldots$

$\theta \approx 135°$

EXAMPLE 8

Find the angle between the vectors $\underset{\sim}{u} = 3\underset{\sim}{i} + 2\underset{\sim}{j} + \underset{\sim}{k}$ and $\underset{\sim}{v} = \underset{\sim}{i} + 2\underset{\sim}{j} - \underset{\sim}{k}$, correct to the nearest minute.

Solution

$\underset{\sim}{u} \cdot \underset{\sim}{v} = 3 \times 1 + 2 \times 2 + 1 \times (-1) = 6$

$|\underset{\sim}{u}| = \sqrt{3^2 + 2^2 + 1^2} = \sqrt{14}$

$|\underset{\sim}{v}| = \sqrt{1^2 + 2^2 + (-1)^2} = \sqrt{6}$

$\cos \theta = \dfrac{6}{\sqrt{14}\sqrt{6}}$

$\approx 0.654\,65\ldots$

$\theta \approx 49°6'$

Exercise 3.03 Angle between vectors

1 Calculate the scalar product of each pair of vectors.

a $\begin{pmatrix} 2 \\ 5 \\ -1 \end{pmatrix}$ and $\begin{pmatrix} 4 \\ 1 \\ 1 \end{pmatrix}$ **b** $2\underset{\sim}{i}$ and $6\underset{\sim}{j}$ **c** $4\underset{\sim}{k}$ and $2\underset{\sim}{i} + \underset{\sim}{k}$

d $\begin{pmatrix} 2 \\ 0 \\ 4 \end{pmatrix}$ and $\begin{pmatrix} -3 \\ 1 \\ 3 \end{pmatrix}$ **e** $2\underset{\sim}{i} + 3\underset{\sim}{k}$ and $6\underset{\sim}{i} + 2\underset{\sim}{j} - 4\underset{\sim}{k}$

2 Which 2 pairs of vectors are perpendicular?

    A $5\underset{\sim}{i} + 2\underset{\sim}{j} + 3\underset{\sim}{k}$ **B** $\underset{\sim}{j} - \underset{\sim}{k}$

    C $-2\underset{\sim}{i} + 2\underset{\sim}{j} + 2\underset{\sim}{k}$ **D** $-\underset{\sim}{i} + 2\underset{\sim}{k}$

3 Find the angle between each pair of vectors, correct to the nearest degree.

a $\begin{pmatrix} 1 \\ 2 \\ 3 \end{pmatrix}$ and $\begin{pmatrix} 4 \\ -1 \\ 0 \end{pmatrix}$ **b** $2\underset{\sim}{i} + \underset{\sim}{j} - 2\underset{\sim}{k}$ and $\underset{\sim}{i} + 5\underset{\sim}{j} - \underset{\sim}{k}$

c $\begin{pmatrix} 0 \\ 5 \\ 1 \end{pmatrix}$ and $\begin{pmatrix} 1 \\ 5 \\ -1 \end{pmatrix}$ **d** $2\underset{\sim}{i} + \underset{\sim}{j} - 2\underset{\sim}{k}$ and $4\underset{\sim}{j}$

4 Find the cosine of the angle between the vectors $2\underset{\sim}{i} + \underset{\sim}{j} - 2\underset{\sim}{k}$ and $3\underset{\sim}{j} + 4\underset{\sim}{k}$.

5 Given $\underset{\sim}{u} = 2\underset{\sim}{i} + 2\underset{\sim}{j} + 2\underset{\sim}{k}$, $\underset{\sim}{v} = 3\underset{\sim}{i} + 2\underset{\sim}{j} - \underset{\sim}{k}$ and $\underset{\sim}{w} = -\underset{\sim}{i} + 4\underset{\sim}{j} + \underset{\sim}{k}$:

    a Show that $\underset{\sim}{u} \cdot \underset{\sim}{v} = \underset{\sim}{u} \cdot \underset{\sim}{w}$.

    b Rearrange $\underset{\sim}{u} \cdot \underset{\sim}{v} = \underset{\sim}{u} \cdot \underset{\sim}{w}$ so that the expression equals zero.
     Which 2 vectors are perpendicular?

6 Explain how you know that $2\underset{\sim}{i} - 3\underset{\sim}{j} + 4\underset{\sim}{k}$ and $6\underset{\sim}{i} + 8\underset{\sim}{j} + 3\underset{\sim}{k}$ are perpendicular vectors.

7 Find a vector that is perpendicular to both $2\underset{\sim}{i} + \underset{\sim}{j} - \underset{\sim}{k}$ and $\underset{\sim}{i} - 2\underset{\sim}{j} + \underset{\sim}{k}$.

8 $ABCD$ is a rectangle, with vector $\overrightarrow{AB} = 3\underset{\sim}{i}$ and $\overrightarrow{AD} = 2\underset{\sim}{j}$.

    a Express the diagonals of the rectangle as vectors in component form.

    b Determine, correct to the nearest minute, the obtuse angle between the diagonals.

9 A parallelogram has all sides equal in length. Show, using vectors, that its diagonals are perpendicular.

3.04 Geometry proofs using vectors

WS

Geometric
proofs using
vectors

Many geometric theorems can be formulated as relations between points directly, without needing coordinates. In many cases the vector approach is simpler and more direct.

In using vectors to solve geometrical problems, we call on the actual properties of the 2D and 3D shapes to allow us to prove results as required. In particular, we use vector results that relate to parallel lines, perpendicular lines and equal intervals.

Vector properties used in geometry proofs

Scalar (dot) product: $u \cdot v = |u||v| \cos \theta$, when the angle between 2 vectors is required

Parallel vectors: A vector parallel to $xi + yj + zk$ is $cxi + cyj + czk$ where c is a scalar: write and solve $u \cdot v = |u||v|$.

Perpendicular vectors: A vector perpendicular to $x_1i + y_1j + z_1k$ is $-y_1i + x_1j + z_1k$: write and solve $u \cdot v = 0$.

Midpoint of vectors:

$\dfrac{u}{2}$, for a single vector

$\dfrac{u + v}{2}$, for the sum of 2 vectors

$\dfrac{u - v}{2}$, for the vector from u to the midpoint between u and v.

EXAMPLE 9

Show that the vectors $u = 2i + 3j - 2k$ and $v = 4i + 6j - 4k$ are parallel.

Solution

Method 1

$v = 4i + 6j - 4k$

$\quad = 2(2i + 3j - 2k)$

$\quad = 2u$

Hence, the first vector is a multiple of the second vector and so they are parallel.

Method 2

Using the scalar product,

$$u \cdot v = |u||v| \cos \theta$$

$$8 + 18 + 8 = \sqrt{17}\ \sqrt{68}\ \cos \theta$$

$$34 = 34 \cos \theta$$

$$\cos \theta = \frac{34}{34} = 1$$

$$\theta = 0°$$

Hence, the vectors are parallel.

Prove that if the opposite sides and opposite angles of a quadrilateral are equal, then the opposite sides are parallel as well.

Solution

In the diagram, opposite sides are equal and $\angle A = \angle C$, so $\overrightarrow{AB} \cdot \overrightarrow{AD} = \overrightarrow{CD} \cdot \overrightarrow{CB}$,

$\overrightarrow{AB} = \underset{\sim}{b} - \underset{\sim}{a}$, $\overrightarrow{AD} = \underset{\sim}{d} - \underset{\sim}{a}$,

$\overrightarrow{CD} = \underset{\sim}{d} - \underset{\sim}{c}$, $\overrightarrow{CB} = \underset{\sim}{b} - \underset{\sim}{c}$

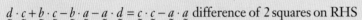

$$\left(\underset{\sim}{b} - \underset{\sim}{a}\right) \cdot \left(\underset{\sim}{d} - \underset{\sim}{a}\right) = \left(\underset{\sim}{d} - \underset{\sim}{c}\right) \cdot \left(\underset{\sim}{b} - \underset{\sim}{c}\right)$$

$$\underset{\sim}{b} \cdot \underset{\sim}{d} - \underset{\sim}{b} \cdot \underset{\sim}{a} - \underset{\sim}{a} \cdot \underset{\sim}{d} + \underset{\sim}{a} \cdot \underset{\sim}{a} = \underset{\sim}{d} \cdot \underset{\sim}{b} - \underset{\sim}{d} \cdot \underset{\sim}{c} - \underset{\sim}{b} \cdot \underset{\sim}{c} + \underset{\sim}{c} \cdot \underset{\sim}{c}$$

$$-\underset{\sim}{b} \cdot \underset{\sim}{a} - \underset{\sim}{a} \cdot \underset{\sim}{d} + \underset{\sim}{a} \cdot \underset{\sim}{a} = -\underset{\sim}{d} \cdot \underset{\sim}{c} - \underset{\sim}{b} \cdot \underset{\sim}{c} + \underset{\sim}{c} \cdot \underset{\sim}{c}$$

$$\underset{\sim}{d} \cdot \underset{\sim}{c} + \underset{\sim}{b} \cdot \underset{\sim}{c} - \underset{\sim}{b} \cdot \underset{\sim}{a} - \underset{\sim}{a} \cdot \underset{\sim}{d} = \underset{\sim}{c} \cdot \underset{\sim}{c} - \underset{\sim}{a} \cdot \underset{\sim}{a} \text{ difference of 2 squares on RHS}$$

$$\underset{\sim}{c} \cdot \left(\underset{\sim}{d} + \underset{\sim}{b}\right) - \underset{\sim}{a} \cdot \left(\underset{\sim}{b} + \underset{\sim}{d}\right) = \left(\underset{\sim}{c} + \underset{\sim}{a}\right) \cdot \left(\underset{\sim}{c} - \underset{\sim}{a}\right)$$

$$\left(\underset{\sim}{d} + \underset{\sim}{b}\right) \cdot \left(\underset{\sim}{c} - \underset{\sim}{a}\right) = \left(\underset{\sim}{c} + \underset{\sim}{a}\right) \cdot \left(\underset{\sim}{c} - \underset{\sim}{a}\right)$$

$$\underset{\sim}{d} + \underset{\sim}{b} = \underset{\sim}{c} + \underset{\sim}{a}$$

$$\underset{\sim}{d} - \underset{\sim}{c} = \underset{\sim}{a} - \underset{\sim}{b} \text{ and } \underset{\sim}{d} - \underset{\sim}{a} = \underset{\sim}{c} - \underset{\sim}{b}$$

$$\overrightarrow{CD} = \overrightarrow{BA} \text{ and } \overrightarrow{AD} = \overrightarrow{BC}$$

$$CD \parallel BA \text{ and } AD \parallel BC$$

Opposite sides are parallel.

Prove that the midpoint of the 2 position vectors $\overrightarrow{OA} = \underset{\sim}{a}$ and $\overrightarrow{OB} = \underset{\sim}{b}$ is $\frac{1}{2}(\underset{\sim}{a} + \underset{\sim}{b})$.

Solution

Let M be the midpoint of A and B.

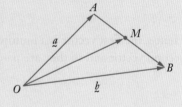

$$\overrightarrow{OM} = \overrightarrow{OA} + \frac{1}{2}\overrightarrow{AB}$$

$$= \underset{\sim}{a} + \frac{1}{2}(\underset{\sim}{b} - \underset{\sim}{a})$$

$$= \frac{1}{2}\underset{\sim}{a} + \frac{1}{2}\underset{\sim}{b}$$

$$= \frac{1}{2}(\underset{\sim}{a} + \underset{\sim}{b})$$

EXAMPLE 12

Prove the cosine rule using vectors.

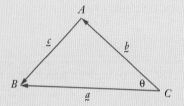

Solution

Since $\underset{\sim}{a} = \underset{\sim}{b} + \underset{\sim}{c}$ then $\underset{\sim}{c} = -\underset{\sim}{b} + \underset{\sim}{a}$

Hence

$$\underset{\sim}{c} \cdot \underset{\sim}{c} = (-\underset{\sim}{b} + \underset{\sim}{a}) \cdot (-\underset{\sim}{b} + \underset{\sim}{a})$$

$$\left|\underset{\sim}{c}\right|^2 = \underset{\sim}{b} \cdot \underset{\sim}{b} - \underset{\sim}{a} \cdot \underset{\sim}{b} - \underset{\sim}{b} \cdot \underset{\sim}{a} + \underset{\sim}{a} \cdot \underset{\sim}{a}$$

$$\left|\underset{\sim}{c}\right|^2 = \left|\underset{\sim}{b}\right|^2 + \left|\underset{\sim}{a}\right|^2 - 2\underset{\sim}{a} \cdot \underset{\sim}{b}$$

Using the scalar product,

$$\left|\underset{\sim}{c}\right|^2 = \left|\underset{\sim}{a}\right|^2 + \left|\underset{\sim}{b}\right|^2 - 2\left|\underset{\sim}{a}\right|\left|\underset{\sim}{b}\right| \cos \theta$$

Hence

$$c^2 = a^2 + b^2 - 2ab \cos \theta$$

EXAMPLE 13

One pair of opposite sides of a quadrilateral is parallel and equal in length.
Show that the quadrilateral is a parallelogram.

Solution

$\overrightarrow{AD}$ is the vector from A to D and is expressed as $\underset{\sim}{d} - \underset{\sim}{a}$.

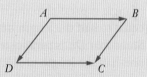

Let the sides AD and BC be parallel and equal, so vectors
$\underset{\sim}{d} - \underset{\sim}{a}$ and $\underset{\sim}{c} - \underset{\sim}{b}$ are in the same direction and equal in
length.

We need to show that the other 2 sides $\underset{\sim}{b} - \underset{\sim}{a}$ and $\underset{\sim}{c} - \underset{\sim}{d}$ are parallel also for $ABCD$ to
be a parallelogram.

Since $\underset{\sim}{d} - \underset{\sim}{a} = \underset{\sim}{c} - \underset{\sim}{b}$, adding $\underset{\sim}{b}$ to both sides gives

$$\underset{\sim}{d} - \underset{\sim}{a} + \underset{\sim}{b} = \underset{\sim}{c} \qquad \text{and subtracting } \underset{\sim}{d} \text{ from both sides gives}$$

$$-\underset{\sim}{a} + \underset{\sim}{b} = \underset{\sim}{c} - \underset{\sim}{d}$$

That is, $\underset{\sim}{b} - \underset{\sim}{a} = \underset{\sim}{c} - \underset{\sim}{d}$.

So vectors $\overrightarrow{AB}$ and $\overrightarrow{DC}$ are equal in size and direction.

Therefore they are parallel and $ABCD$ is a parallelogram.

Exercise 3.04 Geometry proofs using vectors

1 The vector u is drawn. Sketch the vectors represented by $2u, 0.5u$ and $-3u$.

2 ABC is a triangle in which M is the midpoint of BC. Prove that $\overrightarrow{AM} = \frac{1}{2}\left(\overrightarrow{AB} + \overrightarrow{AC} \right)$.

3 Use vectors and the scalar product to prove that if the diagonals of a rectangle are perpendicular then the rectangle is a square.

4 ABC is a triangle in which $\overrightarrow{BC} = a$, $\overrightarrow{CA} = b$ and $\overrightarrow{AB} = c$. Prove that $a + b + c = 0$.

5 Draw the triangle OPQ, where O is the origin, $\overrightarrow{OP} = 4i$ and $\overrightarrow{OQ} = 6j$.

Point M, with position vector $\overrightarrow{OM} = xi + yj$, is at equal distances from O, P and Q. Find the values of x and y.

6 Show that the lines connecting any point on the semicircle of radius 1 to $(1, 0)$ and $(-1, 0)$ are perpendicular.

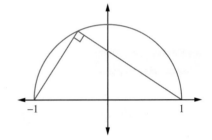

7 The segment that joins the midpoints of 2 sides of a triangle is parallel to the third side and has a length equal to half the length of the third side. Show that DE is parallel to AB and its length is one-half the length of AB.

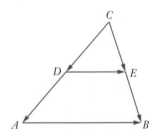

8 Prove using vector methods that the midpoints of the sides of any quadrilateral form a parallelogram.

MATHS IN FOCUS 12. Mathematics Extension 2   ISBN 9780170413435

9 Points A and B have position vectors $\underset{\sim}{a}, \underset{\sim}{b}$ respectively with respect to an origin O. Draw the line AB, and take a point P on that line which divides it in the ratio

$$m : n = AP : PB.$$

What is the position vector of P with respect to O?

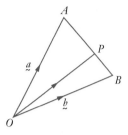

3.05 3D space

In 3D space, every point $P(x, y, z)$, can be determined by the position vector $\overrightarrow{OP}$, which can be represented by the vector $\underset{\sim}{r} = x\underset{\sim}{i} + y\underset{\sim}{j} + z\underset{\sim}{k}$.

EXAMPLE 14

Find the unit vector parallel to $2\underset{\sim}{i} + 3\underset{\sim}{j} - \underset{\sim}{k}$.

Solution

Magnitude $= \sqrt{2^2 + 3^2 + (-1)^2} = \sqrt{14}$

Therefore, the unit vector is $\dfrac{2\underset{\sim}{i}}{\sqrt{14}} + \dfrac{3\underset{\sim}{j}}{\sqrt{14}} - \dfrac{\underset{\sim}{k}}{\sqrt{14}}$ or $\dfrac{1}{\sqrt{14}}(2\underset{\sim}{i} + 3\underset{\sim}{j} - \underset{\sim}{k})$.

The distance between two points A and B in 3D space is the length of the vector $\overrightarrow{AB}$, where $\overrightarrow{AB} = \overrightarrow{OB} - \overrightarrow{OA}$.

EXAMPLE 15

Find the distance between the points $A(2, 3, -1)$ and $B(3, -2, 1)$.

Solution

Distance $=$ length of $\overrightarrow{OB} - \overrightarrow{OA}$

$$= \sqrt{(3-2)^2 + (-2-3)^2 + (1-(-1))^2}$$

$$= \sqrt{30}$$

EXAMPLE 16

Find the angle between the vectors u and v if $u = \begin{pmatrix} 2 \\ 3 \\ -1 \end{pmatrix}$ and $v = \begin{pmatrix} 3 \\ -2 \\ 1 \end{pmatrix}$, correct to one decimal place.

Solution

$$u \cdot v = x_1 x_2 + y_1 y_2 + z_1 z_2 = |u||v| \cos \theta$$

$$|u| = \sqrt{2^2 + 3^2 + (-1)^2} = \sqrt{14}, \quad |v| = \sqrt{3^2 + (-2)^2 + 1^2} = \sqrt{14}$$

$$2(3) + 3(-2) + (-1)1 = \sqrt{14} \sqrt{14} \cos \theta$$

$$-1 = 14 \cos \theta$$

$$\cos \theta = -\frac{1}{14}$$

$$\theta \approx 94.1°$$

Hence, the angle between the vectors is $94.1°$.

Exercise 3.05 3D space

1 Find the magnitude of each vector.

   a $2i - 2j + k$ **b** $3i - 4j + 12k$ **c** $2i + 5j + 14k$

   d $4i + 7j - 32k$ **e** $-3i - 2j + 6k$

2 Find the unit vector for each vector in Question **1**.

3 Find the angle between each pair of vectors correct to 1 decimal place.

   a $2i + j + k$ and $i + j + 2k$

   b $i + 2j + 3k$ and $-i + j + 2k$

   c $2i + 2j + k$ and $i - 2j - 2k$

   d $2i + j + k$ and $i + j - 2k$

   e $-3i - 2j + 6k$ and $i + j + 2k$

4 Find the value of the pronumeral in each vector.

  a $2\underset{\sim}{i} - y\underset{\sim}{j} + 14\underset{\sim}{k}$ has a magnitude of 15 units

  b $2\underset{\sim}{i} + 9\underset{\sim}{j} + z\underset{\sim}{k}$ has a magnitude of 43 units

  c $x\underset{\sim}{i} - 11\underset{\sim}{j} - 110\underset{\sim}{k}$ has a magnitude of 111 units

5 Find the value of m if the vector $2\underset{\sim}{i} - 3\underset{\sim}{j} + m\underset{\sim}{k}$ has a unit vector given by $\dfrac{2\underset{\sim}{i} - 3\underset{\sim}{j} + m\underset{\sim}{k}}{\sqrt{29}}$.

6 Find a vector that is perpendicular to the vector $\begin{pmatrix} 2 \\ -3 \\ 4 \end{pmatrix}$.

7 Find the value of m if the vectors $\begin{pmatrix} 1 \\ m \\ -1 \end{pmatrix}$ and $\begin{pmatrix} 1 \\ -1 \\ 1 \end{pmatrix}$ make an angle of 60°.

DID YOU KNOW?

The bee jive

Honeybees returning to their hive use vectors to communicate information about the location and value of a food source and the amount of energy needed to reach it. This communication takes the form of a 'waggle dance'. On returning to the hive they perform vigorous movements of the abdomen to the left and right while repeatedly moving forward in a straight line. The bees are communicating the locations of important food sources via the direction (direction of line dance) and the distance (the length of the waggle run).

Through the bees' movements, they are indicating 'there is a food smelling of A, requiring an effort B to reach it, in direction C, of economic value D.'

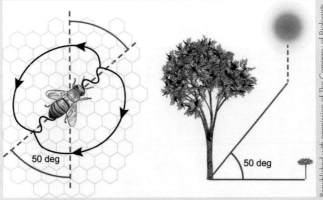

3.06 Vector equation of a curve

A convenient way to describe a line in 3D space is to provide a vector that 'points to' every

point on the line as the **parameter** (t) varies, for example $\begin{pmatrix} 1 \\ 2 \\ 3 \end{pmatrix} + t \begin{pmatrix} 3 \\ -2 \\ 1 \end{pmatrix} = \begin{pmatrix} 1+3t \\ 2-2t \\ 3+t \end{pmatrix}$.

The individual functions of t make up the coordinates of this vector that traces out a

straight line. Any vector with a parameter, for example $\begin{pmatrix} f(t) \\ g(t) \\ h(t) \end{pmatrix}$, will describe some curve in

3 dimensions as t varies through all possible values. We call $\begin{pmatrix} f(t) \\ g(t) \\ h(t) \end{pmatrix}$ a **vector function**.

The equations $x = f(t)$, $y = g(t)$ and $z = h(t)$ are called the
parametric equations for the curve. We first learned
about parametric equations in Year 11 Mathematics
Extension 1, Chapter 7, *Further functions*. We often
think of the parameter t as time and thus the parametric
equations indicate the position of an object in 3D space at
any time.

When graphing lines and curves in 3D space, draw the axes
as shown.

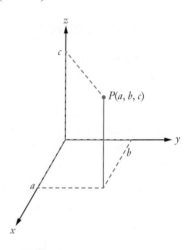

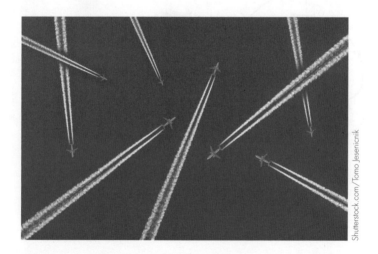

Shutterstock.com/Tomo Jesenicnik

EXAMPLE 17

Graph the line produced by $\begin{pmatrix} t \\ t \\ t \end{pmatrix}$ for $0 \leq t \leq 10$.

Solution

The parametric equations are
$x(t) = t$, $y(t) = t$ and $z(t) = t$.

The points $(0, 0, 0)$ through to
$(10, 10, 10)$ are included in this plot.

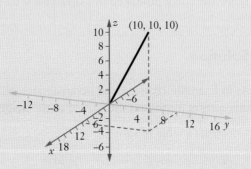

We can use parametric equations to obtain some familiar common curves.

EXAMPLE 18

Graph the curve produced by $\begin{pmatrix} \cos t \\ \sin t \\ 0 \end{pmatrix}$ for all real t.

Solution

The parametric equations are
$x(t) = \cos t$, $y(t) = \sin t$ and $z(t) = 0$.

As t varies and $z(t) = 0$,
the coordinates trace out
the points on the unit
circle on the x–y plane,
starting at $(1, 0)$ when $t = 0$
and proceeding in a
anticlockwise
direction around
the circle as
t increases.

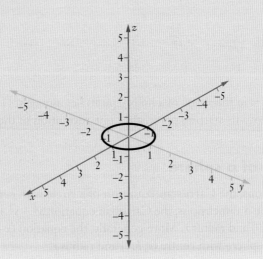

As the relationships between $x(t)$, $y(t)$ and $z(t)$ become more complicated, so too the graph becomes more complex. A simple way to have an idea of the graph of the vector equation is to create a table of values for plotting and identifying the nature of the curve.

EXAMPLE 19

Draw the curve produced by $\begin{pmatrix} \cos t \\ \sin t \\ t \end{pmatrix}$ for $0 \leq t \leq 2\pi$.

Solution

As in Example 18, the x and y coordinates still describe a unit circle, but the z coordinate increases at a constant rate ($z = t$) so the height of the curve increases gradually.

Construct a table of values.

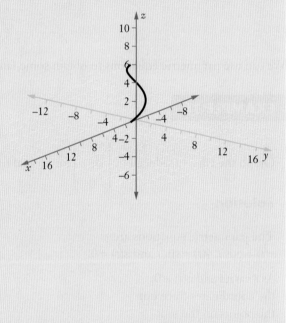

t	$x = \cos t$	$y = \sin t$	$z = t$
0	1	0	0
$\dfrac{\pi}{4}$	$\dfrac{1}{\sqrt{2}}$	$\dfrac{1}{\sqrt{2}}$	$\dfrac{\pi}{4}$
$\dfrac{\pi}{2}$	0	1	$\dfrac{\pi}{2}$
$\dfrac{3\pi}{4}$	$-\dfrac{1}{\sqrt{2}}$	$\dfrac{1}{\sqrt{2}}$	$\dfrac{3\pi}{4}$
π	1	0	π
$\dfrac{5\pi}{4}$	$-\dfrac{1}{\sqrt{2}}$	$-\dfrac{1}{\sqrt{2}}$	$\dfrac{5\pi}{4}$
$\dfrac{3\pi}{2}$	0	-1	$\dfrac{3\pi}{2}$
$\dfrac{7\pi}{4}$	$\dfrac{1}{\sqrt{2}}$	$\dfrac{1}{\sqrt{2}}$	$\dfrac{7\pi}{4}$
2π	1	0	2π

The curve is in the shape of a helix (spiral) beginning at $(1, 0, 0)$ and ending at $(1, 0, 2\pi)$ directly above its starting point, after one revolution of the circle.

Equation of a sphere

In 2 dimensions, the **Cartesian equation** of a circle with centre $(0, 0)$ and radius r is $x^2 + y^2 = r^2$. In 3 dimensions, the Cartesian equation $x^2 + y^2 + z^2 = r^2$ represents a sphere with centre $(0, 0, 0)$ and radius r. More generally, the equation $(x - a)^2 + (y - b)^2 + (z - c)^2 = r^2$ represents a sphere with centre (a, b, c) and radius r.

MATHS IN FOCUS 12. Mathematics Extension 2 ISBN 9780170413435

EXAMPLE 20

Describe the sphere represented by:

a $(x - 1)^2 + (y + 2)^2 + (z - 3)^2 = 25$

b $x^2 + 4x + y^2 + 6y + z^2 - 2z = 2$

Solution

a This is a sphere with centre at $(1, -2, 3)$ and with a radius of $\sqrt{25} = 5$.

b Completing the square for x, y and z:

$$x^2 + 4x + 4 + y^2 + 6y + 9 + z^2 - 2z + 1 = 2 + (4 + 9 + 1)$$

$$(x + 2)^2 + (y + 3)^2 + (z - 1)^2 = 16$$

This is a sphere with centre at $(-2, -3, 1)$ and with a radius of $\sqrt{16} = 4$.

Exercise 3.06 Vector equation of a curve

1 Graph the curve represented by each vector function.

a $\begin{pmatrix} 1 \\ t \\ 1 \end{pmatrix}$ for $0 \leq t \leq 10$ 　　　　**b** $\begin{pmatrix} 1 \\ t \\ t \end{pmatrix}$ for $0 \leq t \leq 5$

c $\begin{pmatrix} t \\ t^2 \\ 1 \end{pmatrix}$ for $0 \leq t \leq 3$ 　　　　**d** $\begin{pmatrix} 0 \\ t \\ t^2 \end{pmatrix}$ for $0 \leq t \leq 3$

e $\begin{pmatrix} \cos t \\ t \\ \sin t \end{pmatrix}$ for $0 \leq t \leq 2\pi$

2 Graph the curve represented by each vector function.

a $x = t, y = 1 - t, z = 1 + t$

b $x = 2t, y = t^2, z = t$

c $x = \cos t, y = \sin t, z = 2$

d $x = \sec t, y = \tan t, z = 3$

e $x = 2 \sin t, y = 3 \cos t, z = 1$

3 Describe the graph of each vector function for $t \geq 0$.

a $\begin{pmatrix} 1 \\ 1 \\ t \end{pmatrix}$ **b** $\begin{pmatrix} \cos t \\ \sin t \\ -t \end{pmatrix}$

c $x = 1, y = 2 \cos t, z = \sin t$ **d** $x = t \cos t, y = t \sin t, z = t$

4 Find a vector function for the curve where $x^2 + y^2 = 9$ and $z + y = 2$ intersect.

5 Plot the curve $\begin{pmatrix} t^2 \\ t - 1 \\ t^2 + 5 \end{pmatrix}$ in 2 dimensions, projected onto the x–y, x–z and y–z planes,

for $0 \leq t \leq 3$.

6 Find a vector function for the curve where $z = x^2 + y^2$ and $y = x$ intersect.

7 Find a vector function that represents the curve of intersection of the cone $z = \sqrt{x^2 + y^2}$ and the plane $z = 1 + y$ and then sketch this curve.

8 Find the radius and centre of the sphere with equation:

a $(x - 1)^2 + (y + 1)^2 + (z - 1)^2 = 1$
b $(x + 2)^2 + (y - 3)^2 + (z - 1)^2 = 4$
c $(x - 3)^2 + (y + 1)^2 + (z + 1)^2 = 9$
d $x^2 + 2x + y^2 + 2y + z^2 - 2z = 6$
e $x^2 - 4x + y^2 - 6y + z^2 + 2z = 11$

9 The spheres with equations $(x + 2)^2 + (y + 3)^2 + (z - 4)^2 = 16$ and $(x + 2)^2 + (y + 3)^2 + (z + 2)^2 = 25$ intersect at a circle. What is the centre and radius of this circle of contact?

TECHNOLOGY

Graphing in 3D space

Using graphing technology, graph some 3D curves and determine the shape of the curve and its projections on each of the $x–y$, $y–z$ and $x–z$ planes.

For example, investigate the shape of the curve given and the projections made by the vector function $\begin{pmatrix} t \\ t^2 \\ t^3 \end{pmatrix}$.

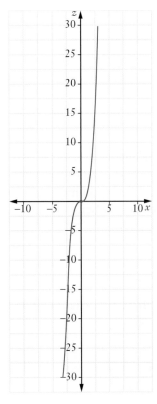

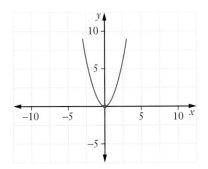

$x–y$ plane
$y = x^2$

$x–z$ plane
$z = x^3$

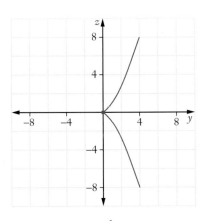

$y–z$ plane
$z = y\sqrt{y}$

Note: $y \geq 0$.

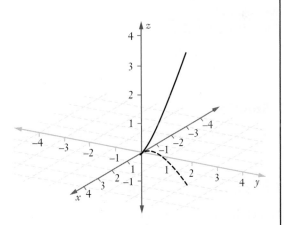

$\begin{pmatrix} t \\ t^2 \\ t^3 \end{pmatrix}$ in 3D

ISBN 9780170413435

WS

Equations
of lines in
space

3.07 Vector equation of a straight line

The equation of a line in 2 dimensions is $ax + by + c = 0$, so you might expect a line in 3 dimensions to be $ax + by + cz + d = 0$, but actually this is the equation of a plane.

To specify a straight line, we need to know 2 things: a point through which the line passes, and the line's direction.

Suppose a line contains the point $A(a_1, a_2, a_3)$ and is parallel to the vector $\underset{\sim}{b} = \begin{pmatrix} b_1 \\ b_2 \\ b_3 \end{pmatrix}$.

By placing these vectors with $\underset{\sim}{a}$ from the origin and $\underset{\sim}{b}$ at its head then any point on the line can be obtained by adding $\underset{\sim}{a}$ and $\lambda\underset{\sim}{b}$, where λ is some real number.

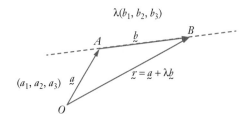

Vector equation of a straight line

The vector equation of a straight line through points A and B is $\underset{\sim}{r} = \underset{\sim}{a} + \lambda\underset{\sim}{b}$, where R is a point on AB, $\underset{\sim}{a} = \overrightarrow{OA}$ is the vector from the origin to a known point, $\underset{\sim}{b} = \overrightarrow{AB}$ is the direction of the required line and λ is a parameter.

Note: If $\lambda < 0$ then the point will lie to the left of A, if $0 \le \lambda \le 1$ the point will lie between A and B, and if $\lambda > 1$ then the point will lie to the right of B.

EXAMPLE 21

What is the vector equation of the line through the point $(2, -1, 0)$ parallel to the vector $-2\underset{\sim}{i} + \underset{\sim}{j} + \underset{\sim}{k}$?

Solution

The line has vector equation $\underset{\sim}{r} = \underset{\sim}{a} + \lambda\underset{\sim}{b}$, that is, $\underset{\sim}{r} = (2\underset{\sim}{i} - \underset{\sim}{j}) + \lambda(-2\underset{\sim}{i} + \underset{\sim}{j} + \underset{\sim}{k})$.

Direction of a vector

Sometimes we do not know the direction of the required vector but are given 2 points through which the vector will pass. In this case, we need to determine the direction of the required vector first.

The direction of the vector joining $A(a_1, b_1, c_1)$ and $B(a_2, b_2, c_2)$ is given by $\begin{pmatrix} a_2 - a_1 \\ b_2 - b_1 \\ c_2 - c_1 \end{pmatrix}$.

MATHS IN FOCUS 12. Mathematics Extension 2 ISBN 9780170413435

EXAMPLE 22

What is the vector equation of the line through the points $(1, -1, 2)$ and $(2, -3, -1)$?

Solution

The required line has direction given by $\begin{pmatrix} 2-1 \\ -3-(-1) \\ -1-2 \end{pmatrix} = \begin{pmatrix} 1 \\ -2 \\ -3 \end{pmatrix}$ and passes through $(1, -1, 2)$. It has the equation

$$\underset{\sim}{r} = (\underset{\sim}{i} - \underset{\sim}{j} + 2\underset{\sim}{k}) + \lambda_1(\underset{\sim}{i} - 2\underset{\sim}{j} - 3\underset{\sim}{k})$$

or, using the other point,

$$\underset{\sim}{r} = (2\underset{\sim}{i} - 3\underset{\sim}{j} - \underset{\sim}{k}) + \lambda_2(\underset{\sim}{i} - 2\underset{\sim}{j} - 3\underset{\sim}{k}).$$

Note that both equations are correct; however, $\lambda_1 \neq \lambda_2$.

EXAMPLE 23

A line passes through the points $A(-2, 1, 5)$ and $B(6, 3, -4)$.

a Write a vector equation of the line.

b Write parametric equations for the line.

c Determine if the point $C(-10, -1, 14)$ lies on the line.

Solution

a $\underset{\sim}{r} = \overrightarrow{OB} - \overrightarrow{OA}$

$= \begin{pmatrix} 6 \\ 3 \\ -4 \end{pmatrix} - \begin{pmatrix} -2 \\ 1 \\ 5 \end{pmatrix}$

$= \begin{pmatrix} 8 \\ 2 \\ -9 \end{pmatrix}$

A vector equation for the line is $\begin{pmatrix} x \\ y \\ z \end{pmatrix} = \begin{pmatrix} -2 \\ 1 \\ 5 \end{pmatrix} + t\begin{pmatrix} 8 \\ 2 \\ -9 \end{pmatrix}$

b The corresponding parametric equations are:

$x = -2 + 8t$

$y = 1 + 2t$

$z = 5 - 9t$

c If the point lies on the line, then:

$-10 = -2 + 8t$, so $t = -1$

$-1 = 1 + 2t$, so $t = -1$

$14 = 5 - 9t$, so $t = -1$

$t = -1$ for $(-10, -1, 14)$.

Therefore, C does lie on the line joining A to B.

Parametric and Cartesian equations for a line

Because λ runs through all possible real values, the vector $\underset{\sim}{r} = \underset{\sim}{a} + \lambda\underset{\sim}{b}$ points to every point on the line when its tail is placed at the origin.

The parametric equations resulting from this expression are:

$$x = a_1 + \lambda b_1, y = a_2 + \lambda b_2, z = a_3 + \lambda b_3$$

If a straight line passes through 2 given points $A(x_1, y_1, z_1)$ and $B(x_2, y_2, z_2)$, then it is parallel to the vector $(x_2 - x_1)\underset{\sim}{i} + (y_2 - y_1)\underset{\sim}{j} + (z_2 - z_1)\underset{\sim}{k}$.

The parametric equations are:

$$x = x_1 + \lambda(x_2 - x_1), y = y_1 + \lambda(y_2 - y_1), z = z_1 + \lambda(z_2 - z_1)$$

The standard Cartesian form of the straight line is found by equating the different expressions for λ:

$$\lambda = \frac{x - x_1}{x_2 - x_1} = \frac{y - y_1}{y_2 - y_1} = \frac{z - z_1}{z_2 - z_1}$$

EXAMPLE 24

Determine whether the lines $\begin{pmatrix} 1 \\ 1 \\ 1 \end{pmatrix} + \lambda_1 \begin{pmatrix} 1 \\ 2 \\ -1 \end{pmatrix}$ and $\begin{pmatrix} 3 \\ 2 \\ 1 \end{pmatrix} + \lambda_2 \begin{pmatrix} -1 \\ -5 \\ 3 \end{pmatrix}$ are parallel, intersect or neither.

Solution

In 3 dimensions, lines that do not intersect might not be parallel (they most likely are skew).

In the 2 equations we can see the direction vectors $\begin{pmatrix} 1 \\ 2 \\ -1 \end{pmatrix}$ and $\begin{pmatrix} -1 \\ -5 \\ 3 \end{pmatrix}$ are different and so they are not parallel.

If they intersect there must be 2 values a and b such that:

$$\begin{pmatrix} 1 \\ 1 \\ 1 \end{pmatrix} + a \begin{pmatrix} 1 \\ 2 \\ -1 \end{pmatrix} = \begin{pmatrix} 3 \\ 2 \\ 1 \end{pmatrix} + b \begin{pmatrix} -1 \\ -5 \\ 3 \end{pmatrix}$$

ISBN 9780170413435

Therefore:

$1 + a = 3 - b$

$1 + 2a = 2 - 5b$

$1 - a = 1 + 3b$

Solving any 2 equations above simultaneously, we get $a = 3$ and $b = -1$.

These values satisfy all 3 equations and so tell us that the 2 lines intersect at $(4, 7, -2)$.

Exercise 3.07 Vector equation of a straight line

1 Write the parametric equations for each vector equation.

a $\begin{pmatrix} x \\ y \\ z \end{pmatrix} = \begin{pmatrix} 1 \\ 1 \\ 0 \end{pmatrix} + \lambda \begin{pmatrix} 2 \\ -3 \\ 0 \end{pmatrix}$

b $\begin{pmatrix} x \\ y \\ z \end{pmatrix} = \begin{pmatrix} 11 \\ 2 \\ 0 \end{pmatrix} + \lambda \begin{pmatrix} 3 \\ 0 \\ 0 \end{pmatrix}$

c $\begin{pmatrix} x \\ y \\ z \end{pmatrix} = \begin{pmatrix} 3 \\ 0 \\ -1 \end{pmatrix} + \lambda \begin{pmatrix} 6 \\ -9 \\ 1 \end{pmatrix}$

d $\begin{pmatrix} x \\ y \\ z \end{pmatrix} = \begin{pmatrix} 5 \\ -2 \\ 1 \end{pmatrix} + \lambda \begin{pmatrix} 7 \\ -4 \\ 2 \end{pmatrix}$

2 a What is the line through the point $(2, -1, 3)$ parallel to the vector $\underset{\sim}{i} + 2\underset{\sim}{j} - \underset{\sim}{k}$?

b What is the line through the point $(-2, 1, -3)$ parallel to the vector $-\underset{\sim}{i} + 2\underset{\sim}{j} + 3\underset{\sim}{k}$?

c What is the line through the point $(1, -1, 1)$ parallel to the vector $2\underset{\sim}{i} - 2\underset{\sim}{j} + \underset{\sim}{k}$?

3 Write a vector equation of the line through each pair of points.

a $(3, -5)$ and $(-2, -8)$ **b** $(6, 2, 5)$ and $(9, 2, 8)$

c $(1, 1, -3)$ and $(1, -1, -5)$ **d** $(1, 0, 3)$ and $(1, 2, 4)$

4 Determine whether the lines in each pair are parallel, intersect or neither.

a $\begin{pmatrix} 1 \\ 3 \\ -1 \end{pmatrix} + \lambda_1 \begin{pmatrix} 1 \\ 1 \\ 0 \end{pmatrix}$ and $\begin{pmatrix} 0 \\ 0 \\ 0 \end{pmatrix} + \lambda_2 \begin{pmatrix} 1 \\ 4 \\ 5 \end{pmatrix}$

b $\begin{pmatrix} 1 \\ 0 \\ 2 \end{pmatrix} + \lambda_1 \begin{pmatrix} -1 \\ -1 \\ 2 \end{pmatrix}$ and $\begin{pmatrix} 4 \\ 4 \\ 2 \end{pmatrix} + \lambda_2 \begin{pmatrix} 2 \\ 2 \\ -4 \end{pmatrix}$

c $\begin{pmatrix} 1 \\ 2 \\ -1 \end{pmatrix} + \lambda_1 \begin{pmatrix} 1 \\ 2 \\ 3 \end{pmatrix}$ and $\begin{pmatrix} 1 \\ 0 \\ 1 \end{pmatrix} + \lambda_2 \begin{pmatrix} \frac{2}{3} \\ 2 \\ \frac{4}{3} \end{pmatrix}$

d $\begin{pmatrix} 1 \\ 1 \\ 2 \end{pmatrix} + \lambda_1 \begin{pmatrix} 1 \\ 2 \\ -3 \end{pmatrix}$ and $\begin{pmatrix} 2 \\ 3 \\ -1 \end{pmatrix} + \lambda_2 \begin{pmatrix} 2 \\ 4 \\ -6 \end{pmatrix}$

5 Find the equation of the line through $(3, 2, 6)$ and $(-1, 0, 4)$ in vector, parametric and Cartesian forms.

6 Find a_1 and a_3 so that $(a_1, 1, a_3)$ lies on the line through $(0, 2, 1)$ and $(2, 7, 4)$.

7 Show that $\begin{pmatrix} 2 \\ 1 \\ 3 \end{pmatrix} + \lambda_1 \begin{pmatrix} 1 \\ 1 \\ 2 \end{pmatrix}$ and $\begin{pmatrix} 3 \\ 2 \\ 5 \end{pmatrix} + \lambda_2 \begin{pmatrix} 2 \\ 2 \\ 4 \end{pmatrix}$ are the same line.

8 The equations $\dfrac{-x + 2}{7} = \dfrac{3y - 1}{5} = \dfrac{2z + 1}{3}$ determine a straight line.
Write the equation of the line in vector form.

9 The x coordinate of a point on the line joining $(2, 2, 1)$ and $(5, 1, -2)$ is 4. Find its z coordinate.

3.08 Parallel and perpendicular lines

A **normal vector** to a line is a vector that is perpendicular to the line.

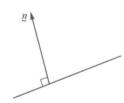

In 2 dimensions, a line has an infinite number of normal vectors, all parallel to each other. However, in 3 dimensions, the normal vectors are not necessarily parallel.

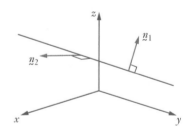

In 2 dimensions, we saw that the scalar product can be used to determine the angle between 2 vectors. We are particularly interested in whether the angle between the vectors is 0° (parallel) or 90° (perpendicular).

A line in 2 dimensions can be defined using its gradient and y-intercept. The gradient indicates the direction of the line and the y-intercept provides its exact position. In 3 dimensions, the gradient of a line is not defined, so a line in 3 dimensions must be defined in a different way.

In 2D, lines can be defined in general form $ax + by + c = 0$, which is a scalar equation. As well, a line can be written in gradient–intercept form $y = mx + c$. Vectors can also be written to define a line in 2 dimensions. We identify the direction vector parallel to the given line and a position vector to establish a unique line.

EXAMPLE 25

Show that $2\underset{\sim}{i} + 3\underset{\sim}{j} - 2\underset{\sim}{k}$ and $\underset{\sim}{i} - 2\underset{\sim}{j} - 2\underset{\sim}{k}$ are perpendicular vectors.

Solution

$$x_1x_2 + y_1y_2 + z_1z_2 = |\underset{\sim}{u}||\underset{\sim}{v}| \cos \theta$$
$$2 \times 1 + 3(-2) + (-2)(-2) = \sqrt{17}\sqrt{9} \cos \theta$$
$$0 = \sqrt{153} \cos \theta$$
$$\cos \theta = 0$$
$$\theta = 90°$$

Hence, the vectors are perpendicular.

EXAMPLE 26

Find the vector equation of the line passing through the point $(3, -1, 1)$ and perpendicular to the 2 lines $\dfrac{x-3}{2} = \dfrac{y+1}{3} = \dfrac{z-1}{-2}$ and $\dfrac{x-3}{2} = \dfrac{y+1}{2} = \dfrac{z-1}{5}$.

Solution

Since the equation that is perpendicular to both has equation $(3, -1, 1) + \lambda(a, b, c)$,

then $2a + 3b - 2c = 0$ **and** $2a + 2b + 5c = 0$.

$2a + 3b - 2c = 0$ [1]

$2a + 2b + 5c = 0$ [2]

$[1] - [2]$:

$b - 7c = 0$ [3]

$2 \times [1] - 3 \times [2]$:

$4a + 6b - 4c = 0$

$6a + 6b + 15c = 0$

$-2a - 19c = 0$ [4]

> Simultaneous equations [3] and [4] have an infinite number of solutions, so we can choose any convenient value for c to work out a and b.

Putting $c = 2$, then $a = -19$ and $b = 14$.

Note: there are many lines perpendicular to both lines but only this one passes through their point of intersection $(3, -1, 1)$.

In this case the perpendicular vector is $(x, y, z) = (3, -1, 1) + \lambda(-19, 14, 2)$

or in Cartesian form $\dfrac{x-3}{-19} = \dfrac{y+1}{14} = \dfrac{z-1}{2}$.

Exercise 3.08 Parallel and perpendicular lines

1 Show that each pair of vectors are parallel.

a $\begin{pmatrix} 1 \\ -1 \\ 1 \end{pmatrix} + \lambda \begin{pmatrix} 2 \\ -1 \\ 1 \end{pmatrix}$ and $\begin{pmatrix} 2 \\ 1 \\ -1 \end{pmatrix} + \lambda \begin{pmatrix} -4 \\ 2 \\ -2 \end{pmatrix}$

b $\left(2\underset{\sim}{i} - \underset{\sim}{j} + \underset{\sim}{k}\right) + \lambda\left(\underset{\sim}{i} + \underset{\sim}{j} - 2\underset{\sim}{k}\right)$ and $\left(\underset{\sim}{i} + \underset{\sim}{j} + \underset{\sim}{k}\right) + \lambda\left(-3\underset{\sim}{i} - 3\underset{\sim}{j} + 6\underset{\sim}{k}\right)$

c $\begin{pmatrix} 3 \\ 1 \\ -1 \end{pmatrix} + \lambda \begin{pmatrix} -2 \\ 1 \\ -2 \end{pmatrix}$ and $\begin{pmatrix} -1 \\ -1 \\ 2 \end{pmatrix} + \lambda \begin{pmatrix} 1 \\ -0.5 \\ 1 \end{pmatrix}$

d $\left(3\underset{\sim}{i} - \underset{\sim}{j} + 2\underset{\sim}{k}\right) + \lambda\left(2\underset{\sim}{i} + 2\underset{\sim}{j} - 3\underset{\sim}{k}\right)$ and $\left(\underset{\sim}{i} + 3\underset{\sim}{j} - 2\underset{\sim}{k}\right) + \lambda\left(4\underset{\sim}{i} + 4\underset{\sim}{j} - 6\underset{\sim}{k}\right)$

2 Show that each pair of vectors are perpendicular.

a $\begin{pmatrix} 1 \\ -1 \\ 1 \end{pmatrix} + \lambda \begin{pmatrix} 1 \\ 0 \\ 1 \end{pmatrix}$ and $\begin{pmatrix} 3 \\ 2 \\ -1 \end{pmatrix} + \lambda \begin{pmatrix} 0 \\ 2 \\ 0 \end{pmatrix}$

b $\left(\underset{\sim}{i} - \underset{\sim}{j} + \underset{\sim}{k}\right) + \lambda\left(2\underset{\sim}{i} - \underset{\sim}{j} + 2\underset{\sim}{k}\right)$ and $\left(3\underset{\sim}{i} + 2\underset{\sim}{j} - \underset{\sim}{k}\right) + \lambda\left(2\underset{\sim}{i} - 2\underset{\sim}{j} - 3\underset{\sim}{k}\right)$

c $\begin{pmatrix} 7 \\ -2 \\ 5 \end{pmatrix} + \lambda \begin{pmatrix} -1 \\ 0 \\ -2 \end{pmatrix}$ and $\begin{pmatrix} 3 \\ 4 \\ 2 \end{pmatrix} + \lambda \begin{pmatrix} -2 \\ 0 \\ 1 \end{pmatrix}$

d $\left(\underset{\sim}{i} - 3\underset{\sim}{j} + 2\underset{\sim}{k}\right) + \lambda\left(-\underset{\sim}{i} + 2\underset{\sim}{j} + 5\underset{\sim}{k}\right)$ and $\left(4\underset{\sim}{i} - 2\underset{\sim}{j} + \underset{\sim}{k}\right) + \lambda\left(-\underset{\sim}{i} - 3\underset{\sim}{j} + \underset{\sim}{k}\right)$

3 Write the equation of a vector that is parallel to each line.

a $\begin{pmatrix} 2 \\ -1 \\ 3 \end{pmatrix}$

b $\begin{pmatrix} -1 \\ 2 \\ 3 \end{pmatrix} + \lambda \begin{pmatrix} 3 \\ 1 \\ -2 \end{pmatrix}$

c $4\underset{\sim}{i} + 3\underset{\sim}{j} - \underset{\sim}{k}$

d $\left(3\underset{\sim}{i} - \underset{\sim}{j} + \underset{\sim}{k}\right) + \lambda\left(2\underset{\sim}{i} + 2\underset{\sim}{j} - \underset{\sim}{k}\right)$

4 Write the equation of a vector that is perpendicular to each line.

a $\begin{pmatrix} 2 \\ -1 \\ 3 \end{pmatrix}$

b $\begin{pmatrix} 1 \\ -2 \\ 3 \end{pmatrix} + \lambda \begin{pmatrix} 3 \\ 1 \\ -2 \end{pmatrix}$

c $4\underset{\sim}{i} + 3\underset{\sim}{j} - \underset{\sim}{k}$

d $\left(3\underset{\sim}{i} - \underset{\sim}{j} + \underset{\sim}{k}\right) + \lambda\left(2\underset{\sim}{i} + 2\underset{\sim}{j} - \underset{\sim}{k}\right)$

5 Find which pairs of lines intersect and find their point of intersection.

a $\begin{pmatrix} 7 \\ -2 \\ 5 \end{pmatrix} + \lambda \begin{pmatrix} 4 \\ 0 \\ -2 \end{pmatrix}$

b $\dfrac{x-1}{6} = \dfrac{y+2}{2} = \dfrac{z-3}{5}$

c $2\underset{\sim}{i} + 4\underset{\sim}{j} - 5\underset{\sim}{k}$

d $\left(4\underset{\sim}{i} - 11\underset{\sim}{j} + 5\underset{\sim}{k}\right) + \lambda\left(\underset{\sim}{i} + 2\underset{\sim}{j} + 2\underset{\sim}{k}\right)$

6 A cube has sides of length L. By giving the vertices coordinates, determine the angle between the diagonal of the cube and one of the edges adjacent to the diagonal correct to one decimal place.

7 a Prove that the lines joining $A(2, -3, 3)$ and $C(-2, 3, -1)$ intersect the line joining $B(-3, 2, 1)$ and $D(3, -2, 1)$.

b Find correct to one decimal place the angle between these 2 lines.

8 Write the parametric equations of the line that goes through the point $(6, -2, 1)$

and is perpendicular to both $\begin{pmatrix} 1 \\ 4 \\ -2 \end{pmatrix} + \lambda_1 \begin{pmatrix} 3 \\ -1 \\ 1 \end{pmatrix}$ and $\begin{pmatrix} 9 \\ 5 \\ -3 \end{pmatrix} + \lambda_2 \begin{pmatrix} 1 \\ -3 \\ 7 \end{pmatrix}$.

9 Given the Cartesian equation $\dfrac{x-4}{8} = \dfrac{y-12}{5} = \dfrac{z-15}{2}$, write the corresponding vector equation.

10 Write the equations of 2 lines that intersect at the point $(2, -1, 3)$ and are perpendicular to each other.

1 Find the scalar product of the vectors $\begin{pmatrix} 2 \\ 1 \end{pmatrix}$ and $\begin{pmatrix} 1 \\ 2 \end{pmatrix}$, then find the angle between these vectors to the nearest minute.

2 Are the vectors $\begin{pmatrix} 2 \\ -3 \end{pmatrix}$ and $\begin{pmatrix} 4.5 \\ 3 \end{pmatrix}$ parallel or perpendicular?

3 Express the 2D vector represented by 8 units of length in a direction of $60°$ in component form and then write the unit vector in component form.

4 Find the exact magnitude and direction to the nearest minute of the vector $-2\underset{\sim}{i} + 5\underset{\sim}{j}$.

5 Write the component vector AB that joins the points $A(-1, 2, -3)$ and $B(0, 3, 1)$.

6 Calculate the scalar product for the vectors $2\underset{\sim}{i} - \underset{\sim}{j} + 4\underset{\sim}{k}$ and $\underset{\sim}{i} - 2\underset{\sim}{j} - \underset{\sim}{k}$. What can you say about these 2 vectors?

7 Write a unit vector that is perpendicular to both $2\underset{\sim}{i} - \underset{\sim}{j} + \underset{\sim}{k}$ and $\underset{\sim}{i} - 2\underset{\sim}{j} - \underset{\sim}{k}$.

8 Given that $\overrightarrow{OX} = 2\underset{\sim}{a} + \underset{\sim}{b}$ and $\overrightarrow{OY} = 3\underset{\sim}{a} + 4\underset{\sim}{b}$, express the vector $\overrightarrow{XY}$ in terms of $\underset{\sim}{a}$ and $\underset{\sim}{b}$.

9 OAB is a triangle. $\overrightarrow{OA} = \underset{\sim}{a}$ and $\overrightarrow{OB} = \underset{\sim}{b}$.

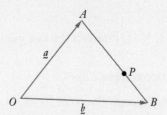

a Find the vector $\overrightarrow{AB}$ in terms of $\underset{\sim}{a}$ and $\underset{\sim}{b}$.

b If P is a point on AB such that $AP : PB = 3 : 2$, show that $\overrightarrow{OP} = \frac{1}{5}(2\underset{\sim}{a} + 3\underset{\sim}{b})$.

10 Find the magnitude of each vector $\underset{\sim}{u} = \begin{pmatrix} 2 \\ -2 \\ 1 \end{pmatrix}$ and $\underset{\sim}{v} = \begin{pmatrix} -3 \\ 1 \\ 2\sqrt{2} \end{pmatrix}$.

Hence, find the angle between them to 2 decimal places.

11 Find a vector that is perpendicular to the vector $\underset{\sim}{i} - 2\underset{\sim}{j} - \underset{\sim}{k}$.

12 Graph the curve represented by the vector $\begin{pmatrix} t \\ 0 \\ t \end{pmatrix}$ for $0 \le t \le 5$.

13 Graph the curve represented by the vector $\begin{pmatrix} 4\cos t \\ 4\sin t \\ t \end{pmatrix}$ for $0 \le t \le 2\pi$.

14 A sphere has equation $x^2 + (y-4)^2 + (z+1)^2 = 5$. What is its radius and centre?

15 For the Cartesian equation $\dfrac{x+3}{2} = \dfrac{y-2}{1} = \dfrac{z+1}{3}$, find the equivalent vector and parametric equations.

16 The y coordinate of a point that lies on the line joining $(1, 5, -2)$ and $(-2, 2, 3)$ is -3. Find the x and z coordinates for this point.

17 Show that $\begin{pmatrix} 1 \\ 0 \\ -2 \end{pmatrix}$ and $\begin{pmatrix} -2 \\ 0 \\ 4 \end{pmatrix}$ are parallel vectors.

18 Show that $\begin{pmatrix} 7 \\ 1 \\ -2 \end{pmatrix}$ and $\begin{pmatrix} 3 \\ -3 \\ 9 \end{pmatrix}$ are perpendicular vectors.

19 Find the angle between the vectors $\begin{pmatrix} -2 \\ 1 \\ 0 \end{pmatrix}$ and $\begin{pmatrix} 0 \\ 1 \\ -2 \end{pmatrix}$ to 1 decimal place.

20 Find the vector equation of the line that passes through $(1, -3, 2)$ and is perpendicular to both $\begin{pmatrix} 2 \\ 3 \\ -1 \end{pmatrix} + \lambda_1 \begin{pmatrix} 1 \\ -2 \\ 3 \end{pmatrix}$ and $\begin{pmatrix} 1 \\ -1 \\ 1 \end{pmatrix} + \lambda_2 \begin{pmatrix} 0 \\ 1 \\ -5 \end{pmatrix}$.

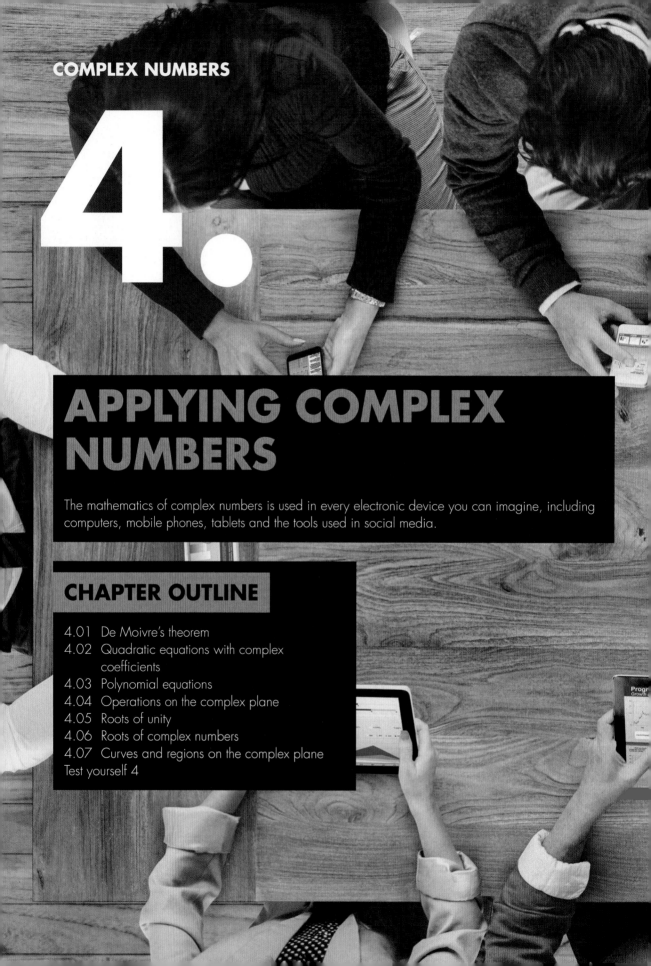

COMPLEX NUMBERS

4.

APPLYING COMPLEX NUMBERS

The mathematics of complex numbers is used in every electronic device you can imagine, including computers, mobile phones, tablets and the tools used in social media.

CHAPTER OUTLINE

IN THIS CHAPTER YOU WILL:

- use De Moivre's theorem with complex numbers in both polar and exponential form
- use De Moivre's theorem to derive trigonometric identities
- find solutions to quadratic equations with real coefficients
- find solutions to polynomial equations with real coefficients
- solve problems using conjugate roots of polynomial equations
- solve quadratic equations with complex coefficients
- revise adding and subtracting complex numbers on the complex plane
- multiply and divide complex numbers on the complex plane using rotation and dilation
- relate a complex number z to the points and vectors $\bar{z}$, iz and cz
- find the nth roots of unity and other complex numbers, and their location on the complex plane
- solve problems using the nth roots of unity and other complex numbers
- sketch curves and regions on the complex plane, the locus of points determined by relations such as $|z-i| \geq 2$, $-\dfrac{\pi}{4} \leq \text{Arg } z \leq \dfrac{\pi}{2}$, $\text{Re}(z) < \text{Im}(z)$ and $|z+1| = 2|z-i|$

TERMINOLOGY

De Moivre's theorem:
$(\cos \theta + i \sin \theta)^n = \cos n\theta + i \sin n\theta = e^{in\theta}$,
$\forall\, n \in \mathbb{Z}$.

locus: A set of points that obey a certain condition. Its graph can be a line, curve or region.

roots of unity: Solutions to the equation $z^n = 1$, $n \in \mathbb{N}$.

Using De Moivre's theorem

Trigonometric identities using De Moivre's theorem

4.01 De Moivre's theorem

We saw in Chapter 1, *Complex numbers*, that if $z = r(\cos \theta + i \sin \theta) = re^{i\theta}$, then

$$z^n = r^n(\cos n\theta + i \sin n\theta) = r^n e^{in\theta}, \text{ where } n \text{ is an integer.}$$

If $r = 1$, then this property simplifies to

$$(\cos \theta + i \sin \theta)^n = \cos n\theta + i \sin n\theta = e^{in\theta}.$$

This theorem is called **De Moivre's theorem**, and can be extended to rational numbers, $\mathbb{Q}$.

De Moivre's theorem

For any non-zero complex number $z = \cos \theta + i \sin \theta = e^{i\theta}$:

- $(\cos \theta + i \sin \theta)^n = \cos n\theta + i \sin n\theta = e^{in\theta}$, $\forall\, n \in \mathbb{Q}$ (polar form)

- $(e^{i\theta})^n = e^{in\theta}$, $\forall\, n \in \mathbb{Q}$ (exponential form)

The statement can be proved to be true for all *real* values of n, not only rational numbers, but this is not part of this course.

We can prove De Moivre's theorem by induction for $n \in \mathbb{N}$, that is, for positive integers only.

Proof

Let $P(n)$ be the proposition that $(\cos \theta + i \sin \theta)^n = \cos n\theta + i \sin n\theta$, $\forall\, n \in \mathbb{N}$.

Prove $P(1)$ is true:

LHS $= (\cos \theta + i \sin \theta)^1$ RHS $= \cos (1\theta) + i \sin (1\theta)$

        $= \cos \theta + i \sin \theta$ $= \cos \theta + i \sin \theta$

$\therefore P(1)$ is true.

Assume $P(k)$ is true; that is, $(\cos \theta + i \sin \theta)^k = \cos k\theta + i \sin k\theta$ for some $k \in \mathbb{N}$.

Prove $P(k + 1)$ is true; that is, $(\cos \theta + i \sin \theta)^{k+1} = \cos (k+1)\theta + i \sin (k+1)\theta$

Proof: Consider the LHS of $P(k + 1)$:

$$\text{LHS} = (\cos \theta + i \sin \theta)^{k+1}$$

$$= (\cos \theta + i \sin \theta)(\cos \theta + i \sin \theta)^k$$

$$= (\cos \theta + i \sin \theta)(\cos k\theta + i \sin k\theta) \text{ using } P(k)$$

RHS of $P(k)$

$$= \cos \theta \cos k\theta + i \cos \theta \sin k\theta + i \sin \theta \cos k\theta - \sin \theta \sin k\theta$$

$$= \cos \theta \cos k\theta - \sin \theta \sin k\theta + i(\cos \theta \sin k\theta + \sin \theta \cos k\theta)$$

$$= \cos (\theta + k\theta) + i \sin (\theta + k\theta)$$

$$= \cos (k + 1)\theta + i \sin (k + 1)\theta$$

$$= \text{RHS of } P(k + 1)$$

$\therefore$ truth of $P(k) \Rightarrow$ truth of $P(k + 1)$

But $P(1)$ is also true.

$\therefore P(n)$ is true for all $n \geq 1$, by mathematical induction.

The exponential form of De Moivre's theorem, $(e^{i\theta})^n = e^{in\theta}$, is obvious from the index laws. It could also be proved by induction. This will be left for you to do as an exercise.

Equating the modulus and argument on both sides of $z^n = r^n(\cos n\theta + i \sin n\theta) = r^n e^{in\theta}$, $\forall\, n \in \mathbb{Z}$, we can see the properties we saw in Chapter 1 follow directly from De Moivre's theorem.

Power of a complex number

$\left| z^n \right| = \left| z \right|^n, \forall\, n \in \mathbb{Z}$

$\arg z^n = n \arg z, \forall\, n \in \mathbb{Z}$

Both of these statements can be proved by induction and will be left as an exercise.

Note that to use De Moivre's theorem, the complex number must be in correct polar form: $z = r(\cos \theta + i \sin \theta) = re^{i\theta}$. If $z = \cos \theta - i \sin \theta$ then we need to use the rule $\cos \theta - i \sin \theta = \cos(-\theta) + i \sin(-\theta) = e^{-i\theta}$ before applying De Moivre's theorem.

EXAMPLE 1

Use De Moivre's theorem to simplify:

a $(2(\cos \theta + i \sin \theta))^8$ **b** $\left(\cos \dfrac{3\pi}{5} - i \sin \dfrac{3\pi}{5}\right)^4$ **c** $\left[8(\cos \pi + i \sin \pi)\right]^{\frac{1}{3}}$

Solution

a $(2(\cos \theta + i \sin \theta))^8 = 2^8(\cos 8\theta + i \sin 8\theta)$

$$= 256(\cos 8\theta + i \sin 8\theta)$$

b First express $\cos \dfrac{3\pi}{5} - i \sin \dfrac{3\pi}{5}$ in the form $\cos \theta + i \sin \theta$.

$$\cos \dfrac{3\pi}{5} - i \sin \dfrac{3\pi}{5} = \cos\left(-\dfrac{3\pi}{5}\right) + i \sin\left(-\dfrac{3\pi}{5}\right)$$

$$\therefore \left(\cos \dfrac{3\pi}{5} - i \sin \dfrac{3\pi}{5}\right)^4 = \left[\cos\left(-\dfrac{3\pi}{5}\right) + i \sin\left(-\dfrac{3\pi}{5}\right)\right]^4$$

$$= \cos\left(-\dfrac{12\pi}{5}\right) + i \sin\left(-\dfrac{12\pi}{5}\right)$$

$$= \cos\left(-\dfrac{2\pi}{5}\right) + i \sin\left(-\dfrac{2\pi}{5}\right)$$

$$= \cos\left(\dfrac{2\pi}{5}\right) - i \sin\left(\dfrac{2\pi}{5}\right)$$

Recall that polar form uses the **principal argument** so we converted $-\dfrac{12\pi}{5}$ to an angle in the interval $(-\pi, \pi]$: $-\dfrac{12\pi}{5} = -\dfrac{2\pi}{5}$

c $\left[8(\cos \pi + i \sin \pi)\right]^{\frac{1}{3}} = 8^{\frac{1}{3}}\left(\cos \dfrac{\pi}{3} + i \sin \dfrac{\pi}{3}\right)$

$$= 2\left(\cos \dfrac{\pi}{3} + i \sin \dfrac{\pi}{3}\right)$$

MATHS IN FOCUS 12. Mathematics Extension 2   ISBN 9780170413435

Proving trigonometric identities

De Moivre's theorem is useful for proving some trigonometric identities.

EXAMPLE 2

Use De Moivre's theorem and the binomial expansion of $(A + B)^3$ to prove each identity.

a $\quad \sin 3\theta = 3 \sin \theta - 4 \sin^3 \theta$
    b $\quad \cos 3\theta = 4 \cos^3 \theta - 3 \cos \theta$

Solution

a $\quad$ Using De Moivre's theorem we have $(\cos \theta + i \sin \theta)^3 = \cos 3\theta + i \sin 3\theta$.

Using the binomial expansion $(A + B)^3 = A^3 + 3A^2B + 3AB^2 + B^3$ we have:
$(\cos \theta + i \sin \theta)^3 = \cos^3 \theta + 3i \cos^2 \theta \sin \theta - 3 \cos \theta \sin^2 \theta - i \sin^3 \theta$

Equating both expansions:

$\cos 3\theta + i \sin 3\theta = \cos^3 \theta + 3i \cos^2 \theta \sin \theta - 3 \cos \theta \sin^2 \theta - i \sin^3 \theta$

Equating imaginary parts we have:

$\sin 3\theta = 3 \cos^2 \theta \sin \theta - \sin^3 \theta$

$\qquad = 3(1 - \sin^2 \theta)\sin \theta - \sin^3 \theta \qquad$ so in terms of sin only

$\qquad = 3\sin \theta - 3 \sin^3 \theta - \sin^3 \theta$

$\qquad = 3\sin \theta - 4 \sin^3 \theta \qquad$ as required.

b $\quad$ Equating real parts from part **a** we have:

$\cos 3\theta = \cos^3 \theta - 3 \cos \theta \sin^2 \theta$

Substituting $\sin^2 \theta = 1 - \cos^2 \theta$:

$\cos 3\theta = \cos^3 \theta - 3 \cos \theta(1 - \cos^2 \theta) \qquad$ so in terms of cos only

$\qquad = \cos^3 \theta - 3 \cos \theta + 3 \cos^3 \theta$

$\qquad = 4 \cos^3 \theta - 3 \cos \theta \qquad$ as required

De Moivre's theorem and binomial expansions can also be used to simplify powers of trigonometric ratios.

These results are useful to express powers of trigonometric ratios as ratios with multiples of θ.

EXAMPLE 3

Use the fact that $z + \dfrac{1}{z} = 2\cos\theta$ and the binomial expansion of $(a + b)^4$ to prove that $\cos^4\theta = \dfrac{1}{8}(\cos 4\theta + 4\cos 2\theta + 3)$.

Solution

Consider the expansion $(2\cos\theta)^4 = \left(z + \dfrac{1}{z}\right)^4$ where $z = \cos\theta + i\sin\theta$.

$$16\cos^4\theta = z^4 + 4z^3\frac{1}{z} + 6z^2\frac{1}{z^2} + 4z\frac{1}{z^3} + \frac{1}{z^4}$$

$$= z^4 + 4z^2 + 6 + 4\frac{1}{z^2} + \frac{1}{z^4}$$

$$= z^4 + \frac{1}{z^4} + 4\left(z^2 + \frac{1}{z^2}\right) + 6$$

$$= 2\cos 4\theta + 4(2\cos 2\theta) + 6 \qquad \text{using } z^n + \frac{1}{z^n} = 2\cos n\theta \text{ twice}$$

$$= 2(\cos 4\theta + 4\cos 2\theta + 3)$$

$$\cos^4\theta = \frac{1}{8}(\cos 4\theta + 4\cos 2\theta + 3)$$

Exercise 4.01 De Moivre's theorem

1 Use De Moivre's theorem to simplify each complex number.

a $(\cos \theta + i \sin \theta)^5$ **b** $(\cos \theta + i \sin \theta)^{-3}$ **c** $(\cos \theta - i \sin \theta)^7$

d $\left(\cos \dfrac{\theta}{2} + i \sin \dfrac{\theta}{2}\right)^{-5}$ **e** $(\cos 4\theta + i \sin 4\theta)^{\frac{3}{4}}$

2 For each value of z, evaluate z^5 in:

     i   polar form   **ii**   Cartesian form

a $z = 2\left(\cos \dfrac{\pi}{6} + i \sin \dfrac{\pi}{6}\right)$ **b** $z = \sqrt{3}\left(\cos \dfrac{\pi}{4} + i \sin \dfrac{\pi}{4}\right)$

c $z = \dfrac{1}{\sqrt{2}}\left(\cos \dfrac{\pi}{3} + i \sin \dfrac{\pi}{3}\right)$ **d** $z = -3\left(\cos \dfrac{7\pi}{10} + i \sin \dfrac{7\pi}{10}\right)$

3 Evaluate $\left[\cos\left(\dfrac{-3\pi}{5}\right) + i \sin\left(\dfrac{-3\pi}{5}\right)\right]^{-6}$, in modulus–argument form.

4 Evaluate each expression by first converting to polar form, giving your answer in Cartesian form.

a $(1 - i)^3$ **b** $\left(1 + i\sqrt{3}\right)^4$ **c** $\left(-\sqrt{2} + i\sqrt{2}\right)^5$

d $\left(\dfrac{1+i}{\sqrt{2}}\right)^{-3}$ **e** $\left(\dfrac{1}{2} + \dfrac{i\sqrt{3}}{2}\right)^{\frac{1}{2}}$

5 Evaluate each expression and give exact answers in modulus–argument form.

a $\dfrac{1}{(3 + 3i)^4}$ **b** $\dfrac{1}{\left(\sqrt{3} - i\right)^9}$

6 Show that $\left(\dfrac{1}{\sqrt{3}} + \dfrac{1}{\sqrt{3}}i\right)^{12}\left(\dfrac{3}{2} + \dfrac{3\sqrt{3}}{2}i\right)^6 = -64$.

7 Simplify each complex number.

a $(\cos \theta + i \sin \theta)^3 \times (\cos \theta + i \sin \theta)^{-7}$

b $(\cos \alpha + i \sin \alpha)^4 \times (\cos \beta - i \sin \beta)^6$

c $\dfrac{\left(\cos 3\delta + i \sin 3\delta\right)^8}{\left(\cos 2\delta + i \sin 2\delta\right)^3}$

d $\dfrac{\left(\cos \beta + i \sin \beta\right)^3 \times \left(\cos 2\beta - i \sin 2\beta\right)^{-2}}{\left(\cos \beta + i \sin \beta\right)^5}$

e $\dfrac{\left(\cos \dfrac{\pi}{2} + i \sin \dfrac{\pi}{2}\right)^{\frac{1}{4}} \times \left(\cos \dfrac{\pi}{2} + i \sin \dfrac{\pi}{2}\right)^{\frac{2}{3}}}{\left(\cos \dfrac{\pi}{2} + i \sin \dfrac{\pi}{2}\right)^{\frac{-1}{12}}}$

---

8 By expanding $(\cos \alpha + i \sin \alpha)^2$ in 2 ways, derive expressions for $\cos 2\alpha$ and $\sin 2\alpha$. Hence find an expression for $\tan 2\alpha$ in terms of $\tan \alpha$.

[Hint: show that $\tan 2\alpha = \dfrac{2 \sin \alpha \cos \alpha}{\cos^2 \alpha - \sin^2 \alpha}$, then divide both sides by $\cos^2 \alpha$.]

9 In Example 2, p.121, we proved that $\cos 3\theta = 4 \cos^3 \theta - 3 \cos \theta$.

   a   Use this fact to find $\displaystyle\int_0^{\frac{\pi}{2}} 4 \cos^3 \theta \, d\theta$.

   b   Use a similar result to evaluate $\displaystyle\int_0^{\frac{\pi}{3}} \sin^3 \theta \, d\theta$.

10 Expand $(\cos \theta + i \sin \theta)^4$ in 2 ways to prove that:

   a $\cos 4\theta = 8 \cos^4 \theta - 8 \cos^2 \theta + 1$

   b $\sin 4\theta = 4 \sin \theta \cos \theta \, (\cos^2 \theta - \sin^2 \theta)$

11 **a**   Complete the statement: $\tan 3\theta = \dfrac{\underline{\quad} 3\theta}{\cos \underline{\quad}}$.

   b   Using the results we derived from $(\cos \theta + i \sin \theta)^3 = \cos 3\theta + i \sin 3\theta$ in Example 2, p.121, show that $\tan 3\theta = \dfrac{3 \sin \theta - 4 \sin^3 \theta}{4 \cos^3 \theta - 3 \cos \theta}$.

   c   Hence prove that $\tan 3\theta = \dfrac{3 \tan \theta - \tan^3 \theta}{1 - 3 \tan^2 \theta}$.

12 Using the fact that $a + ar + ar^2 + ar^3 + \ldots + ar^{n-1} = \dfrac{a(r^n - 1)}{r - 1}$, prove for $n \geq 1$:

$(\cos \theta + i \sin \theta) + (\cos 2\theta + i \sin 2\theta) + (\cos 3\theta + i \sin 3\theta) + \ldots + (\cos n\theta + i \sin n\theta)$

$= \dfrac{(\cos \theta + i \sin \theta)(\cos n\theta + i \sin n\theta - 1)}{\cos \theta + i \sin \theta - 1}$.

13 Using a suitable expansion, find the value of $\displaystyle\int_{\frac{\pi}{3}}^{\frac{2\pi}{3}} \cos^4 x \, dx$.

14 Prove by mathematical induction for all positive integers n:

   a $(\cos \theta + i \sin \theta)^n = \cos n\theta + i \sin n\theta$

   b $\left| z^n \right| = |z|^n$

   c $\arg z^n = n \arg z$

15 If $z = \cos \theta + i \sin \theta$, simplify:

   a $z - \dfrac{1}{z}$ **b** $z^2 - \dfrac{1}{z^2}$ **c** $z^n - \dfrac{1}{z^n}$

16 Simplify each expression.

a $\cos \dfrac{\pi}{12} + i \sin \dfrac{\pi}{12} + \dfrac{1}{\cos \dfrac{\pi}{12} + i \sin \dfrac{\pi}{12}}$

b $\left(\cos \dfrac{\pi}{6} + i \sin \dfrac{\pi}{6} \right)^4 - \dfrac{1}{\left(\cos \dfrac{\pi}{6} + i \sin \dfrac{\pi}{6} \right)^4}$

c $\left(\cos \dfrac{5\pi}{7} + i \sin \dfrac{5\pi}{7} \right)^7 + \left(\cos \dfrac{5\pi}{7} + i \sin \dfrac{5\pi}{7} \right)^{-7}$

17 If $z = \cos \theta + i \sin \theta$, prove that:

a $z^2 + \dfrac{1}{z^2}$ is always real

b $z^3 - \dfrac{1}{z^3}$ is purely imaginary

c $z^n + \dfrac{1}{z^n}$ is always real

18 Let $z = \cos \theta + i \sin \theta$. Use the fact that $z^n - \dfrac{1}{z^n} = 2i \sin n\theta$ and the binomial expansion of $\left(z - \dfrac{1}{z} \right)^3$ to prove that $\sin^3 \theta = \dfrac{3 \sin \theta - \sin 3\theta}{4}$.

4.02 Quadratic equations with complex coefficients

Any quadratic equation of the form $ax^2 + bx + c = 0$ can be solved using the quadratic

formula $x = \dfrac{-b \pm \sqrt{b^2 - 4ac}}{2a}$. We saw in Chapter 1 that if the coefficients a, b and c are real,

then the complex roots always come in conjugate pairs. However, if the coefficients are *complex*, then the roots will not be conjugates.

Shutterstock.com/siambizkit

EXAMPLE 4

Solve each quadratic equation with complex coefficients.

a $z^2 + iz + 2 = 0$

b $w^2 - (5 - 2i)w + 5 - 5i = 0$

c $x^2 + 4i = 0$

Solution

a $z^2 + iz + 2 = 0$

Using the quadratic formula where $a = 1$, $b = i$ and $c = 2$:

$$z = \frac{-b \pm \sqrt{b^2 - 4ac}}{2a}$$

$$= \frac{-i \pm \sqrt{i^2 - 4(1)(2)}}{2(1)}$$

$$= \frac{-i \pm \sqrt{-9}}{2}$$

$$= \frac{-i \pm 3i}{2}$$

$$= \frac{2i}{2} \quad \text{or} \quad \frac{-4i}{2}$$

$$= i \quad \text{or} \quad -2i$$

b $w^2 - (5 - 2i)w + 5 - 5i = 0$

Using the quadratic formula where $a = 1$, $b = -(5 - 2i)$ and $c = 5 - 5i$:

$$w = \frac{(5 - 2i) \pm \sqrt{(5 - 2i)^2 - 4(1)(5 - 5i)}}{2(1)}$$

$$= \frac{(5 - 2i) \pm \sqrt{25 - 20i - 4 - 20 + 20i}}{2}$$

$$= \frac{(5 - 2i) \pm \sqrt{1}}{2}$$

$$= \frac{5 - 2i \pm 1}{2}$$

$$= \frac{6 - 2i}{2} \quad \text{or} \quad \frac{4 - 2i}{2}$$

$$= 3 - i \quad \text{or} \quad 2 - i$$

c $x^2 + 4i = 0$

Method 1

Using the quadratic formula where $a = 1$, $b = 0$ and $c = 4i$

$$x = \frac{-0 \pm \sqrt{0^2 - 4(1)(4i)}}{2(1)}$$

$$= \frac{\pm \sqrt{-16i}}{2}$$

$$= \frac{\pm \sqrt{-16} \times \sqrt{i}}{2}$$

$$= \frac{\pm 4i\sqrt{i}}{2}$$

$$= \pm 2i\sqrt{i}$$

or

$$x^2 + 4i = 0$$

$$x^2 = -4i$$

$$x = \pm\sqrt{-4i}$$

$$= \pm\sqrt{4 \times (-1) \times i}$$

$$= \pm 2i\sqrt{i}$$

Now this solution still contains a square root. Recall we can simplify $\sqrt{i}$ using the following technique.

Let $\sqrt{i} = a + ib$ for some real a and b.

Then $i = (a + ib)^2 = a^2 - b^2 + 2abi$

Equating real and imaginary parts:

$a^2 - b^2 = 0 \Rightarrow a^2 = b^2$ and $1 = 2ab \Rightarrow ab = \dfrac{1}{2}$.

By inspection or solving simultaneously we see that $a = b = \dfrac{1}{\sqrt{2}}$ or $a = b = -\dfrac{1}{\sqrt{2}}$. By convention we take the solution such that $a > 0$.

Then $\sqrt{i} = a + ib = \dfrac{1}{\sqrt{2}} + \dfrac{i}{\sqrt{2}}$.

So the solution to the original equation becomes

$$x = \pm 2i\sqrt{i}$$

$$= \pm 2i\left(\frac{1}{\sqrt{2}} + \frac{i}{\sqrt{2}}\right)$$

$$= \pm\left(i\sqrt{2} - \sqrt{2}\right)$$

$$= -\sqrt{2} + i\sqrt{2} \quad \text{or} \quad \sqrt{2} - i\sqrt{2}$$

Although the equation $x^2 + 4i = 0$ looks harmless, it requires some sophisticated algebra to solve using this approach.

Method 2

An alternative method uses exponential form and Euler's formula:

$x^2 + 4i = 0 \Rightarrow x^2 = -4i$

Let $x = re^{i\theta}$.

Then $x^2 = (re^{i\theta})^2$.

So equating we have $(re^{i\theta})^2 = -4i$.

Then $r^2 e^{2i\theta} = -4i$

So $r^2 e^{2i\theta} = 2^2(-i)$.

We can see that $r = 2$ since $r > 0$ but we need 2 solutions to $e^{2i\theta} = -i$ for the argument θ because $x^2 = -4i$ is a quadratic equation.

Now recall that the domain for the principal argument is $\theta \in (-\pi, \pi]$, so $2\theta \in (-2\pi, 2\pi]$. We can express Arg $(-i)$ as 2 different angles in this domain.

So $\quad e^{2i\theta} = -i$

$\therefore \quad 2\theta = -\dfrac{\pi}{2}$ or $\dfrac{3\pi}{2}$ $\quad$ in the domain

$\therefore \quad \theta = -\dfrac{\pi}{4}$ or $\dfrac{3\pi}{4}$

So the solutions to $x^2 + 4i = 0$ are:

$x = 2e^{-\frac{\pi}{4}i}$ or $x = 2e^{\frac{3\pi}{4}i}$

Converting these to Cartesian form we have:

$$x = 2\left(\cos\left(-\frac{\pi}{4}\right) + i\sin\left(-\frac{\pi}{4}\right)\right) \qquad \text{or} \qquad x = 2\left(\cos\left(\frac{3\pi}{4}\right) + i\sin\left(\frac{3\pi}{4}\right)\right)$$

$$= 2\left(\frac{1}{\sqrt{2}} - \frac{i}{\sqrt{2}}\right) \qquad\qquad\qquad\qquad = 2\left(-\frac{1}{\sqrt{2}} + \frac{i}{\sqrt{2}}\right)$$

$$= \sqrt{2} - i\sqrt{2} \qquad\qquad\qquad\qquad\qquad = -\sqrt{2} + i\sqrt{2}$$

as in Method 1.

Note that in the example we solved the simultaneous equations by inspection rather than by using an algebraic method. The emphasis in this course is to provide elegant and efficient solutions using mathematical insight rather than tedious algebraic manipulations if they can be avoided. Sometimes algebraic techniques are necessary.

Graphical approach to finding the square root of a complex number

De Moivre's theorem is useful for finding a square root.

EXAMPLE 5

Solve the equation $z^2 = -1 + i\sqrt{3}$.

Solution

Express in polar or exponential form:

$$\left|-1+i\sqrt{3}\right| = \sqrt{\left(-1\right)^2 + \left(\sqrt{3}\right)^2}$$
$$= 2$$

$$\arg\left(-1+i\sqrt{3}\right) = \frac{2\pi}{3}$$

So $z^2 = 2\left(\cos\dfrac{2\pi}{3} + i\sin\dfrac{2\pi}{3}\right) = 2e^{\frac{2i\pi}{3}}$.

Since it is a quadratic equation there should be 2 roots.

Using De Moivre's theorem, we can raise both sides to the power of $\dfrac{1}{2}$ to easily calculate one of the roots, z_1:

$$z_1 = \left[2\left(\cos\frac{2\pi}{3} + i\sin\frac{2\pi}{3}\right)\right]^{\frac{1}{2}} = \left(2e^{\frac{2i\pi}{3}}\right)^{\frac{1}{2}}$$

$$= \sqrt{2}\left(\cos\frac{\pi}{3} + i\sin\frac{\pi}{3}\right) = \sqrt{2}e^{\frac{i\pi}{3}}$$

Graphing z^2 and z_1 on an Argand diagram:

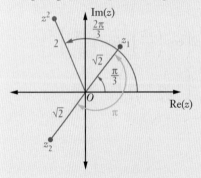

$z^2 = 2\left(\cos\dfrac{2\pi}{3} + i\sin\dfrac{2\pi}{3}\right)$ and $z_1 = \sqrt{2}\left(\cos\dfrac{\pi}{3} + i\sin\dfrac{\pi}{3}\right)$, so we see that z_1 has half the

argument of z^2 and the modulus of z_1 is the square root of mod z^2. The other root z_2 is also graphed, and it is equally spaced around the origin at an angle of π from z_1 with the

same modulus: $z_2 = \sqrt{2}\left[\cos\left(-\dfrac{2\pi}{3}\right) + i\sin\left(-\dfrac{2\pi}{3}\right)\right] = \sqrt{2}e^{-\frac{2i\pi}{3}}$. This will be explained fully

later in this chapter when we examine the roots of complex numbers in more detail.

Using this graphical approach for Example 4c on pages 128–129, we could have seen more easily that $\sqrt{-i} = \left\{ 1\left[\cos\left(-\dfrac{\pi}{2}\right) + i\sin\left(-\dfrac{\pi}{2}\right) \right] \right\}^{\frac{1}{2}} = 1\left[\cos\left(-\dfrac{\pi}{4}\right) + i\sin\left(-\dfrac{\pi}{4}\right) \right]$ without all the algebra.

This is a useful technique to keep in mind.

Exercise 4.02 Quadratic equations with complex coefficients

1 Solve each quadratic equation.

   a $x^2 - 2ix + 3 = 0$ **b** $x^2 + 6ix = 5$

   c $x^2 - (3 + 2i)x + (1 + 3i) = 0$ **d** $3x^2 - 5ix + 2 = 0$

2 Use a graphical approach with De Moivre's theorem to solve each equation in exact form.

   a $z^2 = i$ **b** $z^2 = -9i$ **c** $z^2 = \cos\dfrac{\pi}{3} + i\sin\dfrac{\pi}{3}$

   d $z^2 - 1 + i = 0$ **e** $z^2 = e^{3i\pi}$ **f** $z^2 = 16e^{-\frac{2i\pi}{3}}$

   g $z^2 = \cos 4 + i\sin 4$

3 Show that $5 - 2i$ is a root of $x^2 - 6x + ix + 7 + 3i = 0$. Hence or otherwise, find the other root.

4 Find the quadratic equation that has the roots:

   a $3 - i, 1 + 7i$ **b** $-4i, 3 + 5i$ **c** $\dfrac{2+i}{3}, 1 - i$

5 The quadratic equation $ax^2 + px + q = 0$ has the roots $\dfrac{1}{2} + \dfrac{i}{2}$ and $\dfrac{3}{2} - \dfrac{i}{2}$. Find the values of a, p and q.

6 a Find $\sqrt{8 + 6i}$.

   b Use the quadratic formula to solve the complex quadratic equation $z^2 + 2z + 4iz = 11 + 2i$, expressing the solutions in the form $a + ib$. [Hint: you will need part **a**].

7 Solve each quadratic equation.

   a $x^2 - (1 + i)x + i = 0$

   b $x^2 - 2x + 1 - 2i = 0$

   c $x^2 - 3x + 3ix - 5i = 0$

   d $x^2 - (4 + 3i)x = 2 - 8i$

   e $ix^2 - 3ix - x + 2i + 2 = 0$

The cubic formula

Just as the quadratic formula is used to find the roots of a quadratic equation, there exists a formula to find the roots of a cubic equation. This formula was published in 1545 by an Italian mathematician named Girolamo Cardano (1501–1576). When he was deriving the formula, he was troubled because some of the solutions to a cubic equation required finding the square root of a negative number, such as $\sqrt{-15}$. Mathematicians did not yet know how to handle this.

Cardano then made an astute observation about conjugates. If 2 of the roots were say $5 - \sqrt{-15}$ and $5 + \sqrt{-15}$, he noticed that the product of the 2 roots was real, $\left(5 - \sqrt{-15}\right)\left(5 + \sqrt{-15}\right)$ $= 25 - (-15) = 40$, so the offending $\sqrt{-15}$ disappeared. This was the beginning of complex numbers.

Cardano was not the first to create this formula, only the first to publish it. Mathematicians were very competitive back then. Whoever published first got the fame!

Polynomials with real coefficients

Real and imaginary factors

Complex polynomials

4.03 Polynomial equations

We have seen that if a quadratic equation with real coefficients has complex roots then the roots will be complex conjugate roots.

Complex conjugate root theorem

If a polynomial equation $P(z) = 0$ has real coefficients and if $\alpha = a + ib$, $a, b \in \mathbb{R}$, is a root of $P(z) = 0$, then $\bar{\alpha} = a - ib$ is also a root of $P(z) = 0$.

We use the properties of complex numbers developed in Chapter 1 to prove this theorem.

Proof

Let $P(x) = a_n x^n + a_{n-1} x^{n-1} + a_{n-2} x^{n-2} + \ldots + a_2 x^2 + a_1 x + a_0 = 0$ be a polynomial equation where $a_n, a_{n-1}, a_{n-2}, \ldots, a_2, a_1, a_0 \in \mathbb{R}$.

If α is a root then

$$P(\alpha) = a_n \alpha^n + a_{n-1} \alpha^{n-1} + a_{n-2} \alpha^{n-2} + \ldots + a_2 \alpha^2 + a_1 \alpha + a_0 = 0$$

Using the property that if $\alpha = 0$ then $\bar{\alpha} = 0$, then we can deduce

$$\overline{P(\alpha)} = \overline{a_n \alpha^n + a_{n-1} \alpha^{n-1} + a_{n-2} \alpha^{n-2} + \ldots + a_2 \alpha^2 + a_1 \alpha + a_0} = 0$$

Then, using $\overline{\alpha_1 + \alpha_2} = \overline{\alpha_1} + \overline{\alpha_2}$ it follows that

$$\overline{a_n \alpha^n} + \overline{a_{n-1}\alpha^{n-1}} + \overline{a_{n-2}\alpha^{n-2}} + ... + \overline{a_2 \alpha^2} + \overline{a_1 \alpha} + \overline{a_0} = 0$$

Then, using $\overline{a_1 \alpha_1} = \overline{a_1}\,\overline{\alpha_1}$ we can deduce

$$\overline{a_n}\,\overline{\alpha^n} + \overline{a_{n-1}}\,\overline{\alpha^{n-1}} + \overline{a_{n-2}}\,\overline{\alpha^{n-2}} + ... + \overline{a_2}\,\overline{\alpha^2} + \overline{a_1}\,\overline{\alpha} + \overline{a_0} = 0$$

Then using $\overline{(\alpha_1)^n} = \left(\overline{\alpha_1}\right)^n$ for $n \in \mathbb{N}$, we can say

$$\overline{a_n}\left(\overline{\alpha_n}\right)^n + \overline{a_{n-1}}\left(\overline{\alpha_{n-1}}\right)^n + \overline{a_{n-2}}\left(\overline{\alpha_{n-2}}\right)^n + ... + \overline{a_2}\left(\overline{\alpha_2}\right)^n + \overline{a_1}\left(\overline{\alpha_1}\right)^n + \overline{a_0} = 0$$

But if $a_n, ..., a_0 \in \mathbb{R}$ then $\overline{a_n} = a_n, \overline{a_{n-1}} = a_{n-1}, ..., \overline{a_1} = a_1, \overline{a_0} = a_0$, so

$$a_n\left(\overline{\alpha_n}\right)^n + a_{n-1}\left(\overline{\alpha_{n-1}}\right)^n + a_{n-2}\left(\overline{\alpha_{n-2}}\right)^n + ... + a_2\left(\overline{\alpha_2}\right)^n + a_1\left(\overline{\alpha_1}\right)^n + a_0 = 0$$

Therefore we can deduce that $P(\overline{\alpha}) = 0$

that is, $\overline{\alpha}$ is also a root of $P(x) = a_n x^n + a_{n-1}x^{n-1} + a_{n-2}x^{n-2} + ... + a_2 x^2 + a_1 x + a_0 = 0$.

QED

It follows then that if $x - \alpha$ is a factor of $P(z) = 0$ then $x - \overline{\alpha}$ is also a factor of $P(z) = 0$.

This theorem has implications depending on the degree n of the polynomial $P(z)$:

Real and complex roots

Given a polynomial equation $P(z) = 0$ with real coefficients and of degree n:

- if n is odd, then $P(z) = 0$ has at least one real root and the complex roots will come in conjugate pairs

- if n is even, then $P(z) = 0$ has an even number of real roots or no real roots, and the complex roots will come in conjugate pairs

- if $P(z) = 0$ has complex roots $\alpha = a + ib$ and $\overline{\alpha} = a - ib$ then $P(z) = 0$ will have a quadratic factor of the form $(x - \alpha)(x - \overline{\alpha}) = [x^2 - (\alpha + \overline{\alpha})x + \alpha\overline{\alpha}]$.

As with polynomial equations with real roots, we can use the remainder and factor theorems to solve polynomial equations with complex roots.

EXAMPLE 6

Consider the polynomial $P(x) = x^3 - 4x^2 + 6x - 4$.

a Show that $x = 2$ is a root of $x^3 - 4x^2 + 6x - 4 = 0$.

b Hence solve the equation $x^3 - 4x^2 + 6x - 4 = 0$.

Solution

a Using the factor theorem, show that $P(2) = 0$.

$$P(2) = 2^3 - 4(2)^2 + 6(2) - 4$$

$$= 0$$

Therefore $x = 2$ is a root.

b To solve the equation we can use the fact that $P(x)$ is monic so
$x^3 - 4x^2 + 6x - 4 = (x - 2)(x^2 + bx + c)$ where $b, c \in \mathbb{R}$. We then solve for b and c.

Expand the RHS and equate coefficients:

$-2c = -4$ \qquad Equating the constant terms

$c = 2$

$-4 = -2 + b$ \qquad Equating the coefficients of x^2

$b = -2$

$\therefore x^3 - 4x^2 + 6x - 4 = (x - 2)(x^2 - 2x + 2)$

Now factorise fully or solve the quadratic factor to find all of the roots.

$$(x - 2)(x^2 - 2x + 2) = (x - 2)[(x^2 - 2x + 1) + 1]$$

$$= (x - 2)[(x - 1)^2 - i^2]$$

$$= (x - 2)(x - 1 - i)(x - 1 + i)$$

$$= (x - 2)[x - (1 + i)][(x - (1 - i)]$$

So the roots are $x = 2, 1 + i, 1 - i$. \qquad Note: the complex roots are a conjugate pair.

Creating polynomial equations

Sometimes we are given the roots and need to find the polynomial.

Given that $\alpha = \dfrac{3i}{2}$ and $\beta = 2 + i$ are 2 roots of the polynomial $P(x) = ax^4 + bx^3 + cx^2 + dx + e$ where $a, b, c, d, e \in \mathbb{Z}$, find $P(x)$.

Solution

Using the fact that the coefficients are real, the complex roots must be in conjugate pairs.

Thus $\bar{\alpha} = -\dfrac{3i}{2}$ and $\bar{\beta} = 2 - i$ are also roots.

The structure of $P(x)$ is then

$$a\left(x - \frac{3i}{2}\right)\left(x + \frac{3i}{2}\right)[x - (2 + i)][x - (2 - i)]$$

$$= a\left(x^2 - \frac{9i^2}{4}\right)[x^2 - (2 + i + 2 - i)x + (2 + i)(2 - i)]$$

$$= a\left(x^2 + \frac{9}{4}\right)(x^2 - 4x + 5)$$

Now since all coefficients are integers, let $a = 4$ to eliminate the fraction.

$$ax^4 + bx^3 + cx^2 + dx + e \equiv 4\left(x^2 + \frac{9}{4}\right)(x^2 - 4x + 5)$$

$$= (4x^2 + 9)(x^2 - 4x + 5)$$
$$= 4x^4 - 16x^3 + 20x^2 + 9x^2 - 36x + 45$$
$$= 4x^4 - 16x^3 + 29x^2 - 36x + 45$$

Therefore $P(x) = 4x^4 - 16x^3 + 29x^2 - 36x + 45$.

Note: This is not the only solution for $P(x)$. Any constant multiple of this solution is also a solution.

Exercise 4.03 Polynomial equations

1 Factorise each $P(z)$, showing:

    i   all real factors   **ii**   all complex factors

  a $z^3 + z$ **b** $z^3 - 6z^2 + 10z$

  c $z^3 + 1$ **d** $z^3 - 8$

  e $z^4 + 3z^2 - 4$ **f** $z^4 + 10z^2 + 9$

  g $z^3 + z^2 + z + 1$ **h** $z^3 - z^2 + 2z - 2$

2 Find all the roots of $P(z) = 0$, given:

  a $z = -2$ is a root of $z^3 - 2z^2 - 3z + 10 = 0$

  b $z = i$ is a root of $z^4 - 2z^3 - 2z^2 - 2z - 3 = 0$

  c $z = e^{\frac{i\pi}{4}}$ is a root of $z^3 - \sqrt{2}z^2 - z^2 + z + \sqrt{2}z - 1 = 0$

  d $z^2 - 4z - 5$ is a factor of $z^4 - 6z^3 + 6z^2 - 2z - 15$

  e $z - 2 + i\sqrt{5}$ is a factor of $z^3 - 2z^2 + z + 18$

3 Consider the polynomial $z^4 - 6z^3 + pz^2 + qz + 70 = 0$ where $p, q \in \mathbb{R}$. Given that $1 + 3i$ is a root, find p and q and the other roots of the equation.

4 Find all the roots of each polynomial equation.

  a $z^3 - 3z^2 + z - 3 = 0$

  b $z^4 - z^3 - 3z^2 + 4z - 4 = 0$

5 Find the polynomial equation of minimum degree with integer coefficients that has roots:

  a $z = 1 + i, z = 5$ **b** $z = 1 - i\sqrt{3}, z = -2$

  c $z = 3 + i\sqrt{5}, z = -1 - 4i$ **d** $z = \dfrac{1}{3} + \dfrac{i\sqrt{3}}{3}, z = 4$

  e $z = e^{-\frac{i\pi}{3}}, z = -3$

6 Find the minimum degree of a polynomial that has real coefficients with roots:

  a $x = 4, x = -2, x = -3i$

  b $x = \sqrt{2} + i\sqrt{2}, x = -2 - 5i, x = -1$

4.04 Operations on the complex plane

We saw in Chapter 1, *Complex numbers*, that complex numbers can be expressed as points or vectors on the complex plane. They can also be expressed in Cartesian ($a + ib$), in polar (modulus–argument) or in exponential ($re^{i\theta}$) form.

We now look at multiplication and division of complex numbers on the complex plane.

Multiplying complex numbers

Multiplying complex numbers

If $z_1 = r_1(\cos \theta_1 + i \sin \theta_1)$ and $z_2 = r_2(\cos \theta_2 + i \sin \theta_2)$, then their product is

$$z_1 z_2 = r_1 r_2 [\cos (\theta_1 + \theta_2) + i \sin (\theta_1 + \theta_2)]$$

Geometrically, this means that the product of vectors z and w will have length $|z||w|$ and argument arg z + arg w. Alternatively, if a vector z is multiplied by a vector w then its modulus is increased by the factor $|w|$ and its argument is rotated **anticlockwise** by arg w.

Geometric representation of multiplication

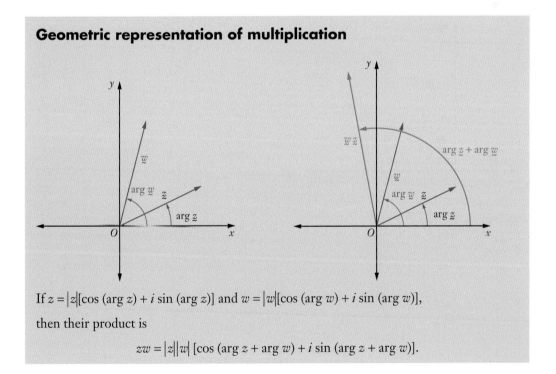

If $z = |z|[\cos (\arg z) + i \sin (\arg z)]$ and $w = |w|[\cos (\arg w) + i \sin (\arg w)]$,

then their product is

$$zw = |z||w| [\cos (\arg z + \arg w) + i \sin (\arg z + \arg w)].$$

If $z = 3\left(\cos\dfrac{\pi}{3} + i\sin\dfrac{\pi}{3}\right)$ and $w = 2\left(\cos\dfrac{\pi}{6} + i\sin\dfrac{\pi}{6}\right)$, plot z and zw on an Argand diagram.

Solution

Multiply the moduli:

$|z||w| = 3 \times 2$

$\qquad = 6$

Add the arguments:

$\text{Arg}\,zw = \dfrac{\pi}{3} + \dfrac{\pi}{6}$

$\qquad = \dfrac{\pi}{2}$

So plotting zw we have dilated ('stretched') the length of z by a factor of 2 and rotated z by an angle of $\dfrac{\pi}{6}$.

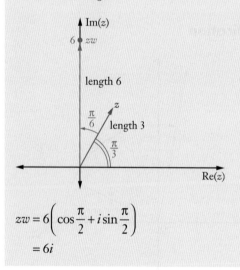

$zw = 6\left(\cos\dfrac{\pi}{2} + i\sin\dfrac{\pi}{2}\right)$

$\quad = 6i$

Dividing complex numbers

Dividing complex numbers

If $z_1 = r_1(\cos \theta_1 + i \sin \theta_1)$ and $z_2 = r_2(\cos \theta_2 + i \sin \theta_2)$, then their quotient is

$$\frac{z_1}{z_2} = \frac{r_1}{r_2}[\cos (\theta_1 - \theta_2) + i \sin (\theta_1 - \theta_2)]$$

Geometrically, this means that the quotient of vectors $\underset{\sim}{z}$ and $\underset{\sim}{w}$ will have length $\dfrac{|\underset{\sim}{z}|}{|\underset{\sim}{w}|}$

and argument $\arg \underset{\sim}{z} - \arg \underset{\sim}{w}$. Alternatively, if a vector $\underset{\sim}{z}$ is divided by a vector $\underset{\sim}{w}$, then its modulus is decreased by the factor $|\underset{\sim}{w}|$ and its argument is rotated **clockwise** by $\arg \underset{\sim}{w}$.

Geometric representation of division

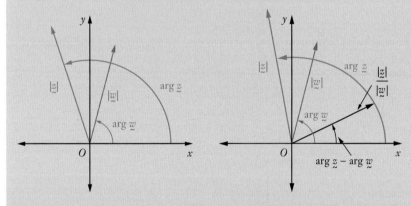

If $z = |z|[\cos (\arg z) + i \sin (\arg z)]$ and $w = |w|[\cos (\arg w) + i \sin (\arg w)]$,

then their quotient is

$$\frac{z}{w} = \frac{|z|}{|w|}[\cos (\arg z - \arg w) + i \sin (\arg z - \arg w)].$$

EXAMPLE 9

If $z = 6\left(\cos \dfrac{5\pi}{6} + i \sin \dfrac{5\pi}{6}\right)$ and $w = 2\left(\cos \dfrac{2\pi}{3} + i \sin \dfrac{2\pi}{3}\right)$, plot z and $\dfrac{z}{w}$ on the complex plane.

Solution

Divide the moduli:

$$\dfrac{|z|}{|w|} = \dfrac{6}{2}$$

$$= 3$$

Subtract the arguments:

$$\text{Arg}\,\dfrac{z}{w} = \dfrac{5\pi}{6} - \dfrac{2\pi}{3}$$

$$= \dfrac{\pi}{6}$$

So plotting $\dfrac{z}{w}$ we have dilated (compressed) the length of z by a factor of $\dfrac{1}{2}$ and rotated z by an angle of $\dfrac{2\pi}{3}$ clockwise.

$$\dfrac{z}{w} = 3\left(\cos \dfrac{\pi}{6} + i \sin \dfrac{\pi}{6}\right)$$

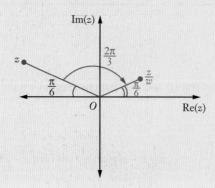

ISBN 9780170413435

Multiplying and dividing by i

Since $i = 1\left(\cos\dfrac{\pi}{2} + i\sin\dfrac{\pi}{2}\right)$, it is easy to see that if we multiply a complex number z by i then this is equivalent to rotating z by $\dfrac{\pi}{2}$ anticlockwise. Similarly, if we divide by i then this is equivalent to rotating z by $\dfrac{\pi}{2}$ clockwise.

Multiplying and dividing by i

Multiplication by i is equivalent to a rotation of $\dfrac{\pi}{2}$ on the complex plane.

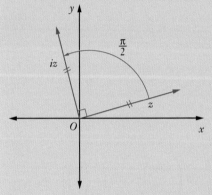

Division by i or multiplication by $-i$ are transformations equivalent to a rotation of $-\dfrac{\pi}{2}$ on the complex plane.

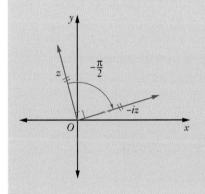
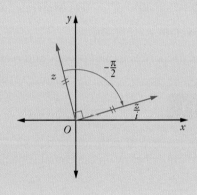

Notice that division by i and multiplication by $-i$ give the same result. We can prove this:

$$\dfrac{z}{i} = \dfrac{z}{i} \times \dfrac{i}{i} = \dfrac{iz}{-1} = -iz.$$

EXAMPLE 10

If $z = \sqrt{3}\left(\cos\dfrac{\pi}{5} + i\sin\dfrac{\pi}{5}\right)$, plot iz and $\dfrac{z}{i}$ on an Argand diagram.

Solution

To find iz, rotate the vector z anticlockwise by $\dfrac{\pi}{2}$ radians. To find $\dfrac{z}{i}$ rotate the vector z clockwise by $\dfrac{\pi}{2}$ radians.

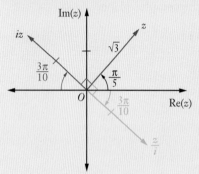

In polar form, $iz = \sqrt{3}\left(\cos\dfrac{7\pi}{10} + i\sin\dfrac{7\pi}{10}\right)$ and $\dfrac{z}{i} = \sqrt{3}\left[\cos\left(-\dfrac{3\pi}{10}\right) + i\sin\left(-\dfrac{3\pi}{10}\right)\right]$.

Recall 2 properties we have met:

- $z^n = r^n(\cos n\theta + i\sin n\theta)$ De Moivre's theorem
- $z^{-1} = \dfrac{1}{z} = \dfrac{1}{r}[\cos(-\theta) + i\sin(-\theta)]$

We can use these properties to plot powers and reciprocals of z on the complex plane.

EXAMPLE 11

Given $z = 1(\cos\theta + i\sin\theta)$ in the diagram, plot each complex number below on the complex plane.

a iz

b $\dfrac{z}{i}$

c z^2

d $\dfrac{1}{z}$

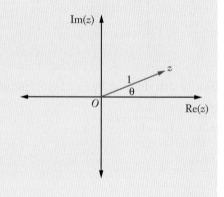

MATHS IN FOCUS 12. Mathematics Extension 2   ISBN 9780170413435

Solution

a $iz = \cos\left(\theta + \dfrac{\pi}{2}\right) + i\sin\left(\theta + \dfrac{\pi}{2}\right)$

b $\dfrac{z}{i} = \cos\left(\theta - \dfrac{\pi}{2}\right) + i\sin\left(\theta - \dfrac{\pi}{2}\right)$

c $z^2 = \cos 2\theta + i\sin 2\theta$

d $\dfrac{1}{z} = \cos(-\theta) + i\sin(-\theta)$

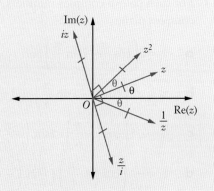

EXAMPLE 12

z and w with corresponding vectors are shown on the complex plane.

Prove that $z^2 + w^2 = 0$.

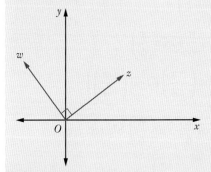

Solution

To find w we rotate z an angle of $90°$ around O. This is equivalent to multiplying by i.

$\therefore w = iz$.

Squaring both sides:

$$w^2 = (iz)^2$$
$$w^2 = -z^2$$
$$\therefore z^2 + w^2 = 0$$

Exercise 4.04 Operations on the complex plane

1 For each pair of complex numbers, find $|z_1||z_2|$ and arg z_1z_2. Hence sketch z_1z_2 on the complex plane.

a $z_1 = 2\left(\cos\dfrac{\pi}{6} + i\sin\dfrac{\pi}{6}\right)$ and $z_2 = 3\left(\cos\dfrac{\pi}{2} + i\sin\dfrac{\pi}{2}\right)$

b $z_1 = \sqrt{2}\left(\cos\dfrac{\pi}{3} + i\sin\dfrac{\pi}{3}\right)$ and $z_2 = \left(\cos\dfrac{2\pi}{3} + i\sin\dfrac{2\pi}{3}\right)$

c $z_1 = \sqrt{5}\left(\cos\left(-\dfrac{\pi}{6}\right) + i\sin\left(-\dfrac{\pi}{6}\right)\right)$ and $z_2 = 2\left(\cos\dfrac{\pi}{3} + i\sin\dfrac{\pi}{3}\right)$

d $z_1 = \dfrac{1}{\sqrt{3}}\left(\cos\left(-\dfrac{3\pi}{4}\right) + i\sin\left(-\dfrac{3\pi}{4}\right)\right)$ and $z_2 = \sqrt{3}\left(\cos\left(-\dfrac{\pi}{2}\right) + i\sin\left(-\dfrac{\pi}{2}\right)\right)$

2 For each z_1 and z_2, find $\dfrac{|z_1|}{|z_2|}$ and arg $\dfrac{z_1}{z_2}$. Hence sketch $\dfrac{z_1}{z_2}$ on the complex plane.

a $z_1 = \left(\cos\dfrac{2\pi}{3} + i\sin\dfrac{2\pi}{3}\right)$ and $z_2 = \left(\cos\dfrac{\pi}{6} + i\sin\dfrac{\pi}{6}\right)$

b $z_1 = 4\left(\cos\dfrac{3\pi}{4} + i\sin\dfrac{3\pi}{4}\right)$ and $z_2 = 2\left(\cos\dfrac{\pi}{2} + i\sin\dfrac{\pi}{2}\right)$

c $z_1 = 2\left(\cos\left(-\dfrac{\pi}{3}\right) + i\sin\left(-\dfrac{\pi}{3}\right)\right)$ and $z_2 = \sqrt{2}\left(\cos\dfrac{\pi}{6} + i\sin\dfrac{\pi}{6}\right)$

d $z_1 = 3\left(\cos\left(-\dfrac{7\pi}{12}\right) + i\sin\left(-\dfrac{7\pi}{12}\right)\right)$ and $z_2 = 6\left(\cos\left(-\dfrac{\pi}{4}\right) + i\sin\left(-\dfrac{\pi}{4}\right)\right)$

3 Find z_1z_2 if $z_1 = 2\left(\cos\dfrac{2\pi}{3} - i\sin\dfrac{2\pi}{3}\right)$ and $z_2 = \sqrt{2}\left(\cos\dfrac{\pi}{6} - i\sin\dfrac{\pi}{6}\right)$. Sketch z_1z_2.

4 Find $\dfrac{z_1}{z_2}$ if $z_1 = 6\left(\cos\dfrac{\pi}{2} - i\sin\dfrac{\pi}{2}\right)$ and $z_2 = 2\left(\cos\dfrac{3\pi}{4} - i\sin\dfrac{3\pi}{4}\right)$. Sketch $\dfrac{z_1}{z_2}$.

5 By first expressing in polar form, sketch $\dfrac{z_1z_2}{z_3}$ on the complex plane if

$z_1 = \sqrt{3}\left(\cos\left(-\dfrac{\pi}{6}\right) + i\sin\left(-\dfrac{\pi}{6}\right)\right), z_2 = \sqrt{6}\left(\cos\dfrac{2\pi}{3} - i\sin\dfrac{2\pi}{3}\right)$ and $z_3 = -3\left(\cos\dfrac{\pi}{3} + i\sin\dfrac{\pi}{3}\right)$.

6 Given the vector $\underset{\sim}{z}$ representing z on the complex plane, copy the diagram and sketch each expression. In each case, state which transformations of z are equivalent.

a iz **b** i^2z **c** i^3z

d $-iz$ **e** $-z$

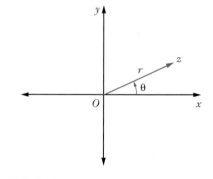

7 Given $\underset{\sim}{z}$ representing z on the complex plane, copy the diagram and sketch each expression. In each case, state which transformations of z are equivalent.

a $\dfrac{z}{i}$

b $\dfrac{z}{i^2}$

c $\dfrac{z}{i^3}$

d $-iz$

e $-z$

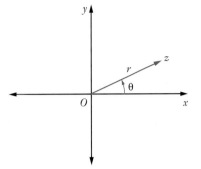

8 Given $\underset{\sim}{z}$ representing $z = \cos\theta + i\sin\theta$ on the complex plane, copy the diagram and sketch each expression.

Explain why $\overline{z} = \dfrac{1}{z}$.

Is this true for all z on the complex plane?

a z^2

b z^3

c $\dfrac{1}{z}$

d $\dfrac{1}{z^2}$

e z^{-3}

f $\overline{z}$

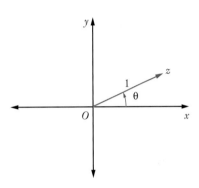

9 Given $|u| = |v|$, express the complex number v in terms of u.

a
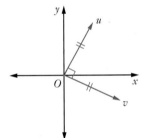

b
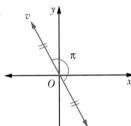

c
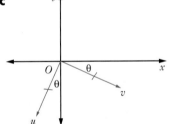

10 Given each z, sketch $\dfrac{1}{z}$.

a $z = 3\operatorname{cis}\dfrac{\pi}{4}$

b $z = \dfrac{1}{\sqrt{2}}\operatorname{cis}\dfrac{2\pi}{3}$

c $z = 4\operatorname{cis}\left(-\dfrac{5\pi}{6}\right)$

11 By first expressing in modulus–argument form, simplify each expression and sketch on an Argand diagram.

a $(1 + i)^3$

b $\left(\dfrac{1}{2} + \dfrac{i}{2}\right)^{-2}$

c $\left(1 - i\sqrt{3}\right)^{-4}$

d $\left(-\dfrac{3}{2} + i\dfrac{\sqrt{3}}{2}\right)^{-3}$

e $\left(\sqrt{2}e^{\frac{i\pi}{4}}\right)^{8}$

f $\left(\dfrac{1}{2}e^{-\frac{i\pi}{6}}\right)^{-3}$

12 Describe each transformation from z to w.

a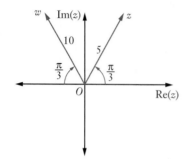

b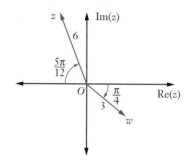

c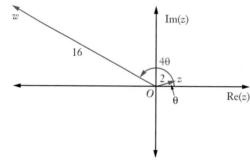

13 Consider the complex numbers z_1, z_2, z_3, O, which correspond to the vertices of a square $ABCO$, where A and C are shown in the diagram.

a Express z_3 in terms of z_1.

b Explain why B is represented by $z_1 - iz_1$.

c The diagonals intersect at M. Find the complex number corresponding to M.

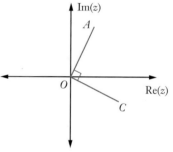

14 Consider the complex numbers α, β, γ, δ, which correspond to the vertices of a rectangle $ABCD$ where $AD = 3AB$.

a Show that $\dfrac{\delta - \gamma}{i} = \dfrac{\beta - \gamma}{3}$.

b Find the complex number corresponding to M, the point of intersection of the diagonals, in terms of δ and β.

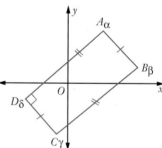

ISBN 9780170413435

15 Consider the points P, Q, R representing w_1, w_2, w_3 on the complex plane as shown.

Prove that $\dfrac{w_2 - w_1}{w_3 - w_1} = i$.

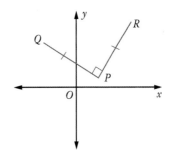

16 Points A and B representing complex numbers z_1, z_2 respectively form a triangle together with O on the complex plane. It is given that $|z_1| = |z_2|$.

    a Plot the information and draw the point C representing $z_1 + z_2$.

    b Describe the quadrilateral $OACB$.

    c Given that $|z_1| = |z_2 - z_1|$, what can you deduce about $\triangle ABC$?

17 Prove that the vectors representing the complex numbers u, v and $\dfrac{u - iv}{1 - i}$ form a right-angled triangle.

INVESTIGATION

FRACTALS AND COMPLEX NUMBERS

Fractals are part of chaos theory. This is a relatively new area in the history of mathematics, the development of which was only possible after the invention of the high-powered IBM computer in the 1970s. Before that, it was difficult to process numerous repetitive calculations quickly. Benoit Mandelbrot coined the term 'fractal', and with his computer he generated what is now known as the Mandelbrot set.

The Mandelbrot set is the set of points that can be generated by multiple iterations using the equation $z_{n+1} = z_n^2 + C$ where $z \in \mathbb{C}$. We can choose the initial points for z_1 and C then iterate a number of times, selecting an arbitrary value of n. This can be done using a spreadsheet. After n iterations, we calculate $|z_n|$. If $|z_n| < 2$ then z_1 is a member of the Mandelbrot set.

For instance, choose $z_1 = 0.5 - 0.5i$ and $C = 0.25 + 0.25i$. Set up a spreadsheet similar to the one below and then determine whether or not $|z_n| < 2$ for some arbitrary value of n (say 50).

ISBN 9780170413435   **4.** Applying complex numbers

n	Re(z_n)	Im(z_n)	Re(C)	Im(C)	Re(z_n^2)	Im(z_n^2)	$|z_n|$
1	0.5	-0.5	0.25	0.25	0	-0.5	0.707107
2	0.25	-0.25	0.25	0.25	0	-0.125	0.353553
3	0.25	0.125	0.25	0.25	0.046875	0.0625	0.279508
4	0.296875	0.3125	0.25	0.25	-0.00952	0.185547	0.431035
5	0.240479	0.435547	0.25	0.25	-0.13187	0.209479	0.497525
6	0.118129	0.459479	0.25	0.25	-0.19717	0.108556	0.474421
7	0.052833	0.358556	0.25	0.25	-0.12577	0.037887	0.362427
8	0.124229	0.287887	0.25	0.25	-0.06745	0.071528	0.313547
9	0.182554	0.321528	0.25	0.25	-0.07005	0.117392	0.369738
10	0.179946	0.367392	0.25	0.25	-0.1026	0.132221	0.409094
11	0.147403	0.382221	0.25	0.25	-0.12437	0.112681	0.409659
12	0.125635	0.362681	0.25	0.25	-0.11575	0.091131	0.383825
13	0.134246	0.341131	0.25	0.25	-0.09835	0.091591	0.366595
14	0.151652	0.341591	0.25	0.25	-0.09369	0.103606	0.373742
15	0.156314	0.353606	0.25	0.25	-0.1006	0.110547	0.386615
16	0.149397	0.360547	0.25	0.25	-0.10767	0.107729	0.390274
17	0.142325	0.357729	0.25	0.25	-0.10771	0.101828	0.385002
18	0.142286	0.351828	0.25	0.25	-0.10354	0.100121	0.37951
19	0.146463	0.350121	0.25	0.25	-0.10113	0.102559	0.37952
20	0.148867	0.352559	0.25	0.25	-0.10214	0.104969	0.3827
21	0.147863	0.354969	0.25	0.25	-0.10414	0.104974	0.384534
22	0.145861	0.354974	0.25	0.25	-0.10473	0.103554	0.383773
23	0.145269	0.353554	0.25	0.25	-0.1039	0.102721	0.382234
24	0.146103	0.352721	0.25	0.25	-0.10307	0.103067	0.381783
25	0.146934	0.353067	0.25	0.25	-0.10307	0.103755	0.382421
26	0.146933	0.353755	0.25	0.25	-0.10355	0.103957	0.383056
27	0.146447	0.353957	0.25	0.25	-0.10384	0.103672	0.383056
28	0.146161	0.353672	0.25	0.25	-0.10372	0.103386	0.382683
29	0.14628	0.353386	0.25	0.25	-0.10348	0.103386	0.382465

If it is the case then $z_1 = 0.5 - 0.5i$ belongs to the Mandelbrot set. We can colour all such points, say, black. The resulting set looks like the image below.

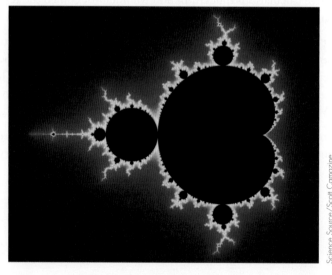

Science Source/Scott Camazine

1 Set up your own spreadsheet and experiment with differing values of z_1 and C.

2 What is the meaning of 'self-similar'?

3 The Mandelbrot set is contained within a circle. What is its radius?

4 What can you say about the area and perimeter of the Mandelbrot set?

5 There are other types of fractals. Find some of them and if they can be generated using an equation. Try to generate them if you can.

6 Where are fractals used now?

7 Find a fractal generator on the Internet and build your own fractal.

4.05 Roots of unity

The solutions for z to an equation in the form $z^n = 1$ are called **roots of unity**, where unity means 1. The equation can be solved by factorising if n is not too large.

EXAMPLE 13

a Show that $a^3 - b^3 = (a - b)(a^2 + ab + b^2)$.

b Use the above identity to find the roots of $z^3 = 1$ and then show them on an Argand diagram.

Solution

a RHS $= (a - b)(a^2 + ab + b^2)$

$\qquad = a^3 + a^2b + ab^2 - a^2b - ab^2 - b^3$

$\qquad = a^3 - b^3$

b $$z^3 = 1$$

$$z^3 - 1 = 0$$

$$(z - 1)(z^2 + 1z + 1^2) = 0 \qquad a = z, b = 1$$

$$(z - 1)(z^2 + z + 1) = 0$$

$$\therefore z = 1 \quad \text{or} \quad z^2 + z + 1 = 0$$

$$\therefore z = 1 \quad \text{or} \quad z = \frac{-1 \pm \sqrt{1^2 - 4(1)(1)}}{2}$$

$$\therefore z = 1 \quad \text{or} \quad z = \frac{-1 \pm \sqrt{-3}}{2}$$

$$\therefore z = 1 \quad \text{or} \quad z = \frac{-1 \pm i\sqrt{3}}{2}$$

In polar form these solutions are

$$z = 1, z = \cos\frac{2\pi}{3} + i\sin\frac{2\pi}{3}, z = \cos\left(-\frac{2\pi}{3}\right) + i\sin\left(-\frac{2\pi}{3}\right).$$

Note that we have one real and 2 complex conjugate solutions that are **equally spaced** on the complex plane around the origin from $z = 1$. Since $z = 1$ satisfies the equation $z^3 = 1$ then $z = 1$ is a solution.

Plotting these on an Argand diagram we have:

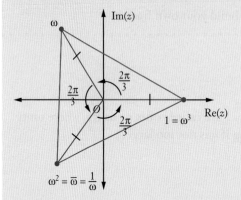

It is conventional to name the complex cube roots of unity using the small Greek letter omega, ω.

If $\omega = \cos\dfrac{2\pi}{3} + i\sin\dfrac{2\pi}{3}$ $\left[\text{or } \cos\left(-\dfrac{2\pi}{3}\right) + i\sin\left(-\dfrac{2\pi}{3}\right)\right]$, then it can be shown that the other 2 roots are ω^2 and ω^3.

Cube roots of unity

The cube roots of unity are the solutions to the equation $z^3 = 1$.

A complex cube root of unity is denoted by $\omega = \cos\dfrac{2\pi}{3} + i\sin\dfrac{2\pi}{3}$

or $\cos\left(-\dfrac{2\pi}{3}\right) + i\sin\left(-\dfrac{2\pi}{3}\right)$

The following properties hold:

- $\omega^2 = \bar{\omega} = \dfrac{1}{\omega}$

- $\omega^2 + \omega + 1 = 0$

- $\omega^3 = 1$

ISBN 9780170413435

EXAMPLE 14

Find the 4 roots of $z^4 = 1$ and show them on the complex plane.

Solution

$$z^4 = 1$$
$$z^4 - 1 = 0$$
$$(z^2 - 1)(z^2 + 1) = 0$$
$$(z - 1)(z + 1)(z - i)(z + i) = 0$$

$\therefore z = 1, -1, i$ or $-i$

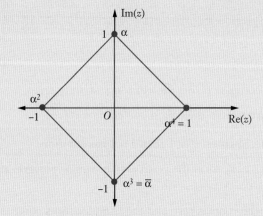

Note that we have 2 real and 2 complex conjugate solutions that are equally spaced on the complex plane around the origin from $z = 1$. Since $z = 1$ satisfies the equation $z^4 = 1$ then $z = 1$ is a solution.

There is no convention for naming higher complex roots of unity so in this example we have named the first one α, so the 4 roots are α, α^2, α^3 and α^4 where

$$\alpha^3 = \bar{\alpha} = \frac{1}{\alpha},$$

$$\alpha^3 + \alpha^2 + \alpha + 1 = 0,$$

$$\alpha^2 = -1$$

and $\alpha^4 = 1$.

The equation $z^n = 1$ cannot be solved by factorising if n is too large. However, notice that when graphed on the complex plane, the roots are equally spaced around the origin with one of the roots at $z = 1$. Since $z = 1$ satisfies the equation $z^n = 1$ then $z = 1$ is one of the solutions.

This observation can be used to solve $z^n = 1$ for any $n \in \mathbb{Z}$.

Roots of unity

If z is a complex number then the equation $z^n = 1$ has n solutions on the complex plane.

The solutions are called the nth roots of unity.

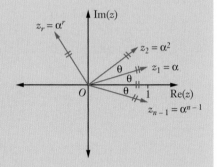

When plotted on the complex plane, the nth roots of unity $z_1, z_2, z_3, z_4, \ldots, z_n$ are equally spaced $\dfrac{2\pi}{n}$ radians apart around the origin, including $z = 1$.

Since $z = 1$ satisfies the equation $z^n = 1$ then $z = 1$ is a solution.

The roots $z_1, z_2, z_3, z_4, \ldots, z_n$ form the vertices of a regular polygon and the vector sum is zero. That is, $z_1 + z_2 + z_3 + z_4 + \ldots + z_n = 0$.

Therefore the sum of the roots is zero and the complex roots come in conjugate pairs.

Note also that if $z_1 = \alpha, z_2 = \alpha^2, z_3 = \alpha^3, \ldots, z_n = \alpha^n = 1$ then $z_1 + z_2 + z_3 + z_4 + \ldots + z_n = 0$ can be written as $1 + \alpha + \alpha^2 + \alpha^3 + \ldots + \alpha^{n-1} = 0$. We can note the conjugate pairs:

$\overline{\alpha} = \alpha^{n-1}, \overline{\alpha^2} = \alpha^{n-2}, \overline{\alpha^3} = \alpha^{n-3}, \ldots$

We can now solve $z^n = 1$ algebraically by first expressing 1 in polar form: $1 = \cos 0 + i \sin 0$.

ISBN 9780170413435

Find the complex roots of $z^7 = 1$ and show them on the complex plane.

Solution

Solving $z^7 = 1$ algebraically in polar form,

Let $z = r(\cos\theta + i\sin\theta)$ for $-\pi < \theta \leq \pi$

LHS $= z^7 = r^7(\cos 7\theta + i\sin 7\theta)$ using De Moivre's theorem

RHS $= 1 = 1(\cos 0 + i\sin 0)$

Then equating we have:

$r^7(\cos 7\theta + i\sin 7\theta) = 1(\cos 0 + i\sin 0)$.

$\therefore r = 1$ and $\cos 7\theta = \cos 0$ for $-7\pi < 7\theta \leq 7\pi$

$\cos 7\theta = 1$

Solving:

$$7\theta = 0, \pm 2\pi, \pm 4\pi, \pm 6\pi$$

$$\therefore \quad \theta = 0, \pm\frac{2\pi}{7}, \pm\frac{4\pi}{7}, \pm\frac{6\pi}{7}$$

For convenience, we will use the shorthand notation $\text{cis}\,\theta = \cos\theta + i\sin\theta$.
Therefore the 7 solutions are:

$$z = \text{cis}\,0, \ \text{cis}\,\frac{2\pi}{7}, \ \text{cis}\,\frac{4\pi}{7}, \ \text{cis}\,\frac{6\pi}{7}, \ \text{cis}\,\frac{-6\pi}{7}, \ \text{cis}\,\frac{-4\pi}{7}, \ \text{cis}\,\frac{-2\pi}{7}$$

$$z = 1, \ \text{cis}\,\frac{2\pi}{7}, \ \text{cis}\,\frac{4\pi}{7}, \ \text{cis}\,\frac{6\pi}{7}, \ \text{cis}\,\frac{-6\pi}{7}, \ \text{cis}\,\frac{-4\pi}{7}, \ \text{cis}\,\frac{-2\pi}{7}$$

Plotting these we see that the roots are equally spaced around the origin on the complex plane, starting at $z = 1$.

Geometrically we could bypass the algebra and use the pattern developed above to plot 7 equally spaced roots $\dfrac{2\pi}{7}$ apart around $z = 1$. Joining the roots forms a regular septagon (polygon with 7 equal sides).

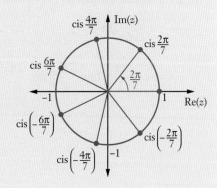

EXAMPLE 16

Find the 5 roots of $z^5 = 1$ and plot them on the complex plane.

a If ρ is a complex root of $z^5 = 1$, explain why $\rho^4 + \rho^3 + \rho^2 + \rho + 1 = 0$.

b Factorise $z^5 - 1$ into one linear and 2 quadratic factors with real coefficients.

c Find the exact value of $\cos\dfrac{2\pi}{5}$.

d Show that $\cos\dfrac{2\pi}{5} - \cos\dfrac{\pi}{5} = -\dfrac{1}{2}$.

Solution

The 5th roots of 1 are equally spaced $\dfrac{2\pi}{5}$ apart from $z = 1$, and the roots form a regular pentagon.

a Since the sum of the roots is zero then
$\rho^4 + \rho^3 + \rho^2 + \rho + 1 = 0$.

b We can factorise using the roots.

$$z^5 - 1 = (z-1)(z-\rho)(z-\rho^2)(z-\rho^3)(z-\rho^4)$$

Now using the conjugates we can pair them to create quadratics with real coefficients.

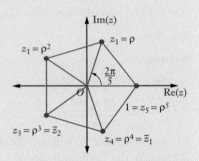

$$z^5 - 1 = (z-1)(z-\rho)(z-\bar{\rho})\left(z-\rho^2\right)\left(z-\overline{\rho^2}\right)$$

$$= (z-1)\left(z^2 - (\rho+\bar{\rho})z + \rho\bar{\rho}\right)\left(z^2 - \left(\rho^2+\overline{\rho^2}\right)z + \rho^2\overline{\rho^2}\right)$$

Now:

$$\rho + \bar{\rho} = \left(\cos\frac{2\pi}{5} + i\sin\frac{2\pi}{5}\right) + \left(\cos\frac{2\pi}{5} - i\sin\frac{2\pi}{5}\right)$$

$$= 2\cos\frac{2\pi}{5}$$

$$\rho\bar{\rho} = \left(\cos\frac{2\pi}{5} + i\sin\frac{2\pi}{5}\right)\left(\cos\frac{2\pi}{5} - i\sin\frac{2\pi}{5}\right)$$

$$= 1$$

Similarly:

$$\rho^2 + \overline{\rho^2} = \left(\cos \frac{4\pi}{5} + i\sin \frac{4\pi}{5} \right) + \left(\cos \frac{4\pi}{5} - i\sin \frac{4\pi}{5} \right)$$

$$= 2\cos \frac{4\pi}{5}$$

$$\rho^2 \overline{\rho^2} = \left(\cos \frac{4\pi}{5} + i\sin \frac{4\pi}{5} \right) \left(\cos \frac{4\pi}{5} - i\sin \frac{4\pi}{5} \right)$$

$$= 1$$

$$\therefore z^5 - 1 = (z-1)\left(z^2 - \left(2\cos \frac{2\pi}{5} \right) z + 1 \right)\left(z^2 - \left(2\cos \frac{4\pi}{5} \right) z + 1 \right)$$

c We know that $\rho^4 + \rho^3 + \rho^2 + \rho + 1 = 0$, or, as conjugates

$$\left(\rho + \overline{\rho} \right) + \left(\rho^2 + \overline{\rho^2} \right) + 1 = 0$$

$$\therefore 2\cos \frac{2\pi}{5} + 2\cos \frac{4\pi}{5} + 1 = 0$$

Using $\cos 2x = 2\cos^2 x - 1$ with $x = \frac{2\pi}{5}$, then we have

$$2\cos \frac{2\pi}{5} + 2\left(2\cos^2 \frac{2\pi}{5} - 1 \right) + 1 = 0$$

$$4\cos^2 \frac{2\pi}{5} + 2\cos \frac{2\pi}{5} - 1 = 0$$

Solving for $\cos \frac{2\pi}{5}$:

$$\cos \frac{2\pi}{5} = \frac{-2 \pm \sqrt{2^2 - 4(4)(-1)}}{2(4)}$$

$$= \frac{-2 \pm \sqrt{20}}{8}$$

$$= \frac{-2 \pm 2\sqrt{5}}{8}$$

$$= \frac{-1 \pm \sqrt{5}}{4}$$

Since $0 < \frac{2\pi}{5} < \frac{\pi}{2}$ then $\cos \frac{2\pi}{5} > 0$

So the exact value of $\cos \frac{2\pi}{5} = \frac{-1 + \sqrt{5}}{4}$.

d We know that $\rho^4 + \rho^3 + \rho^2 + \rho + 1 = 0$.

$$\left(\cos\frac{4\pi}{5} + i\sin\frac{4\pi}{5}\right) + \left(\cos\frac{4\pi}{5} - i\sin\frac{4\pi}{5}\right) + \left(\cos\frac{2\pi}{5} + i\sin\frac{2\pi}{5}\right) + \left(\cos\frac{2\pi}{5} - i\sin\frac{2\pi}{5}\right) = -1$$

$$2\cos\frac{4\pi}{5} + 2\cos\frac{2\pi}{5} = -1$$

$$\cos\frac{4\pi}{5} + \cos\frac{2\pi}{5} = -\frac{1}{2}$$

Now since $\cos\frac{4\pi}{5} = -\cos\left(\pi - \frac{4\pi}{5}\right) = -\cos\frac{\pi}{5}$ using $\cos(\pi - \theta) = -\cos\theta$

then $\cos\frac{2\pi}{5} - \cos\frac{\pi}{5} = -\frac{1}{2}$.

Solving problems using the roots of unity

There are many identities that can be derived and used in solving problems with complex numbers. You saw earlier that we can use De Moivre's theorem to derive results in trigonometry. The relationships between the cube roots of unity such as $\omega^2 = \dfrac{1}{\omega}$ were also demonstrated. Further examples are shown below.

EXAMPLE 17

ω is a complex cube root of unity.

a Show that $\omega^2 + \omega + 1 = 0$.

b Simplify $(1 + 2\omega + 3\omega^2)(1 + 2\omega^2 + 3\omega)$.

Solution

a Using $z^3 = 1$ then

$$z^3 - 1 = 0$$

$(z - 1)(z^2 + z + 1) = 0$

Then either $z - 1 = 0$ or $z^2 + z + 1 = 0$.

Since ω is complex then ω satisfies $z^2 + z + 1 = 0$; therefore $\omega^2 + \omega + 1 = 0$.

ISBN 9780170413435

b Now we can rearrange $\omega^2 + \omega + 1 = 0$ and substitute.

$$(1 + 2\omega + 3\omega^2)(1 + 2\omega^2 + 3\omega) = [1 + 2\omega + 3(-\omega - 1)][1 + 2(-\omega - 1) + 3\omega]$$
$$= (-2 - \omega)(-1 + \omega)$$
$$= 2 - 2\omega + \omega - \omega^2$$
$$= 2 - \omega - \omega^2$$
$$= 2 + 1 \qquad \text{since } -\omega - \omega^2 = 1$$
$$= 3$$

Alternatively we could expand $(1 + 2\omega + 3\omega^2)(1 + 2\omega^2 + 3\omega)$ and use the results $\omega^3 = 1$, $\omega^4 = \omega$, $\omega^5 = \omega^2$ and so on.

Exercise 4.05 Roots of unity

1 Solve each equation on the complex plane and plot the roots on an Argand diagram.

   a $z^3 = 1$ **b** $z^5 - 1 = 0$ **c** $z^8 = 1$ **d** $z^9 - 1 = 0$

2 Verify that each statement is true.

   a $\alpha = \cos\dfrac{\pi}{3} + i\sin\dfrac{\pi}{3}$ is a root of $z^6 = 1$.

   b $\alpha = \cos\left(-\dfrac{4\pi}{7}\right) + i\sin\left(-\dfrac{4\pi}{7}\right)$ is a root of $z^7 = 1$.

   c $\alpha = \cos\dfrac{\pi}{6} + i\sin\dfrac{\pi}{6}$ is a root of $z^{12} = 1$ and $\overline{\alpha^7} = \alpha^5$

   d $\alpha = \cos\left(-\dfrac{2\pi}{3}\right) + i\sin\left(-\dfrac{2\pi}{3}\right)$ is a root of $z^9 = 1$ and $\overline{\alpha^4} = \alpha^{-4}$.

3 List the conjugate root pairs of each equation if α is a complex root.

   a $z^7 - 1 = 0$ **b** $z^{11} = 1$

4 If $\beta = \cos\dfrac{2\pi}{5} + i\sin\dfrac{2\pi}{5}$ is a root of $z^5 - 1 = 0$, explain why $\beta + \beta^2 + \beta^3 + \beta^4 + 1 = 0$.

   Hence show that $\beta^2 + \beta + 1 + \dfrac{1}{\beta} + \dfrac{1}{\beta^2} = 0$.

5 If ω is a complex root of $z^3 = 1$, simplify each expression.

   a $\omega + \omega^2 + \omega^3$ **b** $(\omega + \omega^2)(\omega^2 + \omega^3)(\omega^3 + \omega)$

   c $(6\omega + 1)(6\omega^2 + 1)$ **d** $(1 - \omega - \omega^2)(\omega - \omega^2 - 1)(\omega^2 - 1 - \omega)$

   e $\omega^7 + \omega^8 + \omega^9 + \omega^{10} + \omega^{11} + \omega^{12} + \omega^{13} + \omega^{14}$

6 If α is a complex 9th root of unity, show that:

   a $1 + \alpha + \alpha^2 + \alpha^3 + \alpha^4 + \alpha^5 + \alpha^6 + \alpha^7 + \alpha^8 = 0$

   b $1 + \alpha^3 + \alpha^6 = 0$

   c $\cos\dfrac{2\pi}{9} + \cos\dfrac{4\pi}{9} + \cos\dfrac{6\pi}{9} + \cos\dfrac{8\pi}{9} = -\dfrac{1}{2}$

4.06 Roots of complex numbers

You saw in Example 13 on page 149 that the square roots of a complex number are equally spaced around the origin. This is true for all types of roots and the solution can be shown geometrically on the complex plane.

EXAMPLE 18

Find the square roots of $15 + 8i$.

Solution

Solving $15 + 8i = (a + ib)^2$, $a, b \in \mathbb{R}$ algebraically, we see that the square roots of $15 + 8i$ are $4 + i$ and $-4 - i$.

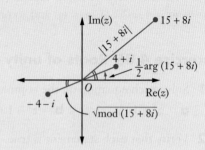

If we plot these on the complex plane we can make the following observations.

1. The first root $4 + i$ has half the argument of $15 + 8i$, that is, $\arg(4 + i) = \dfrac{1}{2}\arg(15 + 8i)$.

2. The modulus is the square root of $|15 + 8i|$, that is, $\operatorname{mod}(4 + i) = \sqrt{\operatorname{mod}(15 + 8i)}$.

3. The 2 square roots of $15 + 8i$ are equally spaced around O on the complex plane (halve 2π to get π apart in this case).

We can use this thinking to solve any complex equation in the form $z^n = a + ib$.

We know that the n solutions to $z^n = a + ib$ are equally spaced around the origin at a spacing of $\dfrac{2\pi}{n}$ apart. We need to find the first solution, which we will call z_1.

First write in polar form:

$z^n = a + ib$

$\therefore r^n(\cos n\theta + i \sin n\theta) = R(\cos \alpha + i \sin \alpha)$

Equating we have

$r^n = R \qquad n\theta = \alpha$

$r = \sqrt[n]{R} \qquad \theta = \dfrac{\alpha}{n}$

So we can plot $z_1 = r(\cos \theta + i \sin \theta)$.

The other solutions to $z^n = a + ib$ are equally spaced around the origin starting at
$z_1 = r(\cos\theta + i\sin\theta)$.

Roots of a complex number

The solutions to $z^n = a + ib$ are equally spaced $\dfrac{2\pi}{n}$ radians apart around the origin, starting at

$z_1 = r(\cos\theta + i\sin\theta)$, where $\theta = \dfrac{\arg(a + ib)}{n}$ and

$r = \sqrt[n]{|a + ib|}$.

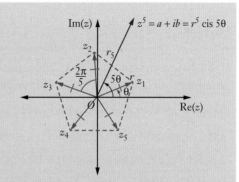

The example of $z^5 = a + ib$ is sketched.
The roots z_1, z_2, z_3, z_4, z_5 form the vertices of a regular pentagon.

Using vector addition of the regular polygon, notice that the sum of the roots is zero; that is, $z_1 + z_2 + z_3 + z_4 + \ldots + z_n = 0$.

However, since $a + ib$ is complex, the roots *do not* come in conjugate pairs.

Note that $a + ib$ is not necessarily a root of the equation $z^n = a + ib$.

Getty Images/Tatiana Dyuvbanova/EyeEm

EXAMPLE 19

Solve the equation $z^4 = -1$.

Solution

Using the technique above, we know the 4 roots z_1, z_2, z_3, z_4 will be equally spaced around the origin starting at z_1, forming the vertices of a square.

Now we need to find the location of z_1.

Let $z_1 = r(\cos \theta + i \sin \theta)$ and note $-1 = 1(\cos \pi + i \sin \pi)$.

Then $z_1^4 = r^4(\cos 4\theta + i \sin 4\theta) = 1(\cos \pi + i \sin \pi)$

So $r = 1$ and $4\theta = \pi$ so $\theta = \dfrac{\pi}{4}$.

Therefore $z_1 = 1\left(\cos \dfrac{\pi}{4} + i \sin \dfrac{\pi}{4} \right)$. The other 3 roots will be equally spaced from

$z_1 = 1\left(\cos \dfrac{\pi}{4} + i \sin \dfrac{\pi}{4} \right)$. Dividing 2π by 4 we have a spacing of $\dfrac{\pi}{2}$. So the other 3 roots are:

$z_2 = 1\left[\cos \left(\dfrac{\pi}{4} + \dfrac{\pi}{2} \right) + i \sin \left(\dfrac{\pi}{4} + \dfrac{\pi}{2} \right) \right] = \cos \dfrac{3\pi}{4} + i \sin \dfrac{3\pi}{4}$

$z_3 = 1\left[\cos \left(\dfrac{\pi}{4} + \pi \right) + i \sin \left(\dfrac{\pi}{4} + \pi \right) \right] = \cos \dfrac{5\pi}{4} + i \sin \dfrac{5\pi}{4} = \cos \left(-\dfrac{3\pi}{4} \right) + i \sin \left(-\dfrac{3\pi}{4} \right)$ using the

principal argument

$z_4 = 1\left[\cos \left(\dfrac{\pi}{4} - \dfrac{\pi}{2} \right) + i \sin \left(\dfrac{\pi}{4} - \dfrac{\pi}{2} \right) \right] = \cos \left(-\dfrac{\pi}{4} \right) + i \sin \left(-\dfrac{\pi}{4} \right)$.

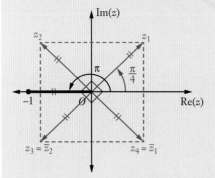

Note that $z = -1$ is not a root of the equation $z^4 = -1$.

Also note that $z_1 + z_2 + z_3 + z_4 = 0$ and that since -1 is real, the roots do come in conjugate pairs.

EXAMPLE 20

Solve the equation $z^5 = -\sqrt{3} + i$.

Solution

The 5 roots, z_1, z_2, z_3, z_4, z_5, will be equally spaced around the origin, starting at z_1. Now we need to find the location of z_1.

Let $z_1 = r(\cos\theta + i\sin\theta)$.

Converting to polar form we have $-\sqrt{3} + i = 2\left(\cos\dfrac{5\pi}{6} + i\sin\dfrac{5\pi}{6}\right)$.

Then, $z_1^{\,5} = r^5(\cos 5\theta + i\sin 5\theta) = 2\left(\cos\dfrac{5\pi}{6} + i\sin\dfrac{5\pi}{6}\right)$

So $r = \sqrt[5]{2}$ and $5\theta = \dfrac{5\pi}{6}$ so $\theta = \dfrac{\pi}{6}$.

Therefore $z_1 = \sqrt[5]{2}\left(\cos\dfrac{\pi}{6} + i\sin\dfrac{\pi}{6}\right)$.

The other 4 roots will be equally spaced from $z_1 = \sqrt[5]{2}\left(\cos\dfrac{\pi}{6} + i\sin\dfrac{\pi}{6}\right)$.

Dividing 2π by 5 we have a spacing of $\dfrac{2\pi}{5}$. So the other 4 roots are:

$$z_2 = \sqrt[5]{2}\left[\cos\left(\dfrac{\pi}{6} + \dfrac{2\pi}{5}\right) + i\sin\left(\dfrac{\pi}{6} + \dfrac{2\pi}{5}\right)\right] = \cos\dfrac{17\pi}{30} + i\sin\dfrac{17\pi}{30}$$

$$z_3 = \sqrt[5]{2}\left[\cos\left(\dfrac{\pi}{6} + \dfrac{4\pi}{5}\right) + i\sin\left(\dfrac{\pi}{6} + \dfrac{4\pi}{5}\right)\right] = \sqrt[5]{2}\left(\cos\dfrac{29\pi}{30} + i\sin\dfrac{29\pi}{30}\right)$$

$$z_4 = \sqrt[5]{2}\left[\cos\left(\dfrac{\pi}{6} + \dfrac{6\pi}{5}\right) + i\sin\left(\dfrac{\pi}{6} + \dfrac{6\pi}{5}\right)\right] = \sqrt[5]{2}\left(\cos\dfrac{41\pi}{30} + i\sin\dfrac{41\pi}{30}\right)$$

$$= \sqrt[5]{2}\left[\cos\left(-\dfrac{19\pi}{30}\right) + i\sin\left(-\dfrac{19\pi}{30}\right)\right] \text{ using the principal argument.}$$

$$z_5 = \sqrt[5]{2}\left[\cos\left(\dfrac{\pi}{6} - \dfrac{2\pi}{5}\right) + i\sin\left(\dfrac{\pi}{6} - \dfrac{2\pi}{5}\right)\right] = \cos\left(-\dfrac{7\pi}{30}\right) + i\sin\left(-\dfrac{7\pi}{30}\right) \text{ using the principal}$$

argument.

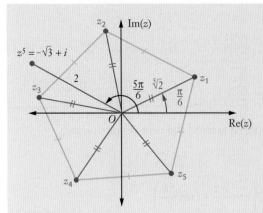

Note that the roots are equally spaced around $z_1 = \sqrt[5]{2}\left(\cos\dfrac{\pi}{6} + i\sin\dfrac{\pi}{6}\right)$.

Also note that $z_1 + z_2 + z_3 + z_4 + z_5 = 0$ and that since $-\sqrt{3} + i$ is not real, the roots do not come in conjugate pairs.

Exercise 4.06 Roots of complex numbers

1 Find the roots of each equation on the complex plane. Sketch your solutions on an Argand diagram.

   a $z^2 = 1 + i\sqrt{3}$ **b** $z^2 = i$ **c** $z^2 = -1 - i\sqrt{3}$ **d** $z^2 = -i$

2 Solve each equation, giving your answer in exact modulus–argument form.

   a $z^3 = -1$ **b** $z^3 = i$ **c** $z^3 = -i$

3 Determine the roots of each equation, answering in modulus–argument form.

   a $z^4 = 16i$ **b** $z^4 = -1 - i\sqrt{3}$ **c** $z^4 = -i$

4 **a** Show that $\cos\left(-\dfrac{3\pi}{5}\right) + i\sin\left(-\dfrac{3\pi}{5}\right)$ is a root of $z^5 + 1 = 0$.

   b Find the other roots of $z^5 + 1 = 0$ and plot them on the complex plane.

   c State which roots are conjugates.

   d Prove that $\cos\dfrac{\pi}{5} + \cos\dfrac{3\pi}{5} = \dfrac{1}{2}$.

5 Find the 7th roots of -1. Hence:

   a explain why the sum of the roots is zero

   b prove that $\cos\dfrac{\pi}{7} + \cos\dfrac{3\pi}{7} + \cos\dfrac{5\pi}{7} = \dfrac{1}{2}$

6 Solve over the complex plane:

   a $z^6 = -1$ **b** $z^8 = -1$ **c** $z^5 = i$

7 **a** Show that $\sqrt{2}\left[\cos\dfrac{3\pi}{4} + i\sin\dfrac{3\pi}{4}\right]$ is a root of $z^3 - 2 - 2i = 0$.

   b Find the other roots of $z^3 - 2 - 2i = 0$ and plot them on an Argand diagram.

4.07 Curves and regions on the complex plane

Complex plane graphs

Curves and regions on the complex plane are sets of points representing complex numbers described by a certain rule or condition placed on the variable complex number z. The set of points is often called the **locus** of z.

We can take a geometric approach or an algebraic approach to solving locus problems. We can graph a locus in the complex plane *algebraically*, by first deriving the Cartesian equation, or *geometrically* by using the definitions of modulus and argument.

Using the algebraic approach we can let $z = x + iy$ and let $\text{Re}(z) = x$ and $\text{Im}(z) = y$. Recall that $|z| = r = \sqrt{x^2 + y^2}$ and Arg z is the principal argument.

Using a geometric approach we can use the definition of $z - z_1$ to mean a vector where z is a variable point and z_1 is a fixed point. This means that $|z - z_1|$ is the distance from z to z_1 and arg $(z - z_1)$ is the angle between the vector $z - z_1$ and the positive x-axis.

Modulus and argument of z – z₁

Given a variable point z and a fixed point z_1 in the complex plane, then:

- $|z - z_1|$ is the distance from z to z_1

- arg $(z - z_1)$ is the angle between the vector $z - z_1$ and the positive x-axis.

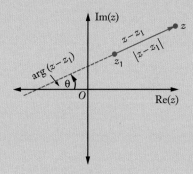

4. Applying complex numbers (163)

We can now develop some basic ideas regarding moduli and arguments.

Modulus and argument on the complex plane

Consider 4 complex numbers z_1, z_2, z_3, z_4 represented by points A, B, C, D respectively on the complex plane.

1 $\left|z_1 - z_2\right| = \left|z_3 - z_4\right|$ means $AB = CD$.

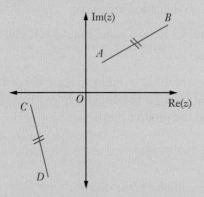

2 $\arg(z_1 - z_2) = \arg(z_3 - z_4)$ means $AB \parallel CD$.

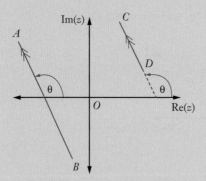

3 $z_1 - z_2 = z_3 - z_4$ means both differences are equal, so their moduli and arguments are equal. Either $ABDC$ forms a parallelogram or A, B, C, D are collinear and $AB = CD$.

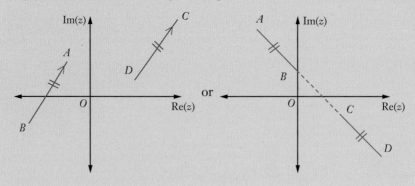

Modulus on the complex plane

EXAMPLE 21

Sketch each curve using:

    i   an algebraic approach   **ii** a geometric approach

a $|z| = 3$ **b** $|z + 1 - 2i| = 1$

Solution

a $|z| = 3$

    i   Let $z = x + iy,\ x, y, \in \mathbb{R}$.

$$|x + iy| = 3$$

$$\sqrt{x^2 + y^2} = 3$$

        Then $x^2 + y^2 = 9$

    ii $|z| = 3$ means $|z - (0 + 0i)| = 3$; that is, the distance of z from O is 3 units.

        In both cases we see that the locus of z is a circle, centre O and radius 3.

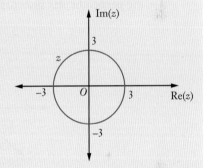

b $|z + 1 - 2i| = 1$

    i   Let $z = x + iy,\ x, y, \in \mathbb{R}$.
        We can write $|z - (-1 + 2i)| = 1$ so

$$|x + iy - (-1 + 2i)| = 1$$

$$\sqrt{(x + 1)^2 + (y - 2)^2} = 1$$

        Then $(x + 1)^2 + (y - 2)^2 = 1$

    ii $|z + 1 - 2i| = 1$ means the distance of z from $(-1 + 2i)$ is 1 unit.

    In both cases we see that the locus of z is a circle, centre $(-1 + 2i)$ and radius 1.

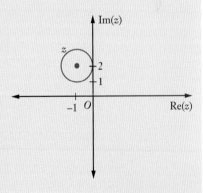

EXAMPLE 22

Sketch $|z - 2| = |z - 2i|$ using:

a an algebraic approach   **b** a geometric approach

Solution

a Let $z = x + iy, x, y, \in \mathbb{R}$. Then $|z - 2| = |z - 2i|$ becomes

$$|(x - 2) + iy| = |x + (y - 2)i|$$
$$\sqrt{(x - 2)^2 + y^2} = \sqrt{x^2 + (y - 2)^2}$$
$$(x - 2)^2 + y^2 = x^2 + (y - 2)^2$$
$$x^2 - 4x + 4 + y^2 = x^2 + y^2 - 4y + 4$$
$$\therefore \qquad\qquad y = x$$

b Geometrically $|z - 2| = |z - 2i|$ means that z is equidistant from both 2 and $2i$; that is, it is the perpendicular bisector, $y = x$.

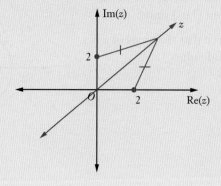

Arguments on the complex plane

EXAMPLE 23

Sketch the graph of each equation.

a $\arg z = \dfrac{\pi}{3}$

b $\arg (z + i) = \dfrac{3\pi}{4}$

c $\arg [z - (1 + i)] = -\dfrac{\pi}{6}$

Solution

a $\arg z = \dfrac{\pi}{3}$

The complex number z is the vector from O at an angle of $\dfrac{\pi}{3}$.

arg 0 is undefined so we draw an open circle there to indicate that it is not part of the graph.

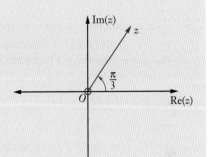

b $\arg (z + i) = \dfrac{3\pi}{4}$

The complex number z is the vector from $-i$ at an angle of $\dfrac{3\pi}{4}$.

arg 0 is undefined so draw an open circle at $-i$.

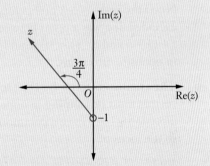

c $\arg [z - (1 + i)] = -\dfrac{\pi}{6}$

The complex number z is the vector from $1 + i$ at an angle of $-\dfrac{\pi}{6}$.

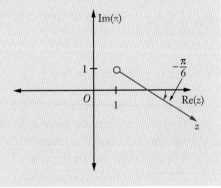

Sketch the graph of each equation.

a $\arg [z - (3 + i)] = \arg [z - (1 + 3i)]$

b $\arg [z - (3 + i)] - \arg [z - (1 + 3i)] = \pm\pi$

c $\arg (z - 3) - \arg (z + 3) = \dfrac{\pi}{2}$

Solution

a The arguments are equal so the vectors $z - (3 + i)$ and $z - (1 + 3i)$ must be running in the same direction. They have a common point z, so the locus of

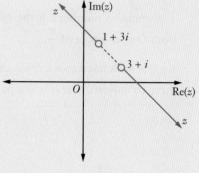

$\arg [z - (3 + i)] = \arg [z - (1 + 3i)]$

must be points on the line through $3 + i$ and $1 + 3i$. The solution has 2 sections, excluding the points between $3 + i$ and $1 + 3i$.

Note: the in-between points must be excluded. If z was a point between $3 + i$ and $1 + 3i$ then the vectors $z - (3 + i)$ and $z - (1 + 3i)$ would be running in opposite directions.

b The arguments differ by 180° so the vectors must be running in opposite directions. They have a common point z, so the locus of

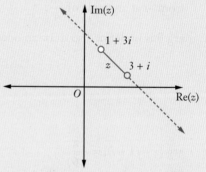

$\arg [z - (3 + i)] - \arg [z - (1 + 3i)] = \pm\pi$

must be points on the line joining $3 + i$ and $1 + 3i$ and must be the points between $3 + i$ and $1 + 3i$.

c $\arg (z - 3) - \arg (z + 3) = \dfrac{\pi}{2}$

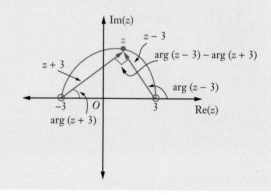

The arguments differ by 90° but they have a common point z. It uses the theorem that in a triangle the exterior angle equals the sum of the two interior opposite angles. The solution is a semicircle with diameter between 3 and −3 because the angle in a semicircle is 90°.

Regions on the complex plane

We will examine more examples of graphing sets of points where the locus is a shaded region rather than a line or curve.

Sketch the graph of each inequality.

a $-\dfrac{\pi}{4} \le \arg z < \dfrac{\pi}{3}$ **b** $\dfrac{1}{2} < |z - 2| \le 1$ **c** $\text{Re}(z) > \text{Im}(z) + 1$

Solution

a $-\dfrac{\pi}{4} \le \arg z < \dfrac{\pi}{3}$ is the region between the two

vectors from O with arguments $-\dfrac{\pi}{4}$ and $\dfrac{\pi}{3}$.

Note the dotted vector since $\arg z = \dfrac{\pi}{3}$ is not included.

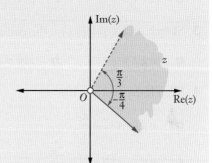

b $\dfrac{1}{2} < |z - 2| \le 1$ represents the region between the

2 circles centred on 2 with radii $\dfrac{1}{2}$ and 1.

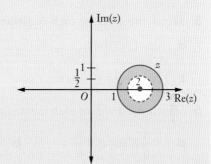

c $\text{Re}(z) > \text{Im}(z) + 1$. Using a Cartesian approach we can say $x > y + 1$ or, rearranging, $y < x - 1$.

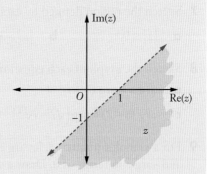

Exercise 4.07 Curves and regions on the complex plane

1 For each equation, draw a sketch and describe it.

a $\arg z = \arg w$ **b** $|z| = |w|$ **c** $\arg z = -\arg w$

d $\arg (z - w) = \arg (u - v)$ **e** $|z - w| = |u - v|$ **f** $z - w = u - v$

g $|z + w| = |z - w|$ **h** $z + w = u + v$ **i** $z - u = i(z + u)$

2 Sketch the circle defined by each equation.

a $|z| = 1$ **b** $|z| = 2$ **c** $|z| = 4$ **d** $|z| = \dfrac{1}{4}$

e $|z - 1| = 3$ **f** $|z + 3| = 1$ **g** $|z - 3i| = 3$ **h** $|z + i| = \dfrac{1}{2}$

3 Sketch each equation on the complex plane.

a $\arg z = \dfrac{\pi}{6}$ **b** $\arg w = \dfrac{3\pi}{4}$ **c** $\arg z = -\dfrac{\pi}{3}$ **d** $\arg u = \pi$

4 Sketch the vector z if:

a $\arg (z - 1) = \dfrac{\pi}{3}$ **b** $\arg (z - i) = \dfrac{\pi}{6}$

c $\arg [z - (1 - i)] = \dfrac{2\pi}{3}$ **d** $\arg (z - 3 - 2i) = -\dfrac{5\pi}{6}$

5 Sketch the graph of each equation.

a $|w - (1 + i)| = 1$ **b** $\left|z - (1 - i\sqrt{3})\right| = 2$

c $|z - 2 - i| = \dfrac{1}{2}$ **d** $|w + 3 - 4i| = 5$

6 Sketch the graph of each inequality.

a $-\dfrac{\pi}{6} \le \arg w \le \dfrac{\pi}{6}$ **b** $\dfrac{\pi}{4} < \arg w \le \dfrac{3\pi}{4}$ **c** $-\pi < \arg w < \dfrac{\pi}{2}$

7 Sketch the region defined by each inequality.

a $|z| \le 9$ **b** $|u| > 3$ **c** $\dfrac{1}{2} < |z| \le 1$ **d** $1 \le |z - 3| < 2$

8 Sketch the graph of each equation.

a $\arg z = \arg (1 + i)$ **b** $\arg z = \arg (1 + i\sqrt{3})$

c $\arg (z - 2) = \arg\left(\sqrt{2} - i\sqrt{2}\right)$ **d** $\arg (z + 3i) = \arg (-\sqrt{3} + i)$

9 For a complex number z, let $\arg z = \theta$. Find a relationship between $\arg z$, $\arg (-z)$ and $-\arg z$ in terms of θ. Draw a sketch.

10 Sketch the graph of z if:

a $\quad \arg z - \arg(-1 - i) = 0$

b $\quad \arg z - \arg(1 - i\sqrt{3}) = 0$

c $\quad \arg z + \arg\left(\dfrac{\sqrt{3}}{2} - \dfrac{i}{2}\right) = 0$

d $\quad \arg z + \arg\left(-\sqrt{2} - i\sqrt{2}\right) = 0$

e $\quad \arg(z - 1) = \arg(-1 + i)$

f $\quad \arg(z - 2i) = \arg(1 + i\sqrt{3})$

11 Sketch the graph of z for:

a $\quad \mathrm{Re}(z) = 3$

b $\quad \mathrm{Im}(z) = 2$

c $\quad \mathrm{Re}(z) = -4$

d $\quad \mathrm{Im}(z) = -1$

12 Find the Cartesian equation of each locus and then sketch it.

a $\quad \mathrm{Re}(z) + \mathrm{Im}(z) = 0$

b $\quad \mathrm{Im}(z) = 2\,\mathrm{Re}(z)$

c $\quad \mathrm{Re}(z) = 2\,\mathrm{Im}(z) - 1$

d $\quad \mathrm{Im}(z) + 3\,\mathrm{Re}(z) = 6$

13 Sketch the locus of w if:

a $\quad \arg w = \arg(w - 2)$

b $\quad \arg w = \arg(w - i)$

c $\quad \arg(w + 2) - \arg w = 0$

d $\quad \arg(w + i) - \arg(w - 1) = 0$

14 Sketch the parabola defined by $\mathrm{Im}(z) = |z - 2i|$.

15 Sketch z defined by each inequality.

a $\quad \mathrm{Re}(z) > 1$

b $\quad \mathrm{Im}(z) < 2$

c $\quad -2 < \mathrm{Re}(z) \leq 3$

d $\quad \mathrm{Im}(z) < -1$ and $\mathrm{Re}(z) \geq -2$

e $\quad |\mathrm{Re}(z)| \leq 3$

f $\quad |\mathrm{Im}(z)| < 1$

g $\quad |\mathrm{Re}(z)| \geq \dfrac{1}{2}$

h $\quad |\mathrm{Re}(z)| > 5$ and $|\mathrm{Im}(z)| \leq 4$

16 Sketch:

a $\quad \arg(z - 2) - \arg z = \pm\pi$

b $\quad \arg(z - 3) - \arg(z + 3) = \dfrac{\pi}{2}$

c $\quad \arg(z - 2i) - \arg(z + 1) = \dfrac{\pi}{2}$

d $\quad \arg(z + 2i) - \arg(z - 2) = \pm\pi$

17 Using the theorem $\arg \dfrac{z}{w} = \arg z - \arg w$, sketch:

a $\quad \arg\left(\dfrac{z}{z - 1}\right) = 0$

b $\quad \arg\left(\dfrac{u - 1}{u + 1}\right) = \dfrac{\pi}{2}$

c $\quad \arg\left(\dfrac{z - 2}{z - 2i}\right) = \pm\pi$

d $\quad \arg\left(\dfrac{z - i}{z + i}\right) = \dfrac{\pi}{2}$

18 Describe each locus and hence sketch the graph.

a $\quad |z| = |z - 4i|$

b $\quad |z - 2| = |z + 2|$

c $\quad |z - 1 + i| = |z + 1 - i|$

d $\quad |z - 5 - i| = |z + 3 + 3i|$

19 Describe the subset of the complex plane defined by each equation and sketch the graph.

a $\arg\left(\dfrac{w-2}{w+2i}\right)=0$

b $\arg\left(\dfrac{u-2i}{u+2}\right)=\dfrac{\pi}{2}$

20 Sketch each region.

a $|z|\le 3$ and $0<\arg z<\dfrac{\pi}{2}$

b $|z-2|\le 1$ and $0<\arg z\le\dfrac{\pi}{3}$

c $|z-i|\ge 1$ and $\dfrac{\pi}{6}\le\arg z<\dfrac{5\pi}{6}$

d $\dfrac{1}{2}\le|z|<3$ and $-\dfrac{\pi}{4}<\arg z\le\dfrac{\pi}{4}$

21 The triangles OXZ and OYW in the diagram are similar. Z and W represent the complex numbers z and w respectively. Let $\arg z=\theta$.

Show that $w|x_1|=iz\,|y_2|$.

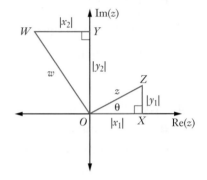

22 PQR is a right-angled isosceles triangle.

If P, Q, R represent the complex numbers p, q, r respectively, prove that:

a $|p-q|=|r-q|$

b $\arg\left(\dfrac{p-q}{r-q}\right)=\dfrac{\pi}{2}$

c $p-q=i(r-q)$

d $(p-q)^2+(r-q)^2=0$

e $|p-q|^2+|r-q|^2=|r-p|^2$

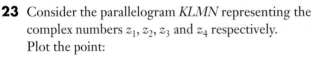

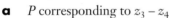

23 Consider the parallelogram $KLMN$ representing the complex numbers z_1, z_2, z_3 and z_4 respectively. Plot the point:

a P corresponding to z_3-z_4

b Q corresponding to z_1-z_3

c R corresponding to $-i(z_2-z_1)$

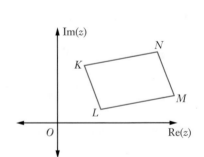

24 Consider the triangle UVW whose vertices represent u, v, w respectively. UV is parallel to the x-axis. Show that:

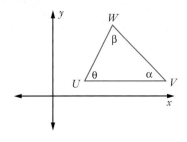

a $\arg(w - u) = \theta$

b $\arg(w - v) = \pi - \alpha$

c $\arg\dfrac{w - v}{w - u} = \beta$

25 Consider the complex numbers a, b, c, d represented by the points A, B, C, D respectively in the complex plane. Express b, c, d in terms of a.

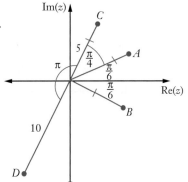

26 Consider the complex numbers p, q and r such that $|p| = |q| = 1$, $\arg p = \theta$ and $\arg q = \alpha$. Draw a sketch and find an expression for r in terms of p and q if:

a $\arg r = \theta - \alpha$ **b** $\arg r = \theta + \alpha$

27 Consider the 5 numbers sketched on the complex plane.

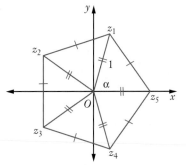

a Find the value of α.

b Write down the complex numbers z_1, z_2, z_3, z_4, z_5.

c Show that $z_1^2 = z_2$.

d Show that $z_2^2 = z_4$.

e Show that $z_3^2 = z_1$.

f Find $z_1^2 + z_2^2 + z_3^2 + z_4^2 + 1$.

28 Sketch:

a $\left|\dfrac{z+4}{z-2}\right| = 1$ **b** $\left|\dfrac{z-2i}{z+2i}\right| \le 1$ **c** $\left|\dfrac{z-3}{2z}\right| = 1$

29 a Show that the equation $|z+3| = 2|z-1|$ describes a circle.

b Find its centre and radius.

30 Find the maximum value of $\arg(z + 1)$ if $|z-1| = 1$.

1 Use De Moivre's theorem to simplify:

a $[\sqrt{3}(\cos \delta + i \sin \delta)]^4$

b $\left(\cos \dfrac{5\pi}{6} - i \sin \dfrac{5\pi}{6} \right)^7$

c $[9(\cos 72° + i \sin 72°)]^{\frac{1}{2}}$

2 Using De Moivre's theorem and the binomial expansion of $(a + b)^5$:

a **i** show that $\sin 5\theta = 5 \cos^4 \theta \sin \theta - 10 \cos^2 \theta \sin^3 \theta + \sin^5 \theta$

ii express $\sin 5\theta$ in terms of $\sin \theta$

b **i** show that $\cos 5\theta = \cos^5 \theta - 10 \cos^3 \theta \sin^2 \theta + 5 \cos \theta \sin^4 \theta$

ii express $\cos 5\theta$ in terms of $\cos \theta$

3 a Given $\tan 5\theta = \dfrac{\sin 5\theta}{\cos 5\theta}$, use your results in Question **2** and divide every term by $\cos^5 \theta$ to find an expression for $\tan 5\theta$ in terms of $\tan \theta$.

b Hence find exact distinct solutions to the equation $x^4 - 10x^2 + 5 = 0$.

4 a Use $z - \dfrac{1}{z} = 2i \sin \theta$ and the binomial expansion of $(a + b)^7$ to prove that:

$$-128i \sin^7 \theta = 2i(\sin 7\theta - 7 \sin 5\theta + 21 \sin 3\theta - 35 \sin \theta)$$

b Hence find $\displaystyle\int 35 \sin \theta - 64 \sin^7 \theta \, d\theta$.

5 Solve each quadratic equation.

a $z^2 + 2iz + 3 = 0$

b $w^2 - (2 - 3i)w - 1 - 3i = 0$

c $ix^2 - 9 = 0$

6 Solve each equation.

a $z^2 = 1 - i\sqrt{3}$

b $z^2 = -\sqrt{2} - i\sqrt{2}$

7 Consider the polynomial $P(x) = x^4 - 4x^3 + 11x^2 - 14x + 12$.

a Show that $x = 1 - i\sqrt{2}$ is a root of $P(x) = 0$.

b Hence solve the equation $P(x) = 0$.

c Express $P(x) = x^4 - 4x^3 + 11x^2 - 14x + 12$ as a product of 2 real quadratic factors.

8 Given that $\alpha = -2 + 3i$ is a root of the polynomial $P(x) = 2x^3 + Bx^2 + Cx + 13$ where $B, C \in \mathbb{Z}$, find $P(x)$ and the other roots.

9 For each value of z and w, plot on an Argand diagram:

    i z and zw **ii** z and $\dfrac{z}{w}$

  a $z = \sqrt{2}\left(\cos \dfrac{\pi}{2} + i \sin \dfrac{\pi}{2} \right)$ and $w = 2\left(\cos \dfrac{\pi}{4} + i \sin \dfrac{\pi}{4} \right)$

  b $z = 2\left(\cos \dfrac{\pi}{3} + i \sin \dfrac{\pi}{3} \right)$ and $w = 3\left(\cos \dfrac{\pi}{6} - i \sin \dfrac{\pi}{6} \right)$

  c $z = (1 + i)^3$ and $w = \left(\dfrac{\sqrt{3} + i}{2} \right)^9$ [Hint: express in polar form first.]

10 If $z = 5\left(\cos \dfrac{\pi}{3} - i \sin \dfrac{\pi}{3} \right)$, plot iz and $\dfrac{z}{i}$ on the complex plane.

11 The complex number $u = 3(\cos 2\beta + i \sin 2\beta)$ is shown in the diagram. Copy the diagram and plot each number below on it.

  a iu **b** $\dfrac{u}{i}$

  c u^2 **d** $\dfrac{1}{u}$

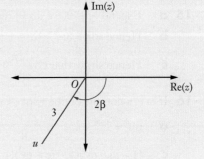

12 Consider the complex numbers u, v and w with corresponding points U, V and W as shown. $OUWV$ forms a square.

  a Find w in terms of u and v.

  b Show that $w = u + iu$.

  c Explain why $u - w = i(v - w)$

  d Prove that $u^2 + v^2 = 0$.

  e Find the vector m corresponding with the point M, the intersection of the diagonals of the square, in terms of u.

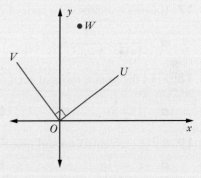

13 For each equation, find the roots of unity, show them on the complex plane and state the conjugate pairs.

   a $z^2 = 1$ **b** $z^3 = 1$ **c** $z^6 = 1$ **d** $z^8 = 1$

14 a   Find the 7 roots of $z^7 = 1$ and plot them on the complex plane.

   b   If α is a complex root, explain in 2 different ways why
$$\alpha^6 + \alpha^5 + \alpha^4 + \alpha^3 + \alpha^2 + \alpha + 1 = 0.$$

   c   Factorise $z^7 - 1$ into one linear and 3 quadratic factors with real coefficients.

   d   Hence show that $z^6 + z^5 + z^4 + z^3 + z^2 + z + 1 =$

$$\left[z^2 + 2\left(\cos \frac{\pi}{7} \right) z + 1 \right]\left[z^2 + 2\left(\cos \frac{3\pi}{7} \right) z + 1 \right]\left[z^2 + 2\left(\cos \frac{5\pi}{7} \right) z + 1 \right]$$

   e   Show that $\cos \dfrac{2\pi}{7} + \cos \dfrac{4\pi}{7} + \cos \dfrac{6\pi}{7} = -\dfrac{1}{2}$.

15 a   Expand $(\cos \theta + i \sin \theta)^3$ in 2 ways to show that $\cos 3\theta = 4 \cos^3 \theta - 3 \cos \theta$.

   b   Hence solve $8x^3 - 6x - 1 = 0$.

   c   Hence show that $\cos \dfrac{2\pi}{9} + \cos \dfrac{4\pi}{9} - \cos \dfrac{\pi}{9} = 0$.

16 If ω is a complex cube root of unity, simplify:

   a $(\omega^2 + 1)^3$ **b** $1 + \dfrac{1}{\omega} + \dfrac{1}{\omega^2}$

   c $(1 - \omega - \omega^2)(1 - \omega + \omega^2)(1 + \omega - \omega^2)$

17 If ω is a complex cube root of unity, prove that:

   a $\dfrac{1}{1+\omega} + \dfrac{1}{1+\omega^2} = 1$ **b** $\dfrac{k + l\omega + m\omega^2}{l + m\omega + k\omega^2} = \omega$

18 Find the square roots of each complex number.

   a $15 - 8i$ **b** $e^{\frac{i\pi}{4}}$ **c** $4\left(\cos \dfrac{\pi}{6} + i \sin \dfrac{\pi}{6} \right)$

19 Solve each equation and plot the roots on the complex plane.

   a $z^4 = -16$ **b** $z^3 = -1 - i$ **c** $z^5 = 32e^{-\frac{i\pi}{2}}$.

20 Solve the equation $z^5 = \dfrac{1+i\sqrt{3}}{2}$, expressing the roots in polar form. Plot them on the complex plane.

21 a Solve $z^9 + 1 = 0$.

    b Factorise $z^9 + 1$.

    c Hence solve $z^6 - z^3 + 1 = 0$.

    d State the real quadratic factors of $z^6 - z^3 + 1$.

22 Sketch each equation using:

    i an algebraic approach   **ii** a geometric approach

    a $|z| = 6$ **b** $|z - 2 - i| = 1$ **c** $|z - 2i| = \text{Im}(z)$

    d $|z| = |z - 2 - 2i|$ **e** $\left|\dfrac{z+6}{z-4i}\right| = 1$

23 Sketch the graph of each equation.

    a $\arg z = -\dfrac{\pi}{3}$ **b** $\arg(z + 1) = \dfrac{\pi}{4}$ **c** $\arg(z + 1 - i\sqrt{3}) = \dfrac{2\pi}{3}$

24 Sketch the graph of:

    a $\arg(z - 2) = \arg(z - 4 - 2i)$ **b** $\arg(z + 1 - i) - \arg(z - 1 + i) = \pi$

    c $\arg(z - 1) - \arg(z + 1) = \dfrac{\pi}{2}$

25 Sketch each region.

    a $-\dfrac{\pi}{3} < \arg(z - 1) \le \dfrac{5\pi}{6}$ **b** $1 \le |z - 2i| \le 2$

    c $\text{Re}(z) + \text{Im}(z) \ge 3$

26 The fixed points Z_1 and Z_2 represent the vectors z_1 and z_2. Sketch the graph of:

a $\left|z - z_1\right| = \left|z - z_2\right|$

b $\arg\left(\dfrac{z - z_1}{z - z_2}\right) = 0$

c $\left|z - z_2\right| = \left|z_1 - z_2\right|$ and $\arg\left(\dfrac{z - z_2}{z_1 - z_2}\right) = \dfrac{\pi}{4}$

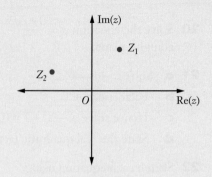

27 The fixed points W_1, W_2 and W_3 represent the vectors w_1, w_2 and w_3 respectively. It is given that $\dfrac{w_2 - w_1}{w_3 - w_1} = \dfrac{w_3 - w_2}{w_1 - w_2}$. Prove that $W_1 W_2 W_3$ forms an equilateral triangle.

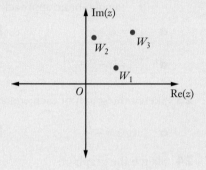

ISBN 9780170413435

Practice set 1

In Questions **1** to **10**, select the correct answer **A**, **B**, **C** or **D**.

1 Jake listed these complex number rules in a summary.
How many are correct?

$$|z_1 z_2| = |z_1||z_2| \qquad \arg z_1 z_2 = \arg z_1 \times \arg z_2 \qquad \left|\frac{z_1}{z_2}\right| = \frac{|z_1|}{|z_2|}$$

$$|z|^2 = z \cdot \bar{z} \qquad \bar{z}_1 \bar{z}_2 = \overline{z_1 z_2}$$

    A   all   **B**   one   **C**   some   **D**   none

2 $(\sin \theta - i \cos \theta)^n =$

    A $\sin n\theta - i \cos n\theta$ **B** $\sin(-n\theta) + i \cos(-n\theta)$

    C $(-i)^n(\cos n\theta + i \sin n\theta)$ **D** $i^n(\cos n\theta + i \sin n\theta)$

3 If ω is a complex root of $z^3 = 1$, which statement is false?

    A $\omega^2 + \omega + 1 = 0$ **B** $\omega^2 = \bar{\omega}$

    C $\omega^4 = \omega$ **D** $\omega^{-2} = \dfrac{1}{\omega}$

4 Consider the statement: 'If there is a stationary point at $x = 3$ then $f'(3) = 0$.'
Which of the following is false?

    A   The converse   **B**   The contrapositive

    C   The negation   **D**   The proposition

5 Which inequality always holds for $a > b$?

    A $\dfrac{1}{a} < \dfrac{1}{b}$ **B** $\dfrac{1}{a^2} < \dfrac{1}{b^2}$ **C** $a^2 > b^2$ **D** $a^3 > b^3$

6 Consider the vectors $\underset{\sim}{u} = \begin{pmatrix} 3 \\ -1 \end{pmatrix}$, $\underset{\sim}{v} = \begin{pmatrix} 6 \\ 2 \end{pmatrix}$, $\underset{\sim}{w} = \begin{pmatrix} -9 \\ 3 \end{pmatrix}$ and $\underset{\sim}{z} = \begin{pmatrix} -1 \\ 3 \end{pmatrix}$.

Which vectors are parallel?

    A $\underset{\sim}{u}$ and $\underset{\sim}{v}$ **B** $\underset{\sim}{u}$ and $\underset{\sim}{w}$

    C $\underset{\sim}{v}$ and $\underset{\sim}{w}$ **D** $\underset{\sim}{w}$ and $\underset{\sim}{z}$

7 What do the set of equations $x = \cos t, y = \sin t, z = t$, where $0 \le t \le 2\pi$ describe?

    A   A circle   **B**   A sphere   **C**   A cylinder   **D**   A helix

8 What is the equation of this circle with centre at $(1, 0)$?

A $|z+1| = 10$ **B** $|z-1| = 10$

C $|z+1| = 9$ **D** $|z-1| = 9$

9 If $v = (1 - 2\cos\alpha) - 2i\sin\alpha$, the real part of v^{-1} is:

A $\dfrac{1-2\cos\alpha}{5-4\cos\alpha}$ **B** $\dfrac{1+2\cos\alpha}{5-4\cos\alpha}$

C $\dfrac{5-4\cos\alpha}{1-2\cos\alpha}$ **D** $\dfrac{5+4\cos\alpha}{1+2\cos\alpha}$

10 What does the equation $|z-3| = |z+3i|$ describe?

A A circle **B** A parabola

C A hyperbola **D** A perpendicular bisector

11 Express each expression in terms of i.

a $\sqrt{-16}$ **b** $\sqrt{\dfrac{-7}{4}}$ **c** $\dfrac{6 \pm \sqrt{-12}}{2}$

12 Simplify:

a i^8 **b** $i^{22} + i^{23} + i^{24} + \ldots + i^{99}$

13 Solve each equation in the complex plane.

a $x^2 + 64 = 0$ **b** $x^2 + 2x + 7 = 0$ **c** $(x-3)^2 + 9 = 0$

14 State the real and imaginary parts of each complex number z.

a $\dfrac{5-2i}{3}$ **b** $\dfrac{x + 2i - yi + 7}{x^2 + y^2}$ where $x, y \in \mathbb{R}$

15 For each complex number z, state the complex conjugate $\bar{z}$.

a $z = 5x - 3iy$, where $x, y \in \mathbb{R}$

b $z = \dfrac{ai + 6b - 2a - ib}{4}$, where $a, b \in \mathbb{R}$

16 If $w = \dfrac{m + in}{m + n}$ where $m, n \in \mathbb{R}$, prove that $w\bar{w}$ is always real.

17 Solve each equation, given that $x, y \in \mathbb{R}$.

a $3x + 2iy - 18 + 6i = 0$ **b** $x + y - i(x - y) = 6 - 2i$

18 Simplify each expression, giving your answer in the form $a + ib$.

a $3 - 4i(5 + 2i) + i$ **b** $(2 - i\sqrt{3})(2 + i\sqrt{3})$ **c** $(1 + 5i)^2 - (1 - 5i)^2$

19 Express each quadratic equation in the form $ax^2 + bx + c = 0$, where $a, b, c \in \mathbb{Z}$.

a $(x - 1 - 2i)(x - 1 + 2i) = 0$

b $\left(x - \dfrac{-1 - i\sqrt{2}}{6}\right)\left(x - \dfrac{-1 + i\sqrt{2}}{6}\right) = 0$

20 Simplify each expression by realising the denominator:

a $\dfrac{2}{1 - i\sqrt{3}}$

b $\dfrac{\sqrt{5} + 2i}{\sqrt{5} - 2i} + \dfrac{\sqrt{5} - 2i}{\sqrt{5} + 2i}$

c $\dfrac{1}{(1 - i)^2}$

21 Find $\sqrt{24 - 10i}$.

22 Find 2 square roots of $-48 + 14i$.

23 Solve the equation $x^2 - (1 + 2i)x + 1 + 7i = 0$.

24 Represent each complex number on an Argand diagram as:

    i a point   **ii** a vector

a $z = 1 - 4i$ **b** $w = 3i$ **c** $u = -2 - 3i$

25 The vector representing the complex number u is sketched below.
Copy the diagram and sketch vectors representing $\bar{u}$, $-u$ and $2u$.

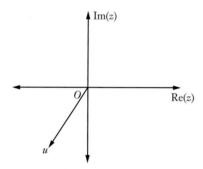

26 The complex numbers z and w are shown.
Copy the diagram and plot the points P and Q
that represent the numbers $z + w$ and $w - z$
respectively.

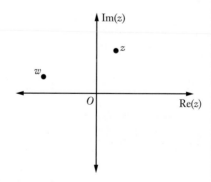

27 For each complex number, find:

    i   the modulus r **ii**   the principal argument Arg z

  a $-1 + i\sqrt{3}$ **b** $2 - 2i$ **c** $\dfrac{-\sqrt{6} - i\sqrt{2}}{2}$

28 Express each complex number in modulus–argument form.

  a $3\left(\cos \dfrac{\pi}{3} - i \sin \dfrac{\pi}{3}\right)$ **b** $\sqrt{2}\left(\sin \dfrac{3\pi}{4} + i \cos \dfrac{3\pi}{4}\right)$

  c $5i$ **d** $-2 + 2i\sqrt{3}$ **e** $\dfrac{1 + i}{3}$

29 Express each complex number u and v graphed below in:

  a polar form   **b** Cartesian form

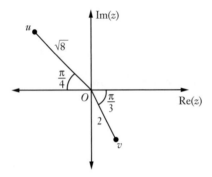

30 Simplify:

  a $r_1(\cos \alpha_1 + i \sin \alpha_1) \times r_2(\cos \alpha_2 + i \sin \alpha_2)$

  b $\dfrac{r_1\left(\cos \alpha_1 + i \sin \alpha_1\right)}{r_2\left(\cos \alpha_2 + i \sin \alpha_2\right)}$

31 Simplify:

  a $\arg (\cos \theta + i \sin \theta)^n$ **b** $\arg (\cos \theta - i \sin \theta)^n$

  c $\arg (\cos \theta + i \sin \theta)^{-n}$ **d** $\arg (\cos \theta - i \sin \theta)^{-n}$

32 If $z_1 = r_1(\cos \alpha_1 + i \sin \alpha_1)$ and $z_2 = r_2(\cos \alpha_2 + i \sin \alpha_2)$, prove $\left|\dfrac{z_1}{z_2}\right| = \dfrac{|z_1|}{|z_2|}$.

33 If $z_1 = 3\left(\cos \dfrac{5\pi}{6} + i \sin \dfrac{5\pi}{6}\right)$ and $z_2 = 2\left(\cos \dfrac{\pi}{3} - i \sin \dfrac{\pi}{3}\right)$, find:

  a $z_1 z_2$ **b** $\dfrac{z_1}{z_2}$ **c** $(z_2)^3$ **d** $(z_1)^{-4}$

MATHS IN FOCUS 12. Mathematics Extension 2   ISBN 9780170413435

34 Express each expression in modulus–argument form and hence find its exact value.

   a $(1-i)^8$ **b** $\dfrac{(1+i\sqrt{3})^2}{\sqrt{2}-i\sqrt{2}}$ **c** $\dfrac{1}{(\sqrt{3}+i)^4}$

35 Simplify $(1+i)(\sqrt{3}+i)$ in 2 different ways. Hence find the exact value of $\sin\dfrac{5\pi}{12}$.

36 Express each complex number in the form $re^{i\theta}$.

   a $\cos 3 + i\sin 3$ **b** $4\left(\cos\dfrac{\pi}{5}-i\sin\dfrac{\pi}{5}\right)$ **c** $-\sqrt{3}-i$

37 Express each complex number in polar form:

   a $2e^{3i\alpha}$ **b** $e^{-\frac{i\pi}{7}}$ **c** $-\dfrac{1}{2}e^{\frac{i\pi}{3}}$

38 Evaluate, expressing your answer in the form $re^{i\theta}$.

   a $2e^{\frac{i}{2}}\times 3e^{2i}$ **b** $\dfrac{e^{i\pi}\times -1}{e^{\frac{i\pi}{5}}}$

39 If $z=e^{i\theta}$ and $w=e^{i\alpha}$, prove that $\arg zw = \arg z + \arg w$.

40 Write the following statement as an implication:

   If it rains, then the dam is full.

41 Write the converse of the statement:

   If there is not enough food, then the people are starving.

42 For each statement write the converse and determine if it is an equivalence.
   a   If a number is even, then it is divisible by 2.
   b   If a number is positive, then its reciprocal is positive.
   c   If a quadrilateral has 4 equal angles, then it is a rectangle.
   d   If an animal is a kangaroo, then it eats grass.

43 Write the negation of each statement.
   a   The dam is full.   **b**   The teacher is good.
   c   All cats are fluffy.   **d**   There is at least one smart politician.
   e   No wine is sweet.   **f**   Some sheep are black.

44 Write the contrapositive of each statement and hence determine if the original statement is true.

   a   If you get a speeding ticket, then you speed.

   b   If you get the old-age pension, then you are over 65.

   c   If a triangle is equilateral, then it has 3 equal sides.

   d   If you go swimming, then you get wet.

45 Write in words: $\forall\, x \in \mathbb{N}, \exists\, y \in \mathbb{N}: y = 2x$.

46 Write the following statement in mathematical notation:

For all natural numbers x such that x is a multiple of 4, there exists a natural number y such that $\sqrt{x} = 2\sqrt{y}$.

47 If $P \Rightarrow Q$ is true, which of the following is always true?

   A $Q \Rightarrow P$ **B** $\neg P \Rightarrow \neg Q$ **C** $\neg Q \Rightarrow \neg P$ **D** $P \Leftrightarrow Q$

48 Give a proof by contradiction to prove that $\sqrt{10}$ is irrational.

49 Find a counter-example to show that the following statement is false.

Given $y = f(x)$ and $f''(p) = 0$ then there is a point of inflection at $x = p$.

50 If $M, N \in \mathbb{N}, M > N$ and:

   a M and N are even, prove that $M^2 - N^2$ is even.

   b M and N are odd, prove that $M^2 - N^2$ is even.

   c M is even and N is odd, prove that $M^2 - N^2$ is odd.

51 a   Prove $\forall\, x, y \in \mathbb{R}: x, y > 0$ that $\dfrac{x^2 + y^2}{2} \geq xy$.

   b   Prove $\forall\, x \in \mathbb{R}: x > 0$ that $x + \dfrac{1}{x} \geq 2$.

   c   Prove $\forall\, a, b, c, d \in \mathbb{R}: a, b, c, d > 0$ that $\dfrac{a + b + c + d}{4} \geq \sqrt[4]{abcd}$.

52 Find the dot product of the vectors $\underset{\sim}{u} = 3\underset{\sim}{i} - 2\underset{\sim}{j}$ and $\underset{\sim}{v} = 4\underset{\sim}{i} + 6\underset{\sim}{j}$.

What can you say about the angle between the vectors?

53 Consider the points $A(1, -2, 4)$ and $B(3, 1, 2)$. Find:

   a   the vector $\overrightarrow{AB}$

   b   the magnitude of $\overrightarrow{AB}$

   c   the unit vector $\hat{\underset{\sim}{u}}$ in the direction of $\overrightarrow{AB}$

MATHS IN FOCUS 12. Mathematics Extension 2   ISBN 9780170413435

54 The vectors $p = 2i + 5j + k$ and $q = -7i + j + nk$ are perpendicular. Find the value of n.

55 Consider the points $A(-2, 3, 4)$, $B(2, 5, 8)$, $C(3, -2, 4)$ and $D(1, -3, 2)$.

   a Show that $\overrightarrow{AB}$ is parallel to $\overrightarrow{CD}$.

   b Find the length of each vector $\overrightarrow{AB}$ and $\overrightarrow{CD}$.

   c What type of quadrilateral is $ABCD$?

56 Find the angle between the vectors $\begin{pmatrix} 5 \\ -1 \\ 3 \end{pmatrix}$ and $\begin{pmatrix} -2 \\ 4 \\ -3 \end{pmatrix}$. Answer to the nearest minute.

57 Find a vector equation of the line through $F(1, 3, -2)$ and $G(4, -2, 7)$.

58 Find a Cartesian equation of the line joining $P(3, -1, 3)$ and $Q(4, 5, 1)$.

59 Determine whether the point $K(-5, 18, 1)$ lies on the line $\begin{pmatrix} x \\ y \\ z \end{pmatrix} = \begin{pmatrix} 1 \\ 4 \\ -2 \end{pmatrix} + \lambda \begin{pmatrix} -3 \\ 7 \\ 1 \end{pmatrix}$.

60 Express the Cartesian equation $\dfrac{x-2}{6} = \dfrac{y-1}{2} = \dfrac{z-4}{3}$ as a vector equation.

61 Plot the vector function $x = 2$, $y = 3 \cos t$, $z = 3 \sin t$ for $0 \le t \le 2\pi$.

62 Prove by mathematical induction: $(\cos\theta + i\sin\theta)^n = \cos n\theta + i\sin n\theta$, $\forall\, n \in \mathbb{N}$.

63 Use De Moivre's theorem to simplify:

   a $[\sqrt{2}(\cos 3\beta + i\sin 3\beta)]^5$ **b** $[512(\cos 144° - i\sin 144°)]^{\frac{1}{9}}$

   c $\left[\dfrac{1}{2}\left(\cos\dfrac{-2\pi}{3} + i\sin\dfrac{-2\pi}{3}\right)\right]^8$

64 a Use De Moivre's theorem and the binomial expansion of $(A + B)^6$ to show that $\cos 6\theta = 32\cos^6\theta - 48\cos^4\theta + 18\cos^2\theta - 1$.

   b Find the roots of $\cos 6\theta = 0$.

   c Hence prove that $\cos 6\theta = 32\left(\cos^2\theta - \cos^2\dfrac{\pi}{12}\right)\left(\cos^2\theta - \dfrac{1}{2}\right)\left(\cos^2\theta - \cos^2\dfrac{5\pi}{12}\right)$

65 Let $z = \cos\theta + i\sin\theta$.

   a Prove that $z - \dfrac{1}{z} = 2i\sin\theta$.

   b Use the expansion of $\left(z - \dfrac{1}{z}\right)^5$ to express $\sin^5\theta$ in the form

      $\sin^5\theta = A\sin 5\theta + B\sin 3\theta + C\sin\theta$, and state the values of A, B and C.

   c Hence evaluate $\displaystyle\int_0^{\frac{\pi}{2}} \sin^5 x \, dx$.

66 Solve the quadratic equation $z^2 - 4iz - 12 = 0$.

67 Solve the equation $z^2 = -1 + i\sqrt{3}$.

68 Consider the polynomial $P(x) = x^5 + 2x^3 - x^2 - 2$.

   a Explain why $P(x)$ has at least one real root.

   b Show that $x = i\sqrt{2}$ is a root of $P(x) = x^5 + 2x^3 - x^2 - 2$.

   c Hence solve the equation $x^5 + 2x^3 - x^2 - 2 = 0$.

   d Express $P(x) = x^5 + 2x^3 - x^2 - 2$ as a product of real factors.

69 The complex number z is shown on an Argand diagram. Copy the diagram and sketch each expression if $w = 2\left(\cos\dfrac{\pi}{3} + i\sin\dfrac{\pi}{3}\right)$.

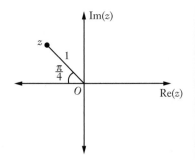

   a zw **b** $\dfrac{z}{w}$

70 The complex numbers z, w, u and v are shown on the Argand diagram. Express w, u and v in terms of z.

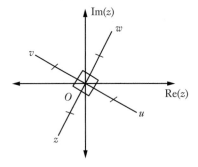

71 Consider the complex numbers a, b and c with corresponding points A, B and C as shown. $ABCD$ forms a parallelogram.

   a Find d in terms of a, b and c.

   b If $\arg\left(\dfrac{a-b}{c-b}\right) = \dfrac{\pi}{4}$, find $\arg\left(\dfrac{d-a}{b-a}\right)$.

   c Find the vector m corresponding with the point M, the intersection of the diagonals of the parallelogram, in terms of a, b and c.

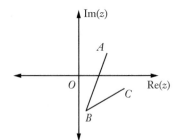

72 a Find the 5th roots of unity for $z^5 = 1$ and show them on an Argand diagram. State the conjugate pairs.

b If α is a complex solution to $z^5 = 1$, show that $\alpha^4 + \alpha^3 + \alpha^2 + \alpha + 1 = 0$.

c Factorise $z^5 - 1$ into one linear and two quadratic factors with real coefficients.

d Show that $\cos\dfrac{2\pi}{5} + \cos\dfrac{4\pi}{5} = -\dfrac{1}{2}$.

73 If ω is a complex cube root of unity, simplify:

a $\omega^2 + \omega + 1$

b $\omega^9 + \omega^8 + \omega^7 + \omega^6 + \omega^5 + \omega^4$

c $(1 - \omega^{-1})(1 - \omega^{-2})$

74 Find both square roots of each complex number.

a $e^{\frac{i\pi}{2}}$

b $9\left(\cos\dfrac{\pi}{3} + i\sin\dfrac{\pi}{3}\right)$

75 Solve each equation and plot its solutions on the complex plane.

a $z^3 = -8$

b $z^4 = -1 + i\sqrt{3}$

76 Sketch each equation on an Argand diagram.

a $|z| = 2$

b $|z - 1 - 2i| = 1$

c $|z + 1| = |z + i|$

d $\mathrm{Re}(z) = 2\,\mathrm{Im}(z)$

77 Sketch each equation.

a $\arg(z - 1) = \dfrac{3\pi}{4}$

b $\arg(z + 1 + i) = \arg(z - 1 - i)$

c $\arg(z + 4i) - \arg(z - 4) = \pi$

78 Sketch each region.

a $\dfrac{\pi}{6} < \arg z \leq \dfrac{\pi}{3}$

b $|z - 3| \leq 3$

c $|\mathrm{Im}(z)| < 1$

79 The points Z_1, Z_2, Z_3 and Z_4 form a quadrilateral and represent the complex numbers z_1, z_2, z_3 and z_4. It is given that $z_1 - z_2 + z_3 - z_4 = 0$. What type of quadrilateral is $Z_1Z_2Z_3Z_4$?

5.

FURTHER MATHEMATICAL INDUCTION

Mathematical induction is used to prove results in series, divisibility, inequality, algebra, calculus, probability, combinatorics and geometry. In Mathematics Extension 1, you learned about the logic and technique of this type of proof. You will now be introduced to its use in a variety of contexts, including a new notation – sigma notation. A further application, recursive formula proofs, will also be discussed.

CHAPTER OUTLINE

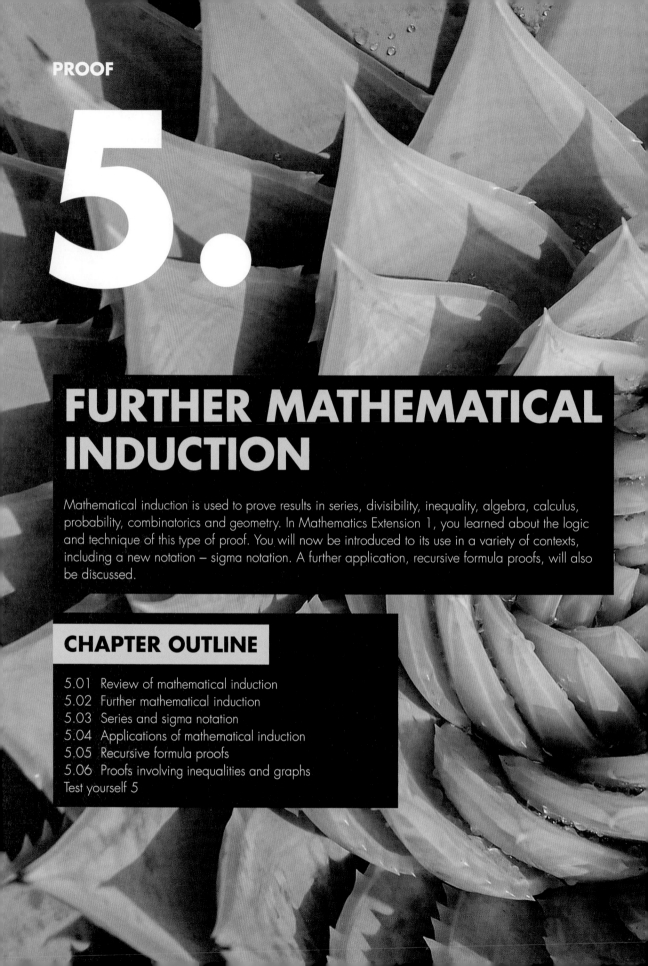

IN THIS CHAPTER YOU WILL:

- review proofs of sums and divisibility by mathematical induction
- prove results for cases other than for *n* being a positive integer
- understand and use sigma notation for sums
- use induction to prove results in inequalities, algebra, calculus, probability, combinatorics and geometry
- use induction to prove recursive formulas

TERMINOLOGY

divisibility: Whether or not a number is divisible by another number.

induction: A method of proof where, based on the truth of earlier statements forming a pattern, pattern is proved true for all statements.

recursive formula: A formula for calculating the next terms of a sequence based on the previous terms.

series: A sum of terms $T_1 + T_2 + T_3 + \dots + T_n$

sigma notation: A shorthand way of writing a series using the Greek letter sigma Σ:

$$\sum_{r=1}^{n} T_r = T_1 + T_2 + T_3 + \dots + T_n.$$

Proof by
mathematical
induction

5.01 Review of mathematical induction

In Mathematics Extension 1, we learn that a proof by mathematical **induction** takes a statement or proposition, proves it is true for an initial value such as $n = 1$, assumes it is true for some value k, then shows it can be proved for the next value $k + 1$ and beyond, using the assumption, thus proving the proposition true for all the defined values.

> ### Proof by mathematical induction
>
> Let $P(n)$ be the proposition $\forall\, n \in \mathbb{N}$, where $\mathbb{N}$ is the set of natural numbers $\{1, 2, 3, \dots\}$.
>
> Step 1: Show that $P(n)$ is true for $n = 1$.
>
> Step 2: Assume the statement is true for some positive integer value $n = k$.
>
> Step 3: Using the assumption, prove that the statement is also true for the next integer $n = k + 1$.
>
> Conclusion: State why the statement is true for all (positive) integers $n \geq 1$.

We will now revise proof by mathematical induction used to prove **series** and divisibility.

Proof of sums

EXAMPLE 1

Use mathematical induction to prove the following proposition is true for all positive integers:

$P(n)$: $2 + 4 + 6 + \dots + 2n = n(n + 1)$.

Solution

Step 1: Prove $P(1)$ is true.

LHS $= 2$ RHS $= 1 \times (1 + 1) = 2$

$\therefore P(1)$ is true.

MATHS IN FOCUS 12. Mathematics Extension 2 ISBN 9780170413435

Step 2: Assume $P(k)$ is true, that is:

$2 + 4 + 6 + \ldots + 2k = k(k + 1)$ for some $k \in \mathbb{N}$

Step 3: Prove $P(k + 1)$ is true, that is:

$2 + 4 + 6 + \ldots + 2k + 2(k + 1) = (k + 1)[(k + 1) + 1] = (k + 1)(k + 2)$

Consider $P(k)$: $2 + 4 + 6 + \ldots + 2k = k(k + 1)$

Adding the $(k + 1)$th term to both sides of $P(k)$:

$2 + 4 + 6 + \ldots + 2k + 2(k + 1) = k(k + 1) + 2(k + 1)$ ◄—— 2(k + 1) is the (k + 1)th term of the series.

$\qquad \therefore \quad$ LHS of $P(k + 1) = (k + 1)(k + 2)$ on factorising

$\qquad\qquad\qquad\qquad = $ RHS of $P(k + 1)$.

$\therefore$ Truth of $P(k)$ implies truth of $P(k + 1)$

Conclusion: But $P(1)$ is also true.

$\therefore$ $P(n)$ is true for all positive integers n by mathematical induction.

EXAMPLE 2

Use mathematical induction to prove that $P(n)$ is true for all positive integers:

$P(n)$: $1 + 2 + 2^2 + 2^3 + \ldots 2^{n-1} = 2^n - 1$.

Solution

Prove $P(1)$ is true:

LHS $= 1 \qquad$ RHS $= 2^1 - 1 = 1 \qquad \therefore P(1)$ is true.

Assume $P(k)$ is true: $1 + 2 + 2^2 + 2^3 + \ldots 2^{k-1} = 2^k - 1$ for some $k \in \mathbb{N}$

Prove $P(k + 1)$ is true: $1 + 2 + 2^2 + 2^3 + \ldots 2^{k-1} + 2^k = 2^{k+1} - 1$

Consider $P(k)$:

$1 + 2 + 2^2 + 2^3 + \ldots 2^{k-1} = 2^k - 1$

Adding the $(k + 1)$th term to both sides of $P(k)$:

$\underbrace{1 + 2 + 2^2 + 2^3 + \ldots 2^{k-1} + 2^k} = 2^k - 1 + 2^k$

$\qquad \therefore \quad$ LHS of $P(k + 1) = 2 \times 2^k - 1$

$\qquad\qquad\qquad\qquad = 2^{k+1} - 1$ on adding the indices

$\qquad\qquad\qquad\qquad = $ RHS of $P(k + 1)$

$\therefore$ Truth of $P(k)$ implies truth of $P(k + 1)$

But $P(1)$ is also true.

$\therefore$ $P(n)$ is true $\forall\, n \in \mathbb{N}$, by mathematical induction.

Proof of divisibility

Note that saying X is divisible by Y is equivalent to saying X is a multiple of Y. You saw in Chapter 2 that we can write $X = pY$ for some $p \in \mathbb{N}$. We can use induction to prove **divisibility**.

EXAMPLE 3

Use mathematical induction to prove the following proposition is true for all positive integers:

$P(n)$: $5^n - 1$ is divisible by 4.

Solution

Prove $P(1)$ is true:

$5^1 - 1 = 5 - 1 = 4$, which is divisible by 4. $\therefore P(1)$ is true.

Assume $P(k)$ is true:

$5^k - 1 = 4Y$, for some $k \in \mathbb{N}$ and some positive integer Y.

$\therefore \quad 5^k = 4Y + 1 \qquad$ [*] $\longleftarrow$ ⟨It will be convenient to make 5^k the subject.⟩

Prove $P(k + 1)$ is true, that is: $5^{k+1} - 1 = 4Z$ for some positive integer Z.

$$
\begin{aligned}
\text{LHS of } P(k+1) &= 5^{k+1} - 1 \\
&= 5 \times 5^k - 1 \\
&= 5 \times (4Y + 1) - 1 \text{ using [*]} \\
&= 5 \times 4Y + 5 - 1 \\
&= 20Y + 4 \\
&= 4(5Y + 1) \\
&= 4Z \text{ for some positive integer } Z
\end{aligned}
$$

which is divisible by 4. Therefore $P(k + 1)$ is true.

$\therefore$ truth of $P(k)$ implies truth of $P(k + 1)$

But $P(1)$ is also true.

$\therefore P(n)$ is true for all positive integers n by mathematical induction.

Exercise 5.01 Review of mathematical induction

1 Prove each proposition $P(n)$, $\forall \, n \in \mathbb{N}$, by mathematical induction.

  a $P(n)$: $1 + 2 + 3 + 4 + \ldots + n = \dfrac{n}{2}(n + 1)$

  b $P(n)$: $1 + 3 + 5 + \ldots + (2n - 1) = n^2$

  c $P(n)$: $1^2 + 2^2 + 3^2 + \ldots + n^2 = \dfrac{n}{6}(n + 1)(2n + 1)$

2 Use mathematical induction to prove each proposition.

  a $P(n)$: $1 + 3 + 3^2 + 3^3 + \ldots + 3^{n-1} = \dfrac{3^n - 1}{2}$, $\forall \, n \in \mathbb{N}$

  b $P(n)$: $1 + 4 + 4^2 + 4^3 + \ldots + 4^{n-1} = \dfrac{(4^n - 1)}{3}$, $\forall \, n \in \mathbb{N}$

  c $P(n)$: $1 + \dfrac{1}{2} + \dfrac{1}{2^2} + \dfrac{1}{2^3} + \ldots + \dfrac{1}{2^{n-1}} = 2 - \dfrac{1}{2^{n-1}}$, $\forall \, n \in \mathbb{N}$

3 Prove each divisibility statement is true for all positive integers n by mathematical induction.

  a $P(n)$: $4^n - 1$ is divisible by 3
  b $P(n)$: $7^n - 1$ is divisible by 6
  c $P(n)$: $3^{2n} - 1$ is divisible by 8

4 Use proof by mathematical induction to prove each proposition $P(n)$, $\forall \, n \in \mathbb{N}$.

  a $P(n)$: $1^3 + 2^3 + 3^3 + 4^3 + \ldots + n^3 = \dfrac{n^2}{4}(n + 1)^2$

  b $P(n)$: $3 + 3^2 + 3^3 + \ldots + 3^n = \dfrac{3(3^n - 1)}{2}$

  c $P(n)$: $\dfrac{1}{1 \times 2} + \dfrac{1}{2 \times 3} + \dfrac{1}{3 \times 4} + \ldots + \dfrac{1}{n \times (n+1)} = \dfrac{n}{n+1}$

5 Write a proof for each proposition for all natural numbers n using mathematical induction.
  a $P(n)$: $9^{n+2} - 4^n$ is divisible by 5
  b $P(n)$: $n(n + 1)$ is divisible by 2
  c $P(n)$: $3^{2n+4} - 2^{2n}$ is divisible by 5
  d $P(n)$: $n(n + 1)(n + 2)$ is divisible by 6
  e $P(n)$: $n^3 + 2n$ is a multiple of 3

6 Prove by mathematical induction that each series formula is true for all positive integers, n.

  a $P(n)$: $a + (a + d) + (a + 2d) + (a + 3d) + \ldots (a + (n - 1)d) = \dfrac{n}{2}(2a + (n - 1)d)$

  b $P(n)$: $a + ar + ar^2 + \ldots + ar^{n-1} = \dfrac{a(r^n - 1)}{r - 1}$, $r \neq 1$

Pascal's induction

The principle of mathematical induction is often used in computer science but the technique was first formulated in a systematic way by **Blaise Pascal**, long before computers were thought of. He was trying to prove the properties of the triangle that came to be known as Pascal's triangle.

5.02 Further mathematical induction

There are some propositions that are only true for integers n such that $n \geq 2$ (or another value) or only for odd n or even n.

EXAMPLE 4

Use proof by mathematical induction to show that $n^2 + 2n$ is a multiple of 8 $\forall\, n \in \mathbb{N}$, where n is even.

Solution

Let $P(n)$ be the proposition that $n^2 + 2n = 8M\ \forall\, n \in \mathbb{N}$, where n is even and $M \in \mathbb{N}$.

Prove $P(2)$ is true:

$\text{LHS} = 2^2 + 2(2)$

$\qquad = 8$

which is a multiple of 8, $\therefore\ P(2)$ is true.

Assume $P(k)$ is true for some even $k \in \mathbb{N}$.

$k^2 + 2k = 8Y$ for some $Y \in \mathbb{N}$.

Required to prove (RTP): $P(k + 2)$ is true,   ←   If k is even, the next even number after k is $(k + 2)$.

$(k + 2)^2 + 2(k + 2) = 8Z$ for some $Z \in \mathbb{N}$.

$\text{LHS of } P(k+2) = (k+2)^2 + 2(k+2)$

$\qquad\qquad\qquad = k^2 + 4k + 4 + 2k + 4$

$\qquad\qquad\qquad = (k^2 + 2k) + 4k + 8$

$\qquad\qquad\qquad = 8Y + 4k + 8 \text{ using } P(k)$

Now since k is even, let $k = 2A$ for some $A \in \mathbb{N}$. Then:

LHS of $P(k + 2) = 8Y + 4(2A) + 8$

$$= 8Y + 8A + 8$$

$$= 8(Y + A + 1)$$

$$= 8Z$$

which is a multiple of 8.

$\therefore P(k + 2)$ is true.

$\therefore$ Truth of $P(k) \Rightarrow$ truth of $P(k + 2)$

But $P(2)$ is also true

$\therefore P(n)$ is true by mathematical induction.

EXAMPLE 5

Use proof by mathematical induction to show that $3^n + 2^n$ is divisible by 5, $\forall\, n \in \mathbb{N}$, where n is odd.

Solution

Let $P(n)$ be the proposition that $3^n + 2^n = 5B\ \forall\, n \in \mathbb{N}$, where n is odd and $B \in \mathbb{N}$.

Prove $P(1)$ is true:

LHS $= 3^1 + 2^1$

$\quad = 5$

which is a multiple of 5, $\therefore P(1)$ is true.

Assume $P(k)$ is true for some odd $k \in \mathbb{N}$.

$3^k + 2^k = 5p$ for some $p \in \mathbb{N}$.

RTP: $P(k + 2)$ is true: $\longleftarrow$ [If k is odd, the next odd number after k is $(k + 2)$.]

$3^{k+2} + 2^{k+2} = 5q$ for some $q \in \mathbb{N}$.

LHS of $P(k + 2) = 3^{k+2} + 2^{k+2}$

$$= 3^2 \times 3^k + 2^2 \times 2^k$$

$$= 9(3^k) + 4(2^k)$$

Now rearranging $P(k)$ to make 3^k the subject:

$3^k + 2^k = 5p$

$\quad 3^k = 5p - 2^k$

Substitute into $P(k + 2)$:

$\text{LHS of } P(k + 2) = 9(5p - 2^k) + 4(2^k)$

$\qquad\qquad\qquad = 45p - 9(2^k) + 4(2^k)$

$\qquad\qquad\qquad = 45p - 5(2^k)$

$\qquad\qquad\qquad = 5(9p - 2^k)$

$\qquad\qquad\qquad = 5q$

which is a multiple of 5.

$\therefore P(k + 2)$ is true.

$\therefore$ Truth of $P(k) \Rightarrow$ truth of $P(k + 2)$

But $P(1)$ is also true.

$\therefore P(n)$ is true by mathematical induction.

Exercise 5.02 Further mathematical induction

1 Prove each proposition by mathematical induction.

    a $P(n)$: $5^n - 1$ is divisible by 8 for all even $n \geq 2$

    b $P(n)$: $3^n - 2^n$ is divisible by 5 for all even $n \geq 2$

    c $P(n)$: $x^n - 1$ is divisible by $x^2 - 1$ for all even $n \geq 2$

2 Prove each divisibility statement is true by mathematical induction.

    a $P(n)$: $5^n + 2^n$ is divisible by 7 for all odd $n \geq 1$

    b $P(n)$: $6^n + 3^n$ is divisible by 9 for all odd $n \geq 1$

    c $P(n)$: $4^{n-2} + 7^{n-2}$ is divisible by 11 for all odd $n \geq 3$

3 Prove each proposition is true by mathematical induction.

    a $P(n)$: $9^n - 8(n - 1) - 9$ is divisible by 64 for all integers $n \geq 2$

    b $P(n)$: $13^{n+1} - 12n - 13$ is divisible by 144 for all natural numbers n

5.03 Series and sigma notation

You have seen that the capital Greek letter *sigma*, written Σ, is used in statistics to represent a sum.

Sigma notation can also be used to describe a **series**, a sum of terms.

The sum of n terms can be written like this:

$$T_1 + T_2 + T_3 + T_4 + \ldots + T_{n-1} + T_n = \sum_{r=1}^{n} T_r.$$

$\sum_{r=1}^{n} T_r$ means the sum of the terms, T_r, starting at $r = 1$ and ending at $r = n$.

For an **infinite** series, we write $T_1 + T_2 + T_3 + T_4 + \ldots + T_{n-1} + T_n + T_{n+1} + \ldots = \sum_{r=1}^{\infty} T_r.$

Sigma notation

$$\sum_{r=1}^{n} T_r \text{ means } T_1 + T_2 + T_3 + T_4 + \ldots T_{n-1} + T_n$$

Note that we can begin and end at any number r we choose.

EXAMPLE 6

a   Write out the series represented by $\sum_{r=1}^{7} r^3$.

b   Evaluate $\sum_{k=3}^{6} \dfrac{1}{3^{k-2}}$.

Solution

a $\sum_{r=1}^{7} r^3 = 1^3 + 2^3 + 3^3 + 4^3 + 5^3 + 6^3 + 7^3$

b $\sum_{k=3}^{6} \dfrac{1}{3^{k-2}} = \dfrac{1}{3^{3-2}} + \dfrac{1}{3^{4-2}} + \dfrac{1}{3^{5-2}} + \dfrac{1}{3^{6-2}}$

$= \dfrac{1}{3^1} + \dfrac{1}{3^2} + \dfrac{1}{3^3} + \dfrac{1}{3^4}$

$= \dfrac{40}{81}$

EXAMPLE 7

Express each sum using sigma notation.

a $\quad 1^2 - 2^2 + 3^2 - 4^2 + \ldots + (-1)^{100-1} \times 100^2$

b $\quad \dfrac{1}{1 \times 2} + \dfrac{1}{2 \times 3} + \dfrac{1}{3 \times 4} + \dfrac{1}{4 \times 5} + \ldots + \dfrac{1}{n(n+1)} + \ldots$

Solution

a $\quad 1^2 - 2^2 + 3^2 - 4^2 + \ldots + (-1)^{100-1} \times 100^2 = \displaystyle\sum_{r=1}^{r=100} (-1)^{r-1} \times r^2$

Note that the value of $(-1)^{r-1}$ alternates between 1 and −1.

b $\quad \dfrac{1}{1 \times 2} + \dfrac{1}{2 \times 3} + \dfrac{1}{3 \times 4} + \dfrac{1}{4 \times 5} + \ldots + \dfrac{1}{n(n+1)} + \ldots = \displaystyle\sum_{k=1}^{\infty} \dfrac{1}{k(k+1)}$

Often sigma notation is used to abbreviate a sum in an induction proof. We will now look at one such example.

EXAMPLE 8

Prove by mathematical induction:

$\displaystyle\sum_{k=1}^{N} \dfrac{1}{(2k+1)(2k-1)} = \dfrac{N}{2N+1}$ where $k \in \mathbb{N}$.

Solution

First write out the series to see the pattern and where the series stops.

$\displaystyle\sum_{k=1}^{N} \dfrac{1}{(2k+1)(2k-1)} = \dfrac{N}{2N+1}$ means

$\dfrac{1}{[2(1)+1][2(1)-1]} + \dfrac{1}{[2(2)+1][2(2)-1]} + \dfrac{1}{[2(3)+2][2(3)-2]} + \ldots + \dfrac{1}{[2N+1][2N-1]} = \dfrac{N}{2N+1}$

Then simplifying we have:

$\dfrac{1}{3 \times 1} + \dfrac{1}{5 \times 3} + \dfrac{1}{7 \times 5} + \ldots + \dfrac{1}{(2N+1)(2N-1)} = \dfrac{N}{2N+1}$

Now we can begin the proof.

Proof

Let $P(N)$ be the proposition that $\dfrac{1}{3\times1}+\dfrac{1}{5\times3}+\dfrac{1}{7\times5}+...+\dfrac{1}{(2N+1)(2N-1)}=\dfrac{N}{2N+1}$, $\forall\,N\in\mathbb{N}$.

Prove true for $P(1)$:

$$\text{LHS}=\dfrac{1}{3\times1}\qquad\text{RHS}=\dfrac{1}{2(1)+1}$$

$$=\dfrac{1}{3}\qquad\qquad\quad=\dfrac{1}{3}$$

Since LHS = RHS then $P(1)$ is true.

Assume $P(k)$ is true for some $k\in\mathbb{N}$.

$$\dfrac{1}{3\times1}+\dfrac{1}{5\times3}+\dfrac{1}{7\times5}+...+\dfrac{1}{(2k+1)(2k-1)}=\dfrac{k}{2k+1}$$

RTP: $P(k+1)$ is true, that is:

$$\dfrac{1}{3\times1}+\dfrac{1}{5\times3}+\dfrac{1}{7\times5}+...+\dfrac{1}{(2k+1)(2k-1)}+\dfrac{1}{(2(k+1)+1)(2(k+1)-1)}=\dfrac{k+1}{2(k+1)+1}$$

which simplifies to:

$$\dfrac{1}{3\times1}+\dfrac{1}{5\times3}+\dfrac{1}{7\times5}+...+\dfrac{1}{(2k+1)(2k-1)}+\dfrac{1}{(2k+3)(2k+1)}=\dfrac{k+1}{2k+3}$$

Consider the LHS of $P(k+1)$:

$$\text{LHS}=\dfrac{1}{3\times1}+\dfrac{1}{5\times3}+\dfrac{1}{7\times5}+...+\dfrac{1}{(2k+1)(2k-1)}+\dfrac{1}{(2k+3)(2k+1)}$$

$$=\dfrac{k}{2k+1}+\dfrac{1}{(2k+3)(2k+1)}\quad\text{using }P(k)$$

$$=\dfrac{k(2k+3)+1}{(2k+3)(2k+1)}$$

$$=\dfrac{2k^2+3k+1}{(2k+3)(2k+1)}$$

$$=\dfrac{(k+1)(2k+1)}{(2k+3)(2k+1)}$$

$$=\dfrac{k+1}{2k+3}$$

$$=\text{RHS of }P(k+1)$$

$\therefore$ Truth of $P(k)\Rightarrow$ truth of $P(k+1)$

But $P(1)$ is also true.

$\therefore$ $P(n)$ is true by mathematical induction.

Exercise 5.03 Series and sigma notation

1 Write out the series represented by each expression.

a $\displaystyle\sum_{r=1}^{10} r^2$

b $\displaystyle\sum_{k=1}^{n} (2k+3)$

c $\displaystyle\sum_{n=1}^{M+1} \frac{1}{n}$

d $\displaystyle\sum_{r=2}^{9} (-1)^{r-1} r$

e $\displaystyle\sum_{r=1}^{\infty} \frac{1}{2^{r-1}}$

2 Evaluate each expression.

a $\displaystyle\sum_{k=1}^{4} (k+2)$

b $\displaystyle\sum_{r=1}^{5} 3^{r-1}$

c $\displaystyle\sum_{j=1}^{3} j(j+1)$

d $\displaystyle\sum_{k=3}^{8} \frac{(-1)^{k-1}}{k^2}$

3 Express each sum in sigma notation.

a $-1^2 + 2^2 - 3^2 + 4^2 + \ldots - 77^2$

b $\dfrac{1}{2} + \dfrac{1}{3} + \dfrac{1}{4} + \dfrac{1}{5} + \ldots + \dfrac{1}{n}$

c $3 + 3^2 + 3^3 + 3^4 + \ldots + 3^{99}$

d $1 - \dfrac{1}{2} + \dfrac{1}{4} - \dfrac{1}{8} + \dfrac{1}{16} - \ldots$

4 Prove each proposition by induction.

a $P(n): \displaystyle\sum_{r=1}^{n} (3r-2) = \frac{n}{2}(3n-1)$

b $P(n): \displaystyle\sum_{r=1}^{n} 6^{r-1} = \frac{6^n - 1}{5}$

c $P(n): \displaystyle\sum_{r=1}^{n} (2r-1) = n^2$

d $P(n): \displaystyle\sum_{n=1}^{N} \frac{1}{2^{n-1}} = 2 - \frac{1}{2^{N-1}}$

e $P(n): \displaystyle\sum_{k=1}^{n} (2k-1)^2 = \frac{n(2n-1)(2n+1)}{3}$

5 $n! = n(n-1)(n-2)(n-3) \times \ldots \times 3 \times 2 \times 1$. For convenience, we define $0! = 1$ and $1! = 1$.

Prove by induction $P(n): \displaystyle\sum_{r=1}^{n} \frac{r}{(r+1)!} = 1 - \frac{1}{(n+1)!}$

6 Prove by mathematical induction $\displaystyle\sum_{r=1}^{n} \log\left(\frac{r+1}{r}\right) = \log(n+1)$.

ISBN 9780170413435

VON KOCH'S SNOWFLAKE

The Von Koch curve is found by taking the limit of a sequence of shapes $T_1, T_2, T_3, ..., T_n$ based on an equilateral triangle as shown below. It resembles a snowflake and it is sometimes referred to as Von Koch's snowflake.

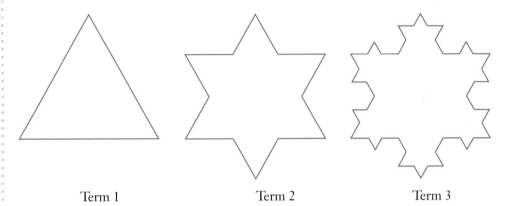

Term 1   Term 2   Term 3

1 Draw the next shape Term 4 in the pattern.

2 If the length of a side in Term 1 is x, find an expression for the perimeter in terms of x when $n = 2, 3, 4$

3 Find an expression for the perimeter if $n = N$.

4 As $n \to \infty$, does the Von Koch curve have finite or infinite perimeter?

5 Is the area finite or infinite?

Getty Images/Cavan Images

5.04 Applications of mathematical induction

In this section we will see applications of proof by mathematical induction in different areas of mathematics.

Proof of inequalities

Recall some important results used in inequality proofs from Chapter 2, *Mathematical proof*.

1 For any two real numbers a and b, $a > b$ if $a - b > 0$.

2 If $a > b$ and $b > c$ then $a > c$.

3 If $a > b$ and $c > d$ then $a + c > b + d$.

4 If $a, b, c > 0$ and $a > b$ then $ac > bc$.

5 If $a > b > 0$ then $\dfrac{1}{a} < \dfrac{1}{b}$.

We can use these simple results to prove further inequalities by induction.

EXAMPLE 9

Prove by mathematical induction: $2^n > n^2$ where $n \in \mathbb{N}$, $n \geq 5$.

Solution

Let $P(n)$ be the proposition that $2^n > n^2$ where $n \in \mathbb{N}$, $n \geq 5$.

Prove true for $P(5)$.

LHS $= 2^5$ RHS $= 5^2$

     $= 32$ $= 25$

Since LHS > RHS, then $P(5)$ is true.

Assume $P(k)$ is true for some $k \in \mathbb{N}$, $k \geq 5$.

$2^k > k^2$

RTP: $P(k + 1)$ is true: $2^{k+1} > (k + 1)^2$

Proof

Consider the LHS of $P(k + 1)$.

$$\text{LHS} = 2^{k+1}$$
$$= 2 \times 2^k$$
$$> 2 \times k^2 \text{ using } P(k)$$
$$= k^2 + k^2$$

Now $k^2 > 2k + 1$ since $k \geq 5$

$$\therefore k^2 + k^2 > k^2 + (2k + 1)$$
$$> (k + 1)^2$$

$\therefore$ Truth of $P(k) \Rightarrow$ truth of $P(k + 1)$

But $P(5)$ is also true.

$\therefore P(n)$ is true by mathematical induction.

Proof of calculus identities

EXAMPLE 10

Prove by mathematical induction:

$$\frac{d}{dx}(x^n) = nx^{n-1}, n \in \mathbb{N}$$

Solution

Let $P(n)$ be the proposition that $\frac{d}{dx}(x^n) = nx^{n-1}, n \in \mathbb{N}$.

Prove true for $P(1)$.

$$\text{LHS} = \frac{d}{dx}(x^1)$$
$$= 1$$

since the gradient of $y = x$ is 1.

$$\text{RHS} = 1 \times x^{1-1}$$
$$= 1x^0$$
$$= 1$$

Since LHS = RHS then $P(1)$ is true.

Assume $P(k)$ is true for some $k \in \mathbb{N}$.

$$\frac{d}{dx}(x^k) = kx^{k-1}$$

RTP: $P(k + 1)$ is true: $\dfrac{d}{dx}(x^{k+1}) = (k + 1)x^k$

Proof

Consider the LHS of $P(k + 1)$.

$$\begin{aligned}
\text{LHS} &= \frac{d}{dx}(x^{k+1}) \\
&= \frac{d}{dx}(x^1 \times x^k) \\
&= x^k \frac{d}{dx}(x) + x\frac{d}{dx}(x^k) \text{ by the product rule} \\
&= x^k \times 1 + x \times kx^{k-1} \text{ using } P(k) \\
&= x^k + kx^k \\
&= (k + 1)x^k \\
&= \text{RHS of } P(k + 1)
\end{aligned}$$

$\therefore$ Truth of $P(k) \Rightarrow$ truth of $P(k + 1)$

But $P(1)$ is also true.

$\therefore$ $P(n)$ is true by mathematical induction.

Proof of binomial theorem

Recall the binomial expansion:

$$(x + a)^n = \sum_{r=0}^{n} {}^nC_r\, x^{n-r}a^r$$

where ${}^nC_r = \begin{pmatrix} n \\ r \end{pmatrix} = \dfrac{n!}{r!(n-r)!}$.

It is possible to prove the binomial theorem by mathematical induction but it will be left for you to do as an exercise below.

Proof of geometry results

The sum S_n of the interior angles of an n-sided polygon is given by the formula

$S_n = (n - 2) \times 180°$, $n \in \mathbb{N}$, $n \geq 3$.

You will find a guided proof of the theorem in Question 5 of the exercise on the next page.

MATHS IN FOCUS 12. Mathematics Extension 2 ISBN 9780170413435

Exercise 5.04 Applications of mathematical induction

1 Prove each inequality by mathematical induction.

   a $3^n > 1 + 2n$ where $n \in \mathbb{N}$

   b $2^n > 1 + n$ where $n \in \mathbb{N}$

   c $3^n > n^3$ where $n \in \mathbb{N}, n \geq 4$

   d $(1 + y)^n \geq 1 + ny$ where $n \in \mathbb{N}$ and $y > -1, y \in \mathbb{R}$

   e $n! > 2^n$ where $n \in \mathbb{N}, n \geq 4$

2 Prove each proposition by mathematical induction.

   a   Let $y = x^M$. $P(n): \dfrac{d^n y}{dx^n} = \dfrac{M! x^{M-n}}{(M-n)!}$ for $n \in \mathbb{N}$, where $M \geq n, M \in \mathbb{N}$.

   b   Let $y = \dfrac{1}{x}, x \neq 0$. $P(n): \dfrac{d^n y}{dx^n} = \dfrac{(-1)^n n!}{x^{n+1}}$ for $n \in \mathbb{N}$.

3 Prove each equation by induction.

   a $(x + a)^n = \displaystyle\sum_{r=0}^{n} {}^nC_r \, x^{n-r} a^r$ where $n \in \mathbb{N}$.

      You will need the identity ${}^nC_r + {}^nC_{r+1} = {}^{n+1}C_{r+1}$ where ${}^nC_r = \dfrac{n!}{r!(n-r)!}$.

   b $n(x + a)^{n-1} = \displaystyle\sum_{r=0}^{n} r \times {}^nC_r \, x^{r-1} a^{n-r}$ where $n \in \mathbb{N}$.

4 Prove each formula by induction.

   a $\sin(n\pi + \theta) = (-1)^n \sin\theta, n \in \mathbb{N}$

   b $\cos(n\pi - \theta) = (-1)^n \cos\theta, n \in \mathbb{N}$

5 Complete in your notebook the blank spaces in the induction proof below.

The sum S_n of the interior angles of an n-sided convex polygon is given by the formula $S_n = (n - 2) \times 180°, n \in \mathbb{N}, n \geq 3$.

Let $P(n)$ be the proposition that $S_n = (n - 2) \times 180°, n \in \mathbb{N}, n \geq 3$.

Prove true for $P(3)$.

LHS $= S_3$ RHS $= (3 - 2)180°$

     $= 180°$ $= 180°$

since the angle sum of a triangle is $180°$.

Since LHS $=$ RHS then $P(3)$ is true.

Assume $P(k)$ is true for some $k \in \mathbb{N}, k \geq 3$.

$$S_k = (k - 2) \times 180°$$

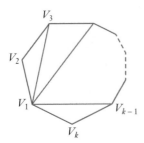

We see from the diagram that there are k vertices labelled $V_1, V_2, V_3, \ldots, V_{k-1}, V_k$. Joining the diagonals from V_1 to the other vertices we create $k - 2$ triangles, with a total interior angle sum of $S_k = (k - 2) \times 180°$.

RTP: $P(k + 1)$ is true; that is $S_{k+1} = $ _____

Proof

Consider the diagram with $(k + 1)$ vertices.

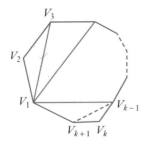

By adding an extra vertex, this has created another triangle.

$\therefore$ LHS = _____

      = _____

      = _____ using $P(k)$

      = _____

      = RHS of $P(k + 1)$

$\therefore$ Truth of $P(k) \Rightarrow$ truth of $P(k + 1)$

But $P(3)$ is also true.

$\therefore$ $P(n)$ is true by mathematical induction.

QED.

MATHS IN FOCUS 12. Mathematics Extension 2

ISBN 9780170413435

6 Consider the number of ways you could cut through a pizza to gain the maximum number of slices, as seen in the diagram below.

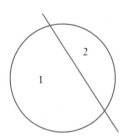 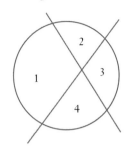 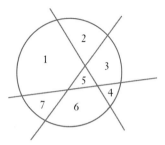

By making a table of values we can find a formula relating the number of cuts n with the number of slices, S_n.

n	1	2	3	...
S_n	2	4	7	...

Find a formula for S_n in terms of n and prove it by mathematical induction.

7 Recall the triangle inequality as shown below for vectors, $z_1, z_2, z_3, \ldots$

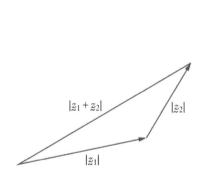 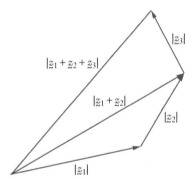

We can see that $\left|z_1 + z_2\right| \le \left|z_1\right| + \left|z_2\right|$ and that $\left|z_1 + z_2 + z_3\right| \le \left|z_1\right| + \left|z_2\right| + \left|z_3\right|$.

Prove by mathematical induction that:

$\left|z_1 + z_2 + z_3 + \ldots + z_n\right| \le \left|z_1\right| + \left|z_2\right| + \left|z_3\right| + \ldots + \left|z_n\right| \ \forall \ n \in \mathbb{N}, n \ge 2.$

5.05 Recursive formula proofs

The sequence $T_1 = 3$, $T_2 = 5$, $T_3 = 7$, $T_4 = 9$, $T_5 = 11$, ... can be described by the **recursive formula** $T_n = T_{n-1} + 2$. We can use recursive formulas to prove by mathematical induction the general formula $T_n = 2n + 1$ for the nth term of the sequence.

EXAMPLE 11

Prove by mathematical induction:

Given $T_1 = 3$ and $T_n = T_{n-1} + 2$, prove by mathematical induction that $T_n = 2n + 1$ is true $\forall\, n \in \mathbb{N}$.

Solution

Let $P(n)$ be the proposition that if $T_1 = 3$ and $T_n = T_{n-1} + 2$, then $T_n = 2n + 1$ is true $\forall\, n \in \mathbb{N}$.

Prove true for $P(1)$.

Given $T_1 = 3$. Using the formula $T_n = 2n + 1$, $T_1 = 2(1) + 1 = 3$ which is consistent.

$P(1)$ is true.

Assume $P(k)$ is true for some $k \in \mathbb{N}$. That is, given $T_k = T_{k-1} + 2$, then $T_k = 2k + 1$.

RTP: $P(k + 1)$ is true; that is, given $T_{k+1} = T_k + 2$, then $T_{k+1} = 2(k + 1) + 1 = 2k + 3$.

Proof

Consider the given $P(k + 1)$ formula:

$$T_{k+1} = T_k + 2$$
$$= (2k + 1) + 2 \text{ using } P(k)$$
$$= 2k + 3 \text{ as required}$$

$\therefore$ Truth of $P(k) \Rightarrow$ truth of $P(k + 1)$

But $P(1)$ is also true.

$\therefore P(n)$ is true by mathematical induction.

MATHS IN FOCUS 12. Mathematics Extension 2

ISBN 9780170413435

EXAMPLE 12

Given $T_1 = 1$, $T_2 = 5$ and $T_n = 5T_{n-1} - 6T_{n-2}$, prove by mathematical induction that $T_n = 3^n - 2^n$ is true $\forall\, n \in \mathbb{N}$, $n \geq 3$.

Solution

Let $P(n)$ be the proposition that if $T_1 = 1$, $T_2 = 5$ and $T_n = 5T_{n-1} - 6T_{n-2}$, then $T_n = 3^n - 2^n$ is true $\forall\, n \in \mathbb{N}$, $n \geq 3$.

Prove true for $P(3)$.

Given $T_1 = 1$, $T_2 = 5$, $T_3 = 5 \times 5 - 6 \times 1 = 19$.

Now using the formula $T_n = 3^n - 2^n$:

$T_3 = 3^3 - 2^3 = 19$, which is consistent.

$P(3)$ is true.

Assume $P(k)$ is true for some $k \in \mathbb{N}$, $k \geq 3$.

That is, given $T_k = 5T_{k-1} - 6T_{k-2}$, then $T_k = 3^k - 2^k$. This also true for $P(k-1)$; that is, $T_{k-1} = 3^{k-1} - 2^{k-1}$.

RTP: $P(k+1)$ is true; that is, given $T_{k+1} = 5T_k - 6T_{k-1}$, then $T_{k+1} = 3^{k+1} - 2^{k+1}$.

Proof

Consider the given $P(k+1)$ formula.

$T_{k+1} = 5T_k - 6T_{k-1}$

Now we can use both $T_k = 3^k - 2^k$ and $T_{k-1} = 3^{k-1} - 2^{k-1}$ to substitute.

$$
\begin{aligned}
T_{k+1} &= 5(3^k - 2^k) - 6(3^{k-1} - 2^{k-1}) \\
&= 5(3^k) - 5(2^k) - 6(3^{k-1}) + 6(2^{k-1}) \\
&= 5(3^k) - 5(2^k) - 2 \times 3 \times 3^{k-1} + 3 \times 2 \times 2^{k-1} \\
&= 5(3^k) - 5(2^k) - 2(3^k) + 3(2^k) \\
&= 3(3^k) - 2(2^k) \\
&= 3^{k+1} - 2^{k+1}
\end{aligned}
$$

which is the required expression.

$\therefore$ Truth of $P(k) \Rightarrow$ truth of $P(k+1)$

But $P(3)$ is also true.

$\therefore P(n)$ is true by mathematical induction.

Exercise 5.05 Recursive formula proofs

1 Prove by mathematical induction:

    a   Given $T_1 = 2$ and $T_n = T_{n-1} + 2$, prove that $T_n = 2n$ is true $\forall\, n \in \mathbb{N}$

    b   Given $T_1 = 2$ and $T_n = 2 \times T_{n-1}$, prove that $T_n = 2^n$ is true $\forall\, n \in \mathbb{N}$

    c   Given $T_1 = 1$ and $T_n = T_{n-1} + 5$, prove that $T_n = 5n - 4$ is true $\forall\, n \in \mathbb{N}$

    d   Given $T_1 = 7$ and $T_n = 3 \times T_{n-1}$, prove that $T_n = 7 \times 3^{n-1}$ is true $\forall\, n \in \mathbb{N}$

2 Prove each result by mathematical induction.

    a   Given $T_1 = 5$, $T_2 = 7$ and $T_n = 3T_{n-1} - 2T_{n-2}$, for $n \geq 3$, prove that $T_n = 2^n + 3$ is true $\forall\, n \in \mathbb{N}$

    b   Given $T_1 = 2$, $T_2 = 16$ and $T_n = 8T_{n-1} - 15T_{n-2}$, for $n \geq 3$, prove that $T_n = 5^n - 3^n$ is true $\forall\, n \in \mathbb{N}$

    c   Given $T_1 = 1$ and $T_n = T_{n-1} + 2n - 1$, for $n \geq 2$, prove that $T_n = n^2$ is true $\forall\, n \in \mathbb{N}$

    d   Given $T_1 = 1$ and $T_n = T_{n-1} + (n-1)(n-1)!$, for $n \geq 2$, prove that $T_n = n!$ is true $\forall\, n \in \mathbb{N}$

3 Probably the most famous recursive formula is that describing the **Fibonacci sequence**:

$$1, 1, 2, 3, 5, 8, 13, \ldots$$

Let $T_1 = 1$, $T_2 = 1$ and $T_k = T_{k-1} + T_{k-2}$, for $n \geq 3$. Rather than expressing the next term as a sum of the 2 previous terms, it has actually been shown that T_n can be calculated using the following formula:

$$T_n = \dfrac{\left(\dfrac{1+\sqrt{5}}{2}\right)^n - \left(\dfrac{1-\sqrt{5}}{2}\right)^n}{\sqrt{5}}, \forall\, n \in \mathbb{N}.\ \text{Prove this by mathematical induction.}$$

DID YOU KNOW?

The Peano axioms

Although the idea of mathematical induction was used informally by a number of mathematicians, the name given to the technique was not stated in a formal manner until an Italian mathematician, **Giuseppe Peano** (1858–1932), produced a work called *Formulaire de mathématiques*. In this publication he attempted to write 5 postulates, known as the **Peano axioms**, that form the basis of arithmetic.

1 Zero is a number.

2 If x is a number, the successor of x is a number.

3 Zero is not the successor of a number.

4 Two numbers of which the successors are equal are themselves equal.

5 If a set S of numbers contains zero and also the successor of every number in S, then every number is in S.

Can you spot the inductive step in the list?

5.06 Proofs involving inequalities and graphs

Most proofs we have seen in this chapter are algebraic. Sometimes results can be proved or problems can be solved using a geometric or graphical approach by comparing concavity, heights or areas.

Graphical solution of inequalities

Often equations and inequalities can be difficult to solve algebraically but are straightforward if we draw a graph.

EXAMPLE 13

Solve $x^2 - 1 > |2x + 2|$.

Solution

Graph $y = x^2 - 1$ and $y = |2x + 2|$ on the same axes.

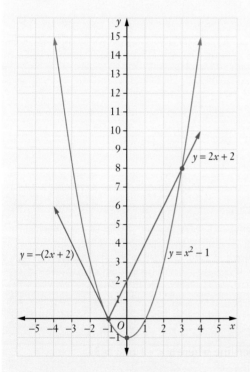

The solutions to the equation $x^2 - 1 = |2x + 2|$ are the points of intersection of the 2 graphs. These can be found by solving simultaneously $y = x^2 - 1$ with the 2 arms of $y = |2x + 2|$, that is, $y = 2x + 2$ and $y = -2x - 2$.

---

5. Further mathematical induction

Solving:

Right arm:

$$x^2 - 1 = 2x + 2$$

$$x^2 - 2x - 3 = 0$$

$$(x - 3)(x + 1) = 0$$

$$x = 3 \text{ or } x = -1$$

Left arm:

$$x^2 - 1 = -2x - 2$$

$$x^2 + 2x + 1 = 0$$

$$(x + 1)^2 = 0$$

$$x = -1$$

From the graph we can see the solution to the right arm is $x = 3$ and to the left arm is $x = -1$.

Now to solve the inequality $x^2 - 1 > |2x + 2|$ we look for the values of x where the parabola is *above* the absolute value graph.

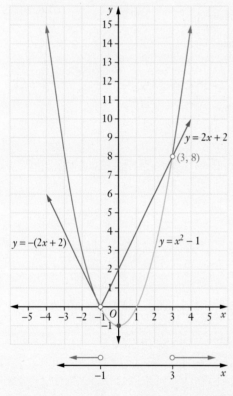

This is equivalent to the solution on the number line where $x < -1$ or $x > 3$.

Therefore the solution to the inequality $x^2 - 1 > |2x + 2|$ is $x < -1$ or $x > 3$.

ISBN 9780170413435

Proofs involving areas

Let us examine some general concepts comparing the area under a curve with approximations using rectangles.

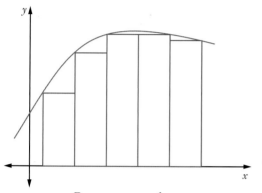

Curve concave down

Area under curve > sum of the rectangles

Curve concave up

Area under curve < sum of the rectangles

We can use these observations to prove various results.

EXAMPLE 14

Consider the areas of the n rectangles each of width $\dfrac{1}{n}$ under the curve $y = \sqrt{1-x^2}$ for $0 \le x \le 1$ as shown.

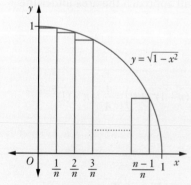

a   Show that the sum A_n of the areas of the rectangles is given by

$$A_n = \frac{1}{n^2}\left(\sqrt{n^2-1} + \sqrt{n^2-2^2} + \sqrt{n^2-3^2} + \ldots + \sqrt{n^2-(n-1)^2}\right).$$

b   Hence show that

$$\lim_{n \to \infty} \frac{1}{n^2}\left(\sqrt{n^2-1} + \sqrt{n^2-2^2} + \sqrt{n^2-3^2} + \ldots\right) = \frac{\pi}{4}.$$

ISBN 9780170413435

Solution

a Let the area of the first rectangle be denoted $A(1)$, the second be $A(2)$, and so on so the last is $A(n)$. Then the height of each rectangle is the y value, calculated using $y = \sqrt{1 - x^2}$.

$$A(1) = \text{height} \times \text{width} \qquad A(2) = \text{height} \times \text{width} \qquad A(n-1) = \text{height} \times \text{width}$$

$$= \sqrt{1 - \left(\frac{1}{n}\right)^2} \times \frac{1}{n} \qquad = \sqrt{1 - \left(\frac{2}{n}\right)^2} \times \frac{1}{n} \qquad = \sqrt{1 - \left(\frac{n-1}{n}\right)^2} \times \frac{1}{n}$$

$$= \frac{1}{n^2}\sqrt{n^2 - 1} \qquad = \frac{1}{n^2}\sqrt{n^2 - 2^2} \qquad = \frac{1}{n^2}\sqrt{n^2 - (n-1)^2}$$

and $A(n) = 0$.

So the sum A_n of the rectangles is given by:

$$A_n = A(1) + A(2) + A(3) + \ldots + A(n-1) + A(n)$$

$$= \frac{1}{n^2}\sqrt{n^2 - 1} + \frac{1}{n^2}\sqrt{n^2 - 2^2} + \frac{1}{n^2}\sqrt{n^2 - 3^2} + \ldots + \frac{1}{n^2}\sqrt{n^2 - (n-1)^2} + 0$$

$$= \frac{1}{n^2}\left(\sqrt{n^2 - 1} + \sqrt{n^2 - 2^2} + \sqrt{n^2 - 3^2} + \ldots + \sqrt{n^2 - (n-1)^2} + 0\right)$$

b From the graph we can see that $y = \sqrt{1 - x^2}$ is concave down so the sum of the rectangles is *less than* the area under the curve; that is, $A_n <$ area under curve.

But as the number of rectangles increases the sum will approach the area under the curve; that is,

$$\lim_{n \to \infty} A_n = \int_0^1 \sqrt{1 - x^2}\, dx \text{ so}$$

$$\lim_{n \to \infty} \frac{1}{n^2}\left(\sqrt{n^2 - 1} + \sqrt{n^2 - 4} + \sqrt{n^2 - 3^2} + \ldots + \sqrt{n^2 - (n-1)^2} + 0\right) = \frac{1}{4} \times \pi \times (1)^2$$

$$\therefore \lim_{n \to \infty} \frac{1}{n^2}\left(\sqrt{n^2 - 1} + \sqrt{n^2 - 2^2} + \sqrt{n^2 - 3^2} + \ldots\right) = \frac{\pi}{4}$$

Exercise 5.06 Proofs involving inequalities and graphs

1 Solve each inequality graphically.

a $|x - 2| > |x|$ **b** $|2x| \le x + 3$ **c** $x^2 < |x|$

d $x^2 - 6 \ge |x|$ **e** $|x| > \dfrac{1}{x}$

2 Explain with the use of a graph why each inequality is true.

a $|2x| > |x| - 1$ for all real values of x

b $|x + 1| < \sqrt{x - 1}$ has no solution

3 Solve $3x^2 - 2x - 2 > |3x|$.

4 a Sketch the graph of $y = xe^{-x}$ and find any stationary points.

   b Hence prove that $x \le e^{x-1}$ for all real x.

5 Consider the secant drawn on $y = \ln x$ between $x = 1$ and $x = 1 + \dfrac{1}{n}$ as shown.

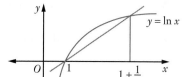

   a Find an expression for the gradient of the secant.

   b Using part **a** and the fact that $\dfrac{d}{dx}(\ln x) = \dfrac{1}{x}$,
   show that $\displaystyle\lim_{n \to \infty} \left(1 + \dfrac{1}{n}\right)^n = e$.

   c Also using the method in part **a**, show that $\left(1 + \dfrac{1}{n+1}\right)^{n+1} > \left(1 + \dfrac{1}{n}\right)^n$.
   Explain with the aid of a sketch.

   d What implication does this have for compound interest?

6 a If $f(x)$ is a continuous function, show with the aid of a diagram the meaning of:

$$\lim_{n \to \infty} \frac{1}{n}\left(f\left(\frac{1}{n}\right) + f\left(\frac{2}{n}\right) + f\left(\frac{3}{n}\right) + \dots + f\left(\frac{n}{n}\right)\right) = \int_0^1 f(x)\,dx$$

   b Hence evaluate $\displaystyle\lim_{n \to \infty} \frac{1}{n}\left(\sin\frac{\pi}{n} + \sin\frac{2\pi}{n} + \sin\frac{3\pi}{n} + \dots + \sin\frac{n\pi}{n}\right)$.

7 a Prove by induction for $a > 0$, $0 < r < 1$ and $n \in \mathbb{N}$ that:

$$a + ar + ar^2 + ar^3 + \dots + ar^{n-1} = \frac{a(r^n - 1)}{r - 1}$$

   b Consider the rays OP and OQ such that $\angle POQ = 30°$ as shown.
   Let OA_0 be a units.

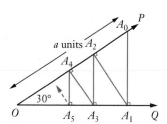

   i Show that the lengths $A_0A_1, A_1A_2, A_2A_3, A_3A_4, \dots$ form a sequence in the form $a, ar, ar^2, ar^3, \dots$ and find the value of r in exact form.

   ii Write down an expression for the length $A_{n-1}A_n$ in exact form.

   iii Show that $A_0A_1 + A_1A_2 + A_2A_3 + A_3A_4 + \dots = a(2 + \sqrt{3})$.

8 Consider the curve $y = \dfrac{1}{x}$, $x > 0$, as shown in the diagram.

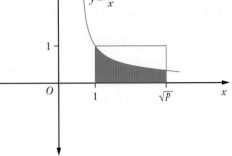

a Explain why $\displaystyle\int_1^{\sqrt{p}} \frac{1}{x}\, dx < \sqrt{p} - 1$ where $p > 1$.

b Hence show that $0 < \ln p < 2\sqrt{p} - 2$.

c Hence deduce that $\displaystyle\lim_{x \to \infty} \frac{\ln x}{x} \to 0$.

9 a Find $\dfrac{d}{dx}(x \ln x)$.

b Hence evaluate exactly the integral $\displaystyle\int_1^n \ln x\, dx$.

c Consider the curve $y = \ln x$.

   i Rectangles each of width 1 unit are drawn below the curve to approximate the area under the curve for $1 \le x \le n$ as shown below.

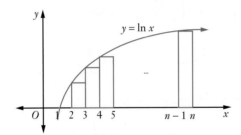

   Show that the sum of the areas of the rectangles is $S_b = \ln[(n-1)!]$

   ii Rectangles each of width 1 unit are now drawn above the curve to approximate the area under the curve for $1 \le x \le n$ as shown below.

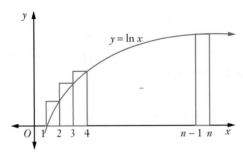

   Find the sum of the areas of the rectangles, S_a.

   iii Hence explain why $\ln[(n-1)!] < n \ln n - n + 1 < \ln(n!)$.

   iv Prove that $(n-1)! < n^n e^{1-n} < n!$

ISBN 9780170413435

1 Prove each series by mathematical induction $\forall\, n \in \mathbb{N}$.

a $5 + 11 + 17 + \ldots + (6n - 1) = 3n^2 + 2n$

b $\dfrac{1}{1 \times 4} + \dfrac{1}{4 \times 7} + \dfrac{1}{7 \times 10} + \ldots + \dfrac{1}{(3n - 2)(3n + 1)} = \dfrac{n}{3n + 1}$

c $a + \dfrac{a}{p} + \dfrac{a}{p^2} + \ldots + \dfrac{a}{p^{n-1}} = \dfrac{a(1 - p^n)}{p^{n-1}(1 - p)},\ p \neq 0,\, 1$

2 Prove by induction:

a $9^n - 1$ is divisible by 8 $\forall\, n \in \mathbb{N}$

b $2^{n+2} + 3^{2n+1}$ is divisible by 7 $\forall\, n \in \mathbb{N}$

3 Prove each series by mathematical induction $\forall\, n \in \mathbb{N}$.

a $\displaystyle\sum_{r=1}^{n} r^3 = \dfrac{1}{4} n^2 (n + 1)^2$ **b** $\displaystyle\sum_{k=1}^{n} k(k!) = (n + 1)! - 1$

c $\displaystyle\sum_{j=1}^{n} \dfrac{1}{j(j + 1)} = \dfrac{n}{n + 1}$

4 Prove by induction:

a $7^n + 13^n + 19^n$ is divisible by 13 if n is odd.

b $n^4 + 4n^2 + 11$ is divisible by 16 if n is odd.

5 Prove each statement by mathematical induction $\forall\, n \in \mathbb{N}$.

a $1 + \dfrac{1}{2^2} + \dfrac{1}{3^2} + \dfrac{1}{4^2} + \ldots + \dfrac{1}{n^2} \leq 2 - \dfrac{1}{n}$

b $\dfrac{1}{x - 1} - \dfrac{1}{x} - \dfrac{1}{x^2} - \dfrac{1}{x^3} - \ldots - \dfrac{1}{x^n} = \dfrac{1}{x^n (x - 1)},\ x \neq 0,\, 1$

6 Prove by induction $(ab)^n = a^n b^n,\ \forall\, n \in \mathbb{N}$.

7 Prove by induction $\dfrac{d}{dx}(x^{-n}) = -nx^{-n-1},\ \forall\, n \in \mathbb{N}$.

8 Prove by induction $\dfrac{d^n}{d\theta^n}(\sin p\theta) = p^n \sin\left(p\theta + \dfrac{n\pi}{2}\right),\ \forall\, n \in \mathbb{N},\, p \in \mathbb{Q}$.

9 Prove by induction $P(n)$: $x^{2n} - y^{2n}$ is divisible by $x^n + y^n$ for all $n \geq 1$.

10 Prove by induction that the expression $(x + 1)^n - nx - 1$ is divisible by x^2 where n is a positive integer.

11 Prove each property by mathematical induction.

a If $m \in \mathbb{N}$ and $m \geq 1$ then $m(m + 3)$ is always even.

b If $n \in \mathbb{N}$ and $n \geq 2$ then $n(n + 1)(n - 1)$ is always divisible by 6.

c If $n \in \mathbb{N}$ and if n is odd, then $n(n + 2) + (n + 2)(n + 4)$ is always even.

12 Given $K > 0$ and $L > 0$ where $K \neq L$ and $n \in \mathbb{N}$:

a find $\dfrac{K^{n+1} - K^n L + L^{n+1} - KL^n}{K - L}$.

b deduce that $K^{n+1} + L^{n+1} \geq K^n L + KL^n$

c hence prove by mathematical induction that $\left(\dfrac{K + L}{2} \right)^n \leq \dfrac{K^n + L^n}{2} \; \forall \, n \in \mathbb{N}$.

13 a Write the expansion for $\cos (A + B)$. Hence prove $\cos 2P = 1 - 2 \sin^2 P$.

b Prove that $\dfrac{\cos Q - \cos (Q + 2P)}{2 \sin P} = \sin (Q + P)$.

c Hence prove by mathematical induction that:

$$\sin P + \sin 3P + \sin 5P + \ldots + \sin (2n - 1)P = \dfrac{1 - \cos 2nP}{2 \sin P}, \; \forall \, n \in \mathbb{N}.$$

14 Use the recursive technique for each proof.

a Given $T_1 = 5$ and $T_n = 2T_{n-1} + 1$, for $n \geq 2$, prove that $T_n = 6(2^{n-1}) - 1$ is true $\forall \, n \in \mathbb{N}$ by mathematical induction.

b i Given $T_1 = 1$ and $T_n = \dfrac{2T_{n-1} - 1}{3}$ for $n \geq 2$, find the values for T_2, T_3 and T_4.

ii Prove that $T_n = 3 \left(\dfrac{2}{3} \right)^n - 1$ is true $\forall \, n \in \mathbb{N}$, by mathematical induction.

c Given $T_1 = 5$, $T_2 = 11$ and $T_n = 4T_{n-1} - 3T_{n-2}$ for $n \geq 3$, prove that $T_n = 3^n + 2$ is true $\forall \, n \in \mathbb{N}$, by mathematical induction.

15 Use mathematical induction to prove that the sum of the exterior angles of an n-sided convex polygon is $360°$.

16 Solve each inequality graphically.

a $|x| \geq x^2 - 2$

b $\dfrac{4}{x - 1} \leq x + 2$

17 Consider the diagram showing an isosceles triangle inscribed in a semicircle inscribed in a rectangle.

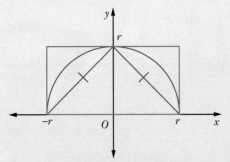

a By considering areas, show that $2 < \pi < 4$.

b Explain how you could find a better approximation for π.

18 Consider the curve $y = \ln x$ and the chord joining $A(a, \ln a)$ and $B(b, \ln b)$ as shown in the diagram. Perpendiculars are drawn from A and B to the x-axis to form a trapezium. Note $a \neq b$.

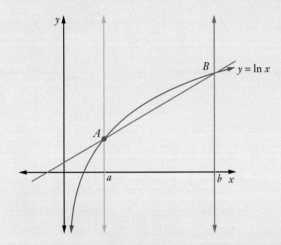

a Explain why the area under the curve between A and B is greater than the area of the trapezium.

b Prove that $e^{2\left(\frac{b-a}{b+a}\right)} < \frac{b}{a}$.

CALCULUS

6.

FURTHER INTEGRATION

One of the first uses of integration was in finding the volumes of wine casks. Because the casks have a curved surface, integration is required to determine area and volume.

Other uses of integration include determining centres of mass, fluid flow and modelling the behaviour of objects under stress. In the real world the equations that model these types of applications are generally not simple nor standard.

In this chapter, we develop a broader range of techniques and strategies to solve more complex problems involving differential equations and integration.

CHAPTER OUTLINE

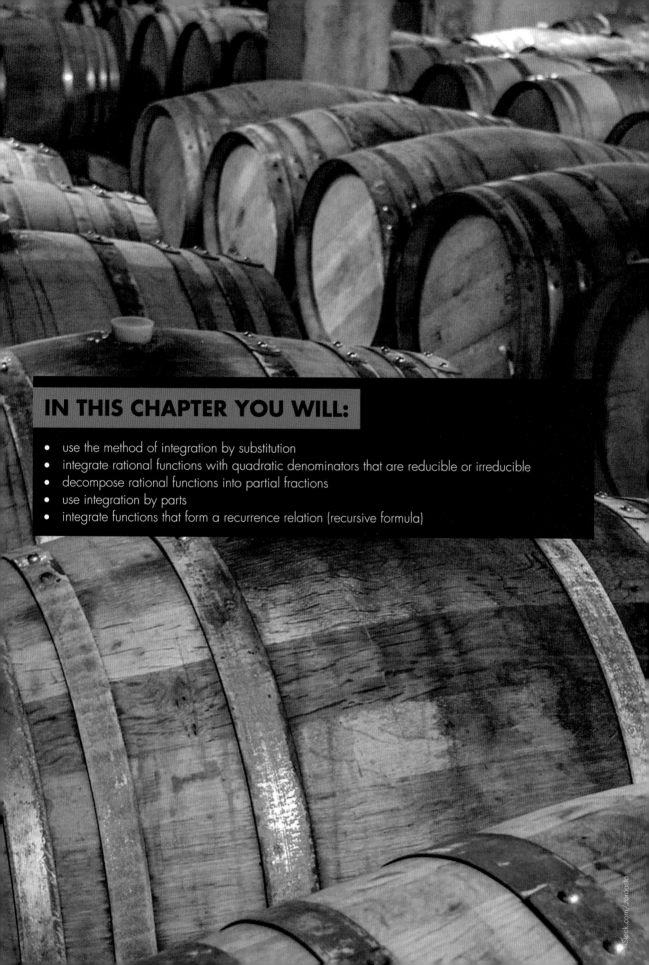

IN THIS CHAPTER YOU WILL:

- use the method of integration by substitution
- integrate rational functions with quadratic denominators that are reducible or irreducible
- decompose rational functions into partial fractions
- use integration by parts
- integrate functions that form a recurrence relation (recursive formula)

iStock.com/zoranlino

TERMINOLOGY

integration by parts: A method of integrating a function by splitting it into one function to be differentiated and one function to be integrated.

partial fractions: A rational function can be expressed as the sum of smaller fractions called partial fractions that are easier to integrate.

rational function: A function that can be expressed as a fraction $\dfrac{f(x)}{g(x)}$, such that both

the numerator and the denominator are polynomials.

recurrence relation or **recursive formula**: A formula or integral that is expressed in terms of itself with a smaller parameter value, for example

$$\int_0^{\frac{\pi}{2}} \sin^n \theta \, d\theta = \frac{n-1}{n} \int_0^{\frac{\pi}{2}} \sin^{n-2} \theta \, d\theta.$$

Integration by substitution

Integration by substitution

6.01 Integration by substitution

In Mathematics Extension 1, Chapter 8, *Further integration*, we learned to use **integration by substitution** for integrals of composite functions involving a function and its derivative.

EXAMPLE 1

Evaluate $\int 6x^2(2x^3 - 1)^4 \, dx$.

Solution

Let $u = 2x^3 - 1$, $\dfrac{du}{dx} = 6x^2$

$du = 6x^2 \, dx$

$$\therefore \int 6x^2(2x^3 - 1)^4 \, dx = \int u^4 \, du$$

$$= \frac{1}{5} u^5 + C$$

$$= \frac{1}{5} (2x^3 - 1)^5 + C$$

In cases like these you will notice that the derivative of 'the function within the function' can be identified as the multiplier.

In many cases, you will be required to find and evaluate indefinite and definite integrals using the method of integration by substitution where the substitution is not given. It is therefore necessary to be able to identify convenient substitutions.

EXAMPLE 2

Evaluate $\displaystyle\int \frac{4x}{\sqrt{2x^2 + 1}} \, dx$.

Solution

Let $u = 2x^2 + 1$, $\dfrac{du}{dx} = 4x$

$du = 4x \, dx$

$$\therefore \int \frac{4x}{\sqrt{2x^2+1}}\,dx = \int \frac{1}{\sqrt{u}}\,du$$

$$= \int u^{-\frac{1}{2}}\,du$$

$$= \frac{1}{\frac{1}{2}}u^{\frac{1}{2}}+C$$

$$= 2\sqrt{u}+C$$

$$= 2\sqrt{2x^2+1}+C$$

EXAMPLE 3

Evaluate $\int \frac{\sin\sqrt{x}}{\sqrt{x}}\,dx$.

Solution

Let $u = \sqrt{x}$, $\dfrac{du}{dx} = \dfrac{1}{2\sqrt{x}}$,

so $dx = 2\sqrt{x}\,du$

$$\int \frac{\sin\sqrt{x}}{\sqrt{x}}\,dx = \int \frac{\sin u}{\sqrt{x}}\,2\sqrt{x}\,du$$

$$= 2\int \sin u\,du$$

$$= -2\cos u + C$$

$$= -2\cos\sqrt{x} + C$$

Many integrations with trigonometric integrands, such as $\int(\sin^n x \cos x)\,dx$ or $\int(\cos^n x \sin x)\,dx$, could use a simple substitution to form a standard integral.

EXAMPLE 4

Evaluate $\int(\sin^9 x \cos x)\,dx$.

Solution

Let $u = \sin x$, noting that $\dfrac{du}{dx} = \cos x$

$du = \cos x\,dx$.

$$\int \sin^9 x \cos x\,dx = \int u^9\,du$$

$$= \frac{1}{10}u^{10} + C$$

$$= \frac{1}{10}\sin^{10} x + C$$

$\sin^2 x$ and $\cos^2 x$

$$\sin^2 x = \frac{1}{2}(1 - \cos 2x)$$

$$\cos^2 x = \frac{1}{2}(1 + \cos 2x)$$

These identities from Mathematics Extension 1, Chapter 8, *Further integration*, allow other similar integrations to be quickly determined.

$$\sin^2 2x = \frac{1}{2}(1 - \cos 4x) \qquad \text{or} \qquad \sin^2 3x = \frac{1}{2}(1 - \cos 6x)$$

$$\cos^2 2x = \frac{1}{2}(1 + \cos 4x) \qquad \qquad \cos^2 3x = \frac{1}{2}(1 + \cos 6x)$$

EXAMPLE 5

Evaluate $\displaystyle\int_0^{\frac{\pi}{4}} \cos^2 2x \, dx$.

Solution

$$\int_0^{\frac{\pi}{4}} \cos^2 2x \, dx = \int_0^{\frac{\pi}{4}} \frac{1}{2}\big[1 + \cos 4x\big] \, dx$$

$$= \frac{1}{2}\left[x + \frac{1}{4}\sin 4x \right]_0^{\frac{\pi}{4}}$$

$$= \frac{1}{2}\left[\left(\frac{\pi}{4} + 0 \right) - (0 + 0) \right]$$

$$= \frac{\pi}{8}$$

EXAMPLE 6

Evaluate $\displaystyle\int_1^4 \frac{1}{(1 + \sqrt{x})^2 \sqrt{x}} \, dx$.

Solution

Substitute $u = 1 + \sqrt{x}$.

Therefore $\dfrac{du}{dx} = \dfrac{1}{2\sqrt{x}}$; that is, $dx = 2\sqrt{x} \, du$

When $x = 1$, $u = 2$, and when $x = 4$, $u = 3$, so we get:

$$\int_1^4 \frac{1}{(1+\sqrt{x})^2 \sqrt{x}} \, dx = \int_2^3 \frac{1}{u^2 \sqrt{x}} \, 2\sqrt{x} \, du$$

$$= 2\int_2^3 u^{-2} \, du$$

$$= 2\left[-\frac{1}{u} \right]_2^3$$

$$= 2\left[-\frac{1}{3} - \left(-\frac{1}{2} \right) \right]$$

$$= 2\left(\frac{1}{6} \right)$$

$$= \frac{1}{3}$$

For integrations of the form $\int \dfrac{dx}{a\cos x + b\sin x}$, $\int \dfrac{dx}{a\cos x + b}$ or $\int \dfrac{dx}{a + b\sin x}$ we can use the t-formulas where $t = \tan \dfrac{x}{2}$:

$\sin x = \dfrac{2t}{1+t^2}$, $\cos x = \dfrac{1-t^2}{1+t^2}$ and $\tan x = \dfrac{2t}{1-t^2}$, along with $dx = \dfrac{2dt}{1+t^2}$ as proven below.

It is easy enough to show that when $t = \tan\left(\dfrac{x}{2} \right)$ then

$$\frac{dt}{dx} = \frac{1}{2}\sec^2 \frac{x}{2}$$

$$= \frac{1}{2}\left[1 + \tan^2\left(\frac{x}{2} \right) \right]$$

$$= \frac{1}{2}(1 + t^2)$$

and therefore: $dx = \dfrac{2dt}{1+t^2}$.

EXAMPLE 7

Use the substitution $t = \tan\dfrac{\theta}{2}$ to show that $\displaystyle\int_{\frac{\pi}{2}}^{\frac{2\pi}{3}} \dfrac{d\theta}{\sin\theta} = \dfrac{1}{2}\log_e 3$.

Solution

When $\theta = \dfrac{2\pi}{3}$, $t = \tan\dfrac{\pi}{3} = \sqrt{3}$, and when $\theta = \dfrac{\pi}{2}$, $t = \tan\dfrac{\pi}{4} = 1$

$$\sin\theta = \frac{2t}{1+t^2}, \ d\theta = \frac{2dt}{1+t^2}$$

$$\int_{\frac{\pi}{2}}^{\frac{2\pi}{3}} \frac{d\theta}{\sin\theta} = \int_1^{\sqrt{3}} \frac{1+t^2}{2t} \frac{2dt}{1+t^2}$$

$$= \int_1^{\sqrt{3}} \frac{1}{t} \, dt$$

$$= \left[\log_e |t|\right]_1^{\sqrt{3}}$$

$$= \log_e \sqrt{3} - \log_e 1$$

$$= \log_e 3^{\frac{1}{2}} - 0$$

$$= \frac{1}{2} \log_e 3 \text{ as required.}$$

When changing the variable x^2 in an integrand it usually involves $\sqrt{x^2}$, which is the same as $|x|$. In this case, we take the positive case for convenience. There are other situations where a similar problem arises, especially in the case of trigonometric functions; again we take the convenient solutions that lie in the domain of the standard inverse trigonometric functions.

When the integrand has a sum or difference of 2 squares:

$$\int \frac{dx}{\sqrt{a^2 + x^2}} \quad \text{use the substitution } x = a \tan\theta$$

$$\int \sqrt{a^2 - x^2} \, dx \quad \text{use the substitution } x = a \sin\theta \text{ or } x = a \cos\theta$$

$$\int \frac{dx}{\sqrt{x^2 - a^2}} \quad \text{use the substitution } x = a \sec\theta$$

These results help to determine what should be the best substitution.

Some standard integrals

$$\int f'(x)[f(x)]^n \, dx = \frac{f(x)^{n+1}}{n+1} + C$$

$$\int f'(x)e^{f(x)} \, dx = e^{f(x)} + C$$

$$\int \frac{f'(x)}{f(x)} \, dx = \ln |f(x)| + C$$

$$\int f'(x) \sin f(x) \, dx = -\cos f(x) + C$$

$$\int f'(x) \cos f(x) \, dx = \sin f(x) + C$$

$$\int f'(x) \sec^2 f(x) \, dx = \tan f(x) + C$$

$$\int \frac{f'(x)}{\sqrt{a^2 - [f(x)]^2}} \, dx = \sin^{-1}\left[\frac{f(x)}{a}\right] + C$$

$$\int \frac{-f'(x)}{\sqrt{a^2 - [f(x)]^2}} \, dx = \cos^{-1}\left[\frac{f(x)}{a}\right] + C$$

$$\int \frac{f'(x)}{a^2 + [f(x)]^2} \, dx = \frac{1}{a} \tan^{-1}\left[\frac{f(x)}{a}\right] + C$$

Exercise 6.01 Integration by substitution

1 Find each integral using the given substitution.

a $\int \dfrac{e^x}{e^x + 1}\, dx,\ u = e^x + 1$

b $\int \dfrac{e^{\frac{x}{2}}}{e^x + 1}\, dx,\ u = e^{\frac{x}{2}}$

c $\int x(1 + x^2)^4\, dx,\ u = 1 + x^2$

d $\int \dfrac{x}{\sqrt{1-x}}\, dx,\ u = 1 - x$

e $\int \dfrac{e^x\, dx}{\sqrt{e^x - 1}},\ x > 0,\ u = e^x - 1$

f $\int \dfrac{dx}{a^2 + x^2},\ x = a \tan \theta$

g $\int x\sqrt{x - 3}\, dx,\ u = x - 3$

h $\int \dfrac{x}{\sqrt{x + 1}}\, dx,\ x = u^2 - 1$

i $\int \dfrac{x}{\sqrt{x - 1}}\, dx,\ x = u^2 + 1$

2 Find each integral using an appropriate substitution.

a $\int \dfrac{e^{\sqrt{x}}}{\sqrt{x}}\, dx$

b $\int \dfrac{x}{e^{x^2}}\, dx$

c $\int \dfrac{x}{\sqrt{x^2 + 1}}\, dx$

d $\int \dfrac{(1 + \ln x)^2}{x}\, dx$

e $\int \dfrac{(\sin^{-1} x + 1)^2}{\sqrt{1 - x^2}}\, dx$

f $\int \dfrac{\tan^{-1}(x + 1)}{(x + 1)^2 + 1}\, dx$

g $\int \dfrac{dx}{x(\ln x)^2}$

h $\int \dfrac{dx}{x \ln x}$

3 Find each integral using an appropriate substitution.

a $\int_0^{\frac{\pi}{2}} \dfrac{\cos x}{1 + \sin x}\, dx$

b $\int_0^{\frac{\pi}{2}} \dfrac{\cos x}{(1 + \sin x)^2}\, dx$

c $\int_0^{\frac{\pi}{2}} \dfrac{\cos x}{(1 + \sin x)^3}\, dx$

d $\int_0^{\frac{\pi}{2}} \dfrac{\cos^3 x}{1 + \sin x}\, dx$

e $\int \sin^2 x \cos^3 x\, dx$

f $\int \sin^3 x \cos^2 x\, dx$

g $\int \sin^4 x \cos^5 x\, dx$

h $\int \sin^5 x \cos^4 x\, dx$

i $\int \dfrac{\sin \theta}{\cos^2 \theta}\, d\theta$

j $\int \dfrac{\sin \theta}{\cos^3 \theta}\, d\theta$

k $\int \dfrac{\sin \theta}{\cos^4 \theta}\, d\theta$

l $\int \dfrac{\cos \theta}{\sin^2 \theta}\, d\theta$

m $\int \dfrac{\cos \theta}{\sin^3 \theta}\, d\theta$

n $\int \dfrac{\cos \theta}{\sin^4 \theta}\, d\theta$

o $\int \dfrac{\sin^3 \theta}{\cos^4 \theta}\, d\theta$

p $\int \dfrac{\sin^3 \theta}{\cos^2 \theta}\, d\theta$

4 Evaluate each definite integral.

a $\int_0^{\frac{\pi}{4}} \tan^4 x \sec^2 x\, dx$

b $\int_0^{\frac{\pi}{4}} \dfrac{e^{\tan \theta}}{\cos^2 \theta}\, d\theta$

c $\int_2^{\sqrt{5}} \dfrac{x}{\sqrt{x^2 - 4}}\, dx$

d $\int_{\ln \frac{\pi}{6}}^{\ln \frac{\pi}{4}} e^x \sin e^x\, dx$

e $\int_0^{\frac{\pi}{4}} \cos^2 x\, dx$

f $\int_0^{\frac{\pi}{4}} \cos^2 4x\, dx$

g $\int_0^{\frac{\pi}{4}} \sin^2 2x\, dx$

5 Evaluate each integral using an appropriate substitution.

a $\displaystyle\int_3^8 \frac{x}{(x+1)\sqrt{x+1}}\, dx$ **b** $\displaystyle\int_0^3 \sqrt{9-x^2}\, dx$ **c** $\displaystyle\int_0^3 \frac{6}{9+x^2}\, dx$

d $\displaystyle\int_0^1 \frac{e^x}{(1+e^x)^2}\, dx$ **e** $\displaystyle\int_0^1 \frac{\sqrt{x}}{1+x}\, dx$

6 Verify the answer to Question **5b** by considering the area the integral represents.

7 Using the substitution $t = \tan\dfrac{x}{2}$:

a evaluate $\displaystyle\int_0^{\frac{\pi}{2}} \frac{1}{1+\sin x}\, dx$ **b** evaluate $\displaystyle\int_{\frac{\pi}{2}}^{\frac{2\pi}{3}} \frac{dx}{\sin x}$

c find $\displaystyle\int \frac{1}{1-\cos x}\, dx$ **d** show that $\displaystyle\int_0^{\frac{\pi}{3}} \frac{1}{1+\sin\theta}\, d\theta = -1+\sqrt{3}$

e show that $\displaystyle\int_0^{\frac{\pi}{2}} \frac{d\theta}{\sin\theta+2} = \frac{\pi}{3\sqrt{3}}$ **f** evaluate $\displaystyle\int_0^{\frac{\pi}{3}} \frac{d\theta}{1+\cos 2\theta}$

8 Find each integral using the given substitution.

a $\displaystyle\int \frac{dx}{x^2+4},\ x=2\tan\theta$ **b** $\displaystyle\int \frac{dx}{\sqrt{9-x^2}},\ x=3\sin\theta$ **c** $\displaystyle\int \frac{dx}{4x^2+9},\ x=\frac{3}{2}\tan\theta$

9 Evaluate each integral using an appropriate substitution.

a $\displaystyle\int_0^1 \frac{dx}{\sqrt{4-x^2}}$ **b** $\displaystyle\int_0^2 \frac{2\,dx}{\sqrt{16-x^2}}$ **c** $\displaystyle\int_0^{\frac{1}{3}} \frac{3\,dx}{1+9x^2}$ **d** $\displaystyle\int_3^4 \frac{2x\,dx}{\sqrt{25-x^2}}$

10 a Using the substitution $u^2 = 9 - x^2$, evaluate $\displaystyle\int_0^3 x^3\sqrt{9-x^2}\, dx$.

b Using the substitution $t = \tan\dfrac{x}{2}$, find $\displaystyle\int \frac{dx}{1+\cos x+\sin x}$.

c Using the substitution $u = \sec x$, or otherwise, find $\displaystyle\int \sec^3 x \tan x\, dx$.

d Use the substitution $x = 2\sin\theta$ to evaluate $\displaystyle\int_0^{\sqrt{2}} \frac{dx}{\sqrt{(4-x^2)^3}}$.

e Find the integral $\displaystyle\int \frac{e^x+e^{2x}}{e^{2x}+1}\, dx$.

f Show that $\displaystyle\int_0^2 \sqrt{x(4-x)}\, dx = \pi$.

g Show that $\displaystyle\int_0^{\frac{2}{3}} \sqrt{4-9x^2}\, dx = \frac{\pi}{3}$.

Length of an arc

Integration can be used to find the length of an arc.
By considering a small portion of a curve and using
Pythagoras' theorem, we get:

$$\delta L^2 \approx \delta x^2 + \delta y^2 = \delta x^2\left[1+\left(\frac{\delta y}{\delta x}\right)^2\right]$$

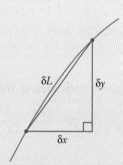

As $\delta x \to 0$, the length of the curve $y = f(x)$ between the
points where $x = a$ and $x = b$ is:

$$L = \int_a^b \sqrt{1+\left(\frac{dy}{dx}\right)^2}\, dx$$

For example, the length of the arc on the circle, in the first quadrant, given by

$x = a\cos t$ and $y = a\sin t$ can be found using this expression:

$$L = \int_0^a \sqrt{1+\left(\frac{dy}{dx}\right)^2}\, dx$$

$$\frac{dy}{dx} = -\frac{x}{y}$$

And the length of the arc $L = \int_0^a \dfrac{a}{\sqrt{a^2 - x^2}}\, dx = \dfrac{1}{2}\pi a$, as expected.

6.02 Rational functions with quadratic denominators

Integration of rational functions with quadratic denominators can be classified into those
with denominators that can be factorised and those with denominators that do not reduce.
Of the 3 integrals below, the first one has a denominator that can be factorised but the others
all have irreducible quadratics so they cannot be factorised.

$$\int \frac{1}{x^2 + 2x + 1}\, dx,\ \int \frac{1}{x^2 + 1}\, dx,\ \int \frac{1}{x^2 + 2x + 5}\, dx$$

Some integrals involve quadratics under a radical (square root) in the denominator.

$$\int \frac{1}{\sqrt{4 - x^2}}\, dx,\ \int \frac{1}{\sqrt{3 - 2x - x^2}}\, dx$$

Other integrands with quadratic denominators also have a linear or quadratic function in the numerator.

$$\int \frac{2x+1}{x^2+x+1}\, dx, \int \frac{2x+1}{x^2+2x+3}\, dx, \int \frac{x^2}{x^2+1}\, dx$$

To integrate these functions, we need to know how to complete the square and recognise standard forms for integration.

Some standard integrals

$$\int (ax+b)^n\, dx = \frac{(ax+b)^{n+1}}{a(n+1)} + C \qquad \int \frac{f'(x)}{f(x)}\, dx = \ln|f(x)| + C$$

$$\int \frac{1}{a^2+x^2}\, dx = \frac{1}{a}\tan^{-1}\left(\frac{x}{a}\right) + C \qquad \int \frac{1}{\sqrt{a^2-x^2}}\, dx = \sin^{-1}\left(\frac{x}{a}\right) + C$$

EXAMPLE 8

Find:

a $\displaystyle\int \frac{1}{x^2+2x+1}\, dx$ **b** $\displaystyle\int \frac{1}{x^2+1}\, dx$ **c** $\displaystyle\int \frac{1}{x^2+2x+5}\, dx$

Solution

a
$$\int \frac{1}{x^2+2x+1}\, dx = \int \frac{1}{(x+1)^2}\, dx$$
$$= \int (x+1)^{-2}\, dx$$
$$= \frac{(x+1)^{-1}}{-1} + C$$
$$= -\frac{1}{x+1} + C$$

b
$$\int \frac{1}{x^2+1}\, dx = \tan^{-1} x + C$$

c Complete the square.
$$\int \frac{1}{x^2+2x+5}\, dx = \int \frac{1}{x^2+2x+1+4}\, dx$$
$$= \int \frac{1}{(x+1)^2+4}\, dx$$
$$= \frac{1}{2}\tan^{-1}\left(\frac{x+1}{2}\right) + C$$

EXAMPLE 9

Find:

a $\int \dfrac{1}{\sqrt{4-x^2}}\,dx$

b $\int \dfrac{1}{\sqrt{3-2x-x^2}}\,dx$

Solution

a $\int \dfrac{1}{\sqrt{4-x^2}}\,dx = \sin^{-1}\dfrac{x}{2} + C$

b Complete the square.

$$\int \dfrac{1}{\sqrt{3-2x-x^2}}\,dx = \int \dfrac{1}{\sqrt{-x^2-2x-1+4}}\,dx$$

$$= \int \dfrac{1}{\sqrt{4-(x+1)^2}}\,dx$$

$$= \sin^{-1}\left(\dfrac{x+1}{2}\right) + C$$

EXAMPLE 10

Find:

a $\int \dfrac{2x+1}{x^2+x+1}\,dx$

b $\int \dfrac{2x+1}{x^2+2x+3}\,dx$

Solution

a Recognise that the numerator is the derivative of the denominator.

$$\int \dfrac{2x+1}{x^2+x+1}\,dx = \ln\left|x^2+x+1\right| + C$$

b Recognise that the numerator is almost the derivative of the denominator.

$$\int \dfrac{2x+1}{x^2+2x+3}\,dx = \int \dfrac{2x+2}{x^2+2x+3} - \dfrac{1}{x^2+2x+3}\,dx$$

$$= \ln\left|x^2+2x+3\right| - \int \dfrac{1}{(x+1)^2+2}\,dx$$

$$= \ln\left|x^2+2x+3\right| - \dfrac{1}{\sqrt{2}}\tan^{-1}\left(\dfrac{x+1}{\sqrt{2}}\right) + C$$

This is a combination of recognising the log result, splitting the numerator, completing the square and recognising the $y = \tan^{-1} x$ result.

Exercise 6.02 Rational functions with quadratic denominators

1 Find the integral of each function.

a $\dfrac{1}{(x+3)^2}$

b $\dfrac{1}{(x-4)^2}$

c $\dfrac{-2}{x^2+4x+4}$

d $\dfrac{1}{x^2+1}$

e $\dfrac{1}{x^2+9}$

f $\dfrac{1}{x^2+3}$

g $\dfrac{1}{x^2+5}$

h $\dfrac{1}{x^2-4x+4}$

2 Find:

a $\displaystyle\int \dfrac{1}{\sqrt{1-x^2}}\,dx$

b $\displaystyle\int \dfrac{1}{\sqrt{9-x^2}}\,dx$

c $\displaystyle\int \dfrac{1}{\sqrt{4-x^2}}\,dx$

d $\displaystyle\int \dfrac{1}{\sqrt{9-4x^2}}\,dx$

e $\displaystyle\int \dfrac{1}{\sqrt{4x-x^2}}\,dx$

f $\displaystyle\int \dfrac{1}{\sqrt{-9x^2+12x}}\,dx$

3 a Find $\displaystyle\int \dfrac{1}{x^2+2x+2}\,dx$ by first showing that $x^2+2x+2=(x+1)^2+1$.

b Find $\displaystyle\int \dfrac{1}{\sqrt{2x-x^2}}\,dx$ by first showing that $2x-x^2=1-(x-1)^2$.

c Show that $\dfrac{x^2+2}{x^2+1}=1+\dfrac{1}{x^2+1}$. Hence, find $\displaystyle\int \dfrac{x^2+2}{x^2+1}\,dx$.

d Show that $x^2-2x+5=4+(x-1)^2$. Hence, find $\displaystyle\int \dfrac{1}{x^2-2x+5}\,dx$.

4 Find:

a $\displaystyle\int \dfrac{x^2}{x^2+1}\,dx$

b $\displaystyle\int \dfrac{x^2-1}{x^2+1}\,dx$

c $\displaystyle\int \dfrac{x^2+4}{x^2+2}\,dx$

d $\displaystyle\int \dfrac{x^2}{x^2+9}\,dx$

e $\displaystyle\int \dfrac{(x+1)^2}{x^2+1}\,dx$

f $\displaystyle\int \dfrac{(x-1)^2}{x^2+2}\,dx$

5 Find:

a $\displaystyle\int \dfrac{2x+1}{x^2+4x+5}\,dx$

b $\displaystyle\int \dfrac{4x+3}{x^2+1}\,dx$

c $\displaystyle\int \dfrac{x}{x^2+1}\,dx$

d $\displaystyle\int \dfrac{x-1}{x^2+1}\,dx$

e $\displaystyle\int \dfrac{x}{x^2-2x+2}\,dx$

f $\displaystyle\int \dfrac{x+1}{x^2-1}\,dx$

ISBN 9780170413435

6 Find each indefinite integral.

a $\displaystyle\int \frac{-dx}{\sqrt{1-x^2}}$

b $\displaystyle\int \frac{dx}{\sqrt{1-4x^2}}$

c $\displaystyle\int \frac{dx}{1+4x^2}$

d $\displaystyle\int \frac{dx}{x^2+4x+8}$

e $\displaystyle\int \frac{dx}{4x^2+4x+10}$

f $\displaystyle\int \frac{2}{x^2-6x+13}\,dx$

7 a The graphs of $y = \dfrac{x^2-1}{x^2+1}$ and $y = 1$ are shown in the diagram. Evaluate the area shaded. Leave your answer in terms of π.

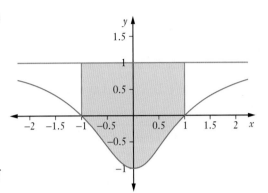

b The area shaded above is rotated about the y-axis. Find the volume of the figure generated. Leave your answer as an exact value.

c Calculate the volume of the solid of revolution formed when

$$f(x) = \frac{1}{\sqrt{x^2+4x+8}} \text{ is rotated about the } x\text{-axis between}$$

$x = -1$ and $x = 2$. Leave your answer in terms of π.

6.03 Partial fractions

WS

Partial fractions

A function $\dfrac{f(x)}{g(x)}$, where $f(x)$ and $g(x)$ are polynomials, is called a **rational function**.

If the degree of $f(x)$ is greater than the degree of $g(x)$, the function is called an **improper function**. If the degree $f(x)$ is less than the degree of $g(x)$, the function is called a **proper function**.

Integral calculus

An improper function can be expressed as the sum of a polynomial and a proper rational function. For example, $\dfrac{x^3}{x^2+1} = x - \dfrac{x}{x^2+1}$

A proper function can be expressed as a sum of simpler fractions called **partial fractions**. This process is also known as **decomposing** the function into partial fractions.

EXAMPLE 11

If $\dfrac{6}{(x-1)(x+1)(x+2)} = \dfrac{A}{x-1} + \dfrac{B}{x+1} + \dfrac{C}{x+2}$, find the values of A, B and C and hence write

$\dfrac{6}{(x-1)(x+1)(x+2)}$ as the sum of partial fractions.

Solution

Multiply both sides by $(x-1)(x+1)(x+2)$:

$6 = A(x+1)(x+2) + B(x-1)(x+2) + C(x-1)(x+1)$

Expand and equate coefficients of x:

$6 = A(x^2 + 3x + 2) + B(x^2 + x - 2) + C(x^2 - 1)$

$\quad = Ax^2 + 3Ax + 2A + Bx^2 + Bx - 2B + Cx^2 - C$

$\quad = (A + B + C)x^2 + (3A + B)x + (2A - 2B - C)$

Hence:

$$A + B + C = 0 \qquad\qquad [1]$$

$$3A + B = 0 \qquad\qquad [2]$$

$$2A - 2B - C = 6 \qquad\qquad [3]$$

From [2]: $\qquad B = -3A \qquad\qquad [4]$

[1] + [3]: $3A - B = 6 \qquad\qquad [5]$

Substitute [4] into [5]:

$$3A - (-3A) = 6$$

$$6A = 6$$

$$A = 1$$

Substitute into [4]:

$$B = -3(1)$$

$$\quad = -3$$

Substitute into [1]:

$$1 - 3 + C = 0$$

$$-2 + C = 0$$

$$C = 2$$

$\therefore A = 1$, $B = -3$ and $C = 2$

$\therefore \dfrac{6}{(x-1)(x+1)(x+2)} = \dfrac{1}{x-1} - \dfrac{3}{x+1} + \dfrac{2}{x+2}$

MATHS IN FOCUS 12. Mathematics Extension 2 ISBN 9780170413435

EXAMPLE 12

If $\dfrac{4}{(x-1)(x+1)^2} = \dfrac{A}{x-1} + \dfrac{B}{x+1} + \dfrac{C}{(x+1)^2}$, find the values of A, B and C and hence write as the sum of partial fractions.

Solution

Multiply both sides by $(x-1)(x+1)^2$:

$$4 = A(x+1)^2 + B(x-1)(x+1) + C(x-1) \qquad [*]$$

Expand and equate coefficients of x:

$$
\begin{aligned}
4 &= A(x+1)^2 + B(x-1)(x+1) + C(x-1) \\
&= A(x^2 + 2x + 1) + B(x^2 - 1) + Cx - C \\
&= Ax^2 + 2Ax + A + Bx^2 - B + Cx - C \\
&= (A+B)x^2 + (2A+C)x + (A - B - C)
\end{aligned}
$$

Hence:
$$A + B = 0 \qquad [1]$$
$$2A + C = 0 \qquad [2]$$
$$A - B - C = 4 \qquad [3]$$

From [1]: $\qquad B = -A \qquad [4]$

From [2]: $\qquad C = -2A \qquad [5]$

Substitute into [3]:
$$A - (-A) - (-2A) = 4$$
$$4A = 4$$
$$A = 1$$

Substitute into [4] and [5]:
$$B = -1$$
$$C = -2(1)$$
$$= -2$$

$\therefore A = 1, B = -1$ and $C = -2$

$$\therefore \frac{4}{(x-1)(x+1)^2} = \frac{1}{x-1} - \frac{1}{x+1} - \frac{2}{(x+1)^2}$$

Alternatively, using [*] above, substituting $x = 1$, we obtain $A = 1$ rather quickly, and by substituting $x = -1$, we can get $C = -2$, again fairly easily. Last, we can choose x to be any other number, say $x = 0$, to find that $A - B - C = 4$ and therefore $B = -1$.

Repeated linear factors in denominator

To each linear factor $ax + b$ occurring n times in the denominator of a proper function, there corresponds a sum of n partial fractions of the form $\dfrac{A_1}{ax+b} + \dfrac{A_2}{(ax+b)^2} + \ldots + \dfrac{A_n}{(ax+b)^n}$.

For example: $\dfrac{4}{(x-1)(x+1)^2} = \dfrac{1}{x-1} - \dfrac{1}{x+1} - \dfrac{2}{(x+1)^2}$

MATHS IN FOCUS 12. Mathematics Extension 2 ISBN 9780170413435

EXAMPLE 13

Find A, B and C such that $\dfrac{5x+2}{(x-1)(x^2+2x+4)} = \dfrac{A}{x-1} + \dfrac{Bx+C}{x^2+2x+4}$.

Solution

Note that when a partial fraction has a **quadratic denominator**, then it has a **linear numerator** $Bx + C$.

Multiply both sides by $(x-1)(x^2+2x+4)$:

$5x + 2 = A(x^2 + 2x + 4) + (Bx + C)(x - 1)$

$\qquad = Ax^2 + 2Ax + 4A + Bx^2 - Bx + Cx - C$

$\qquad = (A + B)x^2 + (2A - B + C)x + (4A - C)$

Hence:

$$A + B = 0 \qquad\qquad [1]$$
$$2A - B + C = 5 \qquad\qquad [2]$$
$$4A - C = 2 \qquad\qquad [3]$$

From [1]: $\qquad B = -A \qquad\qquad [4]$

From [3]: $\qquad C = 4A - 2 \qquad\qquad [5]$

Substitute into [2]:

$$2A - (-A) + 4A - 2 = 5$$
$$7A - 2 = 5$$
$$7A = 7$$
$$A = 1$$

Substitute into [4] and [5]:

$$B = -1$$
$$C = 4(1) - 2$$
$$\quad = 2$$

$\therefore A = 1$, $B = -1$ and $C = 2$

Quadratic factors in denominator

To each irreducible quadratic factor $ax^2 + bx + c$ occurring once in the denominator of a proper function, there corresponds a single partial fraction of the form $\dfrac{Ax + B}{ax^2 + bx + c}$.

For example: $\dfrac{5x+2}{(x-1)(x^2+2x+4)} = \dfrac{1}{x-1} + \dfrac{-x+2}{x^2+2x+4}$

EXAMPLE 14

Find A, B, C, D and E such that $\dfrac{1}{x(x^2+1)^2} = \dfrac{A}{x} + \dfrac{Bx+C}{x^2+1} + \dfrac{Dx+E}{(x^2+1)^2}$.

Solution

Multiply both sides by $x(x^2+1)^2$:

$$1 = A(x^2+1)^2 + (Bx+C)(x^2+1)x + (Dx+E)x$$
$$= A(x^4+2x^2+1) + (Bx+C)(x^3+x) + (Dx^2+Ex)$$
$$= Ax^4 + 2Ax^2 + A + Bx^4 + Bx^2 + Cx^3 + Cx + Dx^2 + Ex$$
$$= (A+B)x^4 + Cx^3 + (2A+B+D)x^2 + (C+E)x + A$$

Hence:

$$A + B = 0 \qquad [1]$$
$$C = 0 \qquad [2]$$
$$2A + B + D = 0 \qquad [3]$$
$$C + E = 0 \qquad [4]$$
$$A = 1 \qquad [5]$$

Substitute $A = 1$ into [1]:

$$1 + B = 0$$
$$B = -1$$

Substitute $C = 0$ into [4]:

$$0 + E = 0$$
$$E = 0$$

Substitute $A = 1, B = -1$ into [3]:

$$2(1) - 1 + D = 0$$
$$1 + D = 0$$
$$D = -1$$

$\therefore A = 1, B = -1, C = 0, D = -1, E = 0$.

Repeated quadratic factors in denominator

To each irreducible quadratic factor $ax^2 + bx + c$ occurring n times in the denominator of a proper function, there is a sum of n partial fractions of the form

$$\dfrac{A_1 x + B_1}{ax^2+bx+c} + \dfrac{A_2 x + B_2}{(ax^2+bx+c)^2} + \cdots + \dfrac{A_n x + B_n}{(ax^2+bx+c)^n}.$$

For example: $\dfrac{1}{x(x^2+1)^2} = \dfrac{A}{x} + \dfrac{Bx+C}{x^2+1} + \dfrac{Dx+E}{(x^2+1)^2} = \dfrac{1}{x(x^2+1)^2} = \dfrac{1}{x} - \dfrac{x}{x^2+1} - \dfrac{x}{(x^2+1)^2}$

Once the rational functions are distributed as partial fractions they become far easier to integrate using common integration strategies.

EXAMPLE 15

Find $\int \dfrac{9x-2}{(2x-1)(x-3)}\,dx$.

Solution

Using partial fractions:

$$\dfrac{9x-2}{(2x-1)(x-3)} = \dfrac{A}{2x-1} + \dfrac{B}{x-3}$$

$$9x - 2 = A(x-3) + B(2x-1)$$

Substituting $x = 3$, we get $B = 5$.

And so by putting $x = 0$, or any other number, $A = -1$.

Hence:

$$\int \dfrac{9x-2}{(2x-1)(x-3)}\,dx = \int \dfrac{-1}{2x-1} + \dfrac{5}{x-3}\,dx$$

Integrating,

$$\int \dfrac{9x-2}{(2x-1)(x-3)}\,dx = -\dfrac{1}{2}\ln|2x-1| + 5\ln|x-3| + C$$

EXAMPLE 16

Find $\int \dfrac{3x^2 - 2x + 1}{(x^2+1)(x^2+2)}\,dx$.

Solution

Using partial fractions:

$$\dfrac{3x^2 - 2x + 1}{(x^2+1)(x^2+2)} = \dfrac{Ax+B}{x^2+1} + \dfrac{Cx+D}{x^2+2}$$

$$\begin{aligned}
3x^2 - 2x + 1 &= (Ax+B)(x^2+2) + (Cx+D)(x^2+1) \\
&= Ax^3 + 2Ax + Bx^2 + 2B + Cx^3 + Cx + Dx^2 + D \\
&= (A+C)x^3 + (B+D)x^2 + (2A+C)x + 2B + D
\end{aligned}$$

Equating coefficients:

$A + C = 0$ [1]

$B + D = 3$ [2]

$2A + C = -2$ [3]

$2B + D = 1$ [4]

$[3] - [1]: A = -2$

Substitute into [1]:

$-2 + C = 0$

$C = 2$

$[4] - [2]: B = -2$

Substitute into [2]:

$-2 + D = 3$

$D = 5$

Hence:

$$\int \frac{3x^2 - 2x + 1}{\left(x^2 + 1\right)\left(x^2 + 2\right)} dx = \int \frac{-2x - 2}{x^2 + 1} + \frac{2x + 5}{x^2 + 2} dx$$

$$= \int \frac{-2x}{x^2 + 1} - \frac{2}{x^2 + 1} + \frac{2x}{x^2 + 2} + \frac{5}{x^2 + 2} dx$$

$$= -\ln\left(x^2 + 1\right) - 2\tan^{-1} x + \ln\left(x^2 + 2\right) + \frac{5}{\sqrt{2}} \tan^{-1}\left(\frac{x}{\sqrt{2}}\right) + C$$

EXAMPLE 17

The expression $\dfrac{2x^2 + 5x + 3}{(x - 1)^2 (x^2 + 4)}$ can be written as $\dfrac{A}{x - 1} + \dfrac{B}{(x - 1)^2} + \dfrac{Cx + D}{x^2 + 4}$ where A, B, C and D are real numbers.

a Find A, B, C and D.

b Hence find $\displaystyle\int \frac{2x^2 + 5x + 3}{(x - 1)^2 (x^2 + 4)} \, dx$.

Solution

a Using partial fractions:

$$\frac{2x^2 + 5x + 3}{(x-1)^2(x^2+4)} = \frac{A}{x-1} + \frac{B}{(x-1)^2} + \frac{Cx+D}{x^2+4}$$

On multiplying both sides by $(x-1)^2(x^2+4)$:

$$2x^2 + 5x + 3 = A(x-1)(x^2+4) + B(x^2+4) + (Cx+D)(x-1)^2$$

Substituting $x = 1$ we get $10 = 5B$, so $B = 2$.

Substituting $x = -1$, we get $0 = -10A + 10 - 4C + 4D$

$$-5 = -5A - 2C + 2D \qquad\qquad [1]$$

Substituting $x = 0$, we get $3 = -4A + 8 + D$

$$-5 = -4A + D \qquad\qquad [2]$$

Substituting $x = 2$, we get $21 = 8A + 16 + 2C + D$

$$5 = 8A + 2C + D \qquad\qquad [3]$$

Solving simultaneously, $A = 1$, $B = 2$, $C = -1$ and $D = -1$.

b Hence:

$$\int \frac{2x^2 + 5x + 3}{(x-1)^2(x^2+4)}\, dx = \int \frac{1}{x-1} + \frac{2}{(x-1)^2} - \frac{x+1}{x^2+4}\, dx$$

Integrating:

$$\int \frac{2x^2 + 5x + 3}{(x-1)^2(x^2+4)}\, dx = \ln|x-1| - \frac{2}{x-1} - \frac{1}{2}\ln(x^2+4) - \frac{1}{2}\tan^{-1}\left(\frac{x}{2}\right) + C$$

Exercise 6.03 Partial fractions

1 Find each integral using partial fractions.

a $\displaystyle\int \frac{3x+1}{(x-3)(x+2)}\, dx$

b $\displaystyle\int \frac{5x+8}{(x+3)(2x-1)}\, dx$

c $\displaystyle\int \frac{3x+1}{(x+3)(x+2)}\, dx$

d $\displaystyle\int \frac{3x+7}{(x-3)(x+5)}\, dx$

e $\displaystyle\int \frac{3+x}{(1+2x)(1-3x)}\, dx$

f $\displaystyle\int \frac{(a-b)x}{(x-a)(x-b)}\, dx,\ a > b$

2 Find each integral.

a $\displaystyle\int \frac{2x}{(x-2)(x+2)} \, dx$

b $\displaystyle\int \frac{11}{6x^2+5x-4} \, dx$

c $\displaystyle\int \frac{x-7}{2x^2-3x-2} \, dx$

d $\displaystyle\int \frac{3x^2-12x+11}{(x-1)(x-2)(x-3)} \, dx$

e $\displaystyle\int \frac{5x^2+9x+6}{(x^2-1)(2x+3)} \, dx$

f $\displaystyle\int \frac{x-1}{x^2-7x+6} \, dx$

3 By writing each integrand in the form $\dfrac{A}{x+a} + \dfrac{B}{(x+a)^2} + \dfrac{C}{x+b}$, determine each integral.

a $\displaystyle\int \frac{4+7x}{(x+1)^2(2+3x)} \, dx$

b $\displaystyle\int \frac{1}{(x+1)^2(x-1)} \, dx$

c $\displaystyle\int \frac{x^3-6x^2+25}{(x+1)^2(x-2)^2} \, dx$

4 By decomposing into partial fractions, determine each integral.

a $\displaystyle\int \frac{x-2}{(x^2+1)(x+1)} \, dx$

b $\displaystyle\int \frac{x}{(x^2+1)(x-1)} \, dx$

c $\displaystyle\int \frac{x+3}{x^3+3x^2+x+3} \, dx$

5 Using the expansions for the sum and difference of 2 cubes, $a^3+b^3 = (a+b)(a^2-ab+b^2)$ and $a^3-b^3 = (a-b)(a^2+ab+b^2)$, find:

a $\displaystyle\int \frac{12}{x^3+8} \, dx$

b $\displaystyle\int \frac{9x+6}{x^3-8} \, dx$

6 a Find the real numbers A, B and C such that $\dfrac{2}{(1+x)(1+x^2)} = \dfrac{A}{1+x} + \dfrac{Bx+C}{1+x^2}$.

b Hence, find $\displaystyle\int \frac{2}{(1+x)(1+x^2)} \, dx$.

7 a Find real values for A and B such that $\dfrac{1}{(x+3)(x+1)} = \dfrac{A}{x+3} + \dfrac{B}{x+1}$.

b Hence evaluate $\displaystyle\int_0^1 \frac{1}{(x+3)(x+1)} \, dx$.

8 Write $\dfrac{2x^2+5x+3}{(x-1)^2(x^2+1)}$ in the form $\dfrac{A}{x-1} + \dfrac{B}{(x-1)^2} + \dfrac{Cx+D}{x^2+1}$ and then find the integral $\displaystyle\int \frac{2x^2+5x+3}{(x-1)^2(x^2+1)} \, dx$.

9 Write $\dfrac{2x^2-x-7}{(x+2)^2(x^2+x+1)}$ as a partial fractions $\dfrac{A}{x+2} + \dfrac{B}{(x+2)^2} + \dfrac{Cx+D}{x^2+x+1}$ and then evaluate $\displaystyle\int_0^1 \frac{2x^2-x-7}{(x+2)^2(x^2+x+1)} \, dx$.

10 a Find the values of the constants A, B, C and D if $\dfrac{(x+1)^3}{x^2} = Ax + B + \dfrac{C}{x} + \dfrac{D}{x^2}$.

b Hence find $\displaystyle\int_1^3 \frac{(x+1)^3}{x^2} \, dx$.

11 a Show that $\dfrac{x}{x^2 + 2x + 3} = \dfrac{1}{2}\left[\dfrac{2x + 2}{x^2 + 2x + 3}\right] - \dfrac{1}{x^2 + 2x + 3}$.

b Hence find $\displaystyle\int \dfrac{x}{x^2 + 2x + 3}\ dx$.

12 a Show that $\dfrac{1}{(x^2 + a^2)(x^2 + b^2)} = \dfrac{1}{a^2 - b^2}\left(\dfrac{1}{x^2 + b^2} - \dfrac{1}{x^2 + a^2}\right)$.

b Hence evaluate $\displaystyle\int_0^\infty \dfrac{1}{(x^2 + a^2)(x^2 + b^2)}\ dx$.

THE SHAPE OF A GOBLET

Mathematics can be used to model many real-life situations. In this task you are modelling the shape of a goblet using mathematics and technology to ascertain the volume of liquid the goblet can hold.

6 cm

12 cm

5 cm

iStock.com/gedzun

Task 1

Determine an equation, of the form $y = a\ \text{cosec}\ \dfrac{1}{2}(x - \alpha) - b$ to model the shape of a goblet. Find suitable values for a, b and α, given that $(0, 0)$ and $(3, 7)$ are on the curve that models the goblet. Show that the function is symmetrical.

Task 2

Find the cross-sectional area of the goblet.

Support your answer with reasoning and calculations.

Task 3

Find the volume of liquid that can fill the goblet to the very top.

Support your answer with reasoning and calculations.

Discuss the accuracy, reasonableness and limitations of your results for this investigation.

6.04 Integration by parts

There are 2 main methods for simplifying integration of a function. The substitution method is based on the chain rule, while **integration by parts** is based on the product rule.

Integration by parts allows us to integrate functions such as $\ln x$, $\tan^{-1} x$ and xe^x; that is, find $\int \ln x \, dx$, $\int \tan^{-1} x \, dx$ and $\int xe^x \, dx$.

The product rule for differentiation is:

$\dfrac{d}{dx} uv = v\dfrac{du}{dx} + u\dfrac{dv}{dx}$ where u and v are functions of x.

Integrating both sides:

$uv = \int v\dfrac{du}{dx} \, dx + \int u\dfrac{dv}{dx} \, dx$

$\therefore \int u\dfrac{dv}{dx} \, dx = uv - \int v\dfrac{du}{dx} \, dx$

Integration by parts

If u and v are both functions of x, then:

$\int u\dfrac{dv}{dx} dx = uv - \int v\dfrac{du}{dx} dx$

or $\int uv' \, dx = uv - \int vu' \, dx$

This expression often allows us to convert a difficult integral into more manageable integral parts. The key to using integration by parts lies in choosing the u and $\dfrac{dv}{dx}$ to split the function into. One part, u, needs to be differentiated, while the other part, $\dfrac{dv}{dx}$, needs to be integrated.

EXAMPLE 18

Find $\int xe^x \, dx$.

Solution

Let $u = x$ and $v' = e^x$

So $u' = 1$ and $v = e^x$

Using integration by parts:

$\int uv' \, dx = uv - \int vu' \, dx$

$\int xe^x \, dx = xe^x - \int e^x \times 1 \, dx$

$\qquad\quad = xe^x - \int e^x \, dx$

$\qquad\quad = xe^x - e^x + C$

EXAMPLE 19

Find $\int x \sin x \, dx$.

Solution

Let $u = x$ and $v' = \sin x$

So $u' = 1$ and $v = -\cos x$

Using integration by parts:

$\int uv' \, dx = uv - \int vu' \, dx$

$\int x \sin x \, dx = -x \cos x - \int -\cos x \times 1 \, dx$

$\qquad\qquad = -x \cos x + \int \cos x \, dx$

$\qquad\qquad = -x \cos x + \sin x + C$

For integrands such as $\ln x$ and $\sin^{-1} x$ that cannot be integrated easily but which can be differentiated, let u be the integrand and $v' = 1$.

EXAMPLE 20

Find $\int \ln x \, dx$.

Solution

Let $u = \ln x$ and $v' = 1$

So $u' = \dfrac{1}{x}$ and $v = x$

Using integration by parts:

$\int uv' \, dx = uv - \int vu' \, dx$

$\int 1 \times \ln x \, dx = (\ln x) \times x - \int \dfrac{1}{x} \times x \, dx$

$\int \ln x \, dx = x \ln x - \int 1 \, dx$

$\qquad\qquad = x \ln x - x + C$

The LIATE rule

For integrating by parts, LIATE is a handy guide to choosing which function should be u to differentiate.

L   logarithmic function: $\ln x, \log_a x$

I   inverse trigonometric functions: $\tan^{-1} x, \sin^{-1} x$

A   algebraic functions: $x^2, 2x^{10}$

T   trigonometric functions: $\sin x, \tan x$

E   exponential functions: $e^x, 5^x$

EXAMPLE 21

Evaluate $\int_1^2 x \ln x \, dx$.

Solution

According to the LIATE rule, L for logarithms comes first, so let $u = \ln x$.

Let $u = \ln x$ and $v' = x$

So $u' = \dfrac{1}{x}$ and $v = \dfrac{1}{2}x^2$

Using integration by parts:

$\int uv' \, dx = uv - \int vu' \, dx$

$$\int_1^2 (\ln x) x \, dx = \left[(\ln x) \frac{1}{2} x^2 \right]_1^2 - \int_1^2 \frac{1}{x} \times \frac{1}{2} x^2 \, dx$$

$$\int_1^2 x \ln x \, dx = \left[\frac{1}{2} x^2 \ln x \right]_1^2 - \frac{1}{2} \int_1^2 x \, dx$$

$$= \left(\frac{1}{2} \times 2^2 \ln 2 - \frac{1}{2} \times 1^2 \ln 1 \right) - \frac{1}{2} \left[\frac{1}{2} x^2 \right]_1^2$$

$$= (2 \ln 2 - 0) - \frac{1}{2} \left(\frac{1}{2} \times 2^2 - \frac{1}{2} \times 1^2 \right)$$

$$= 2 \ln 2 - \frac{1}{2} \left(\frac{3}{2} \right)$$

$$= 2 \ln 2 - \frac{3}{4}$$

DIFFERENTIATE, HENCE INTEGRATE

Because integration by parts is the reversal of the product rule for differentiation, many problems requiring integration by parts can also be approached using differentiation.

For example, $\dfrac{d}{dx} x \sin x = \sin x + x \cos x$ and so we can deduce that

$\int x \cos x \, dx = x \sin x + \cos x$.

Differentiate the given functions to find following integrals.

1 Use the derivative of $x \ln x$ to find $\int \ln x \, dx$.

2 Use the derivative of xe^x to find $\int xe^x \, dx$.

3 Use the derivative of $x \cos x$ to find $\int x \sin x \, dx$.

4 Use the derivative of $x^n e^x$ to find $\int x^n e^x \, dx$.

EXAMPLE 22

Find $\int x^n \ln x \, dx$.

Solution

Let $u = \ln x$ $\qquad$ and $\qquad$ $v' = x^n$

So $u' = \dfrac{1}{x}$ $\qquad$ and $\qquad$ $v = \dfrac{1}{n+1} x^{n+1}$

Using integration by parts:

$\int uv' \, dx = uv - \int vu' \, dx$

$\int (\ln x) x^n \, dx = (\ln x) \dfrac{1}{n+1} x^{n+1} - \int \dfrac{1}{n+1} x^{n+1} \dfrac{1}{x} \, dx$

$\int x^n \ln x \, dx = \dfrac{x^{n+1}}{n+1} (\ln x) - \dfrac{1}{n+1} \int x^n \, dx$

$\qquad = \dfrac{x^{n+1}}{n+1} (\ln x) - \dfrac{1}{n+1} \left(\dfrac{1}{n+1} x^{n+1} \right) + C$

$\qquad = \dfrac{x^{n+1}}{n+1} \left(\ln x - \dfrac{1}{n+1} \right) + C$

The table method

Integration by parts can be done using the **table method**. It works best when applied to certain functions in the form $f(x) = g(x)h(x)$ where one of $g(x)$ or $h(x)$ can be differentiated multiple times easily, while the other function can be integrated multiple times easily.

There are 2 types of tabular integrations. The first type is when one of the factors of $f(x)$ goes to 0 when differentiated multiple times. The second type is when *neither* of the factors of $f(x)$ goes to 0 when differentiated multiple times.

Type 1

Find $\int x^3 \cos x \, dx$. $\qquad\qquad$ Let $u = x^3$ and $v' = \cos x$.

Sign	Derivatives of u		Integrals of v
+	x^3		$\cos x$
−	$3x^2$		$\sin x$
+	$6x^1$		$-\cos x$
−	6		$-\sin x$
+	0		$\cos x$

The integral is then found by multiplying the derivative in line 1 by the integral in line 2, and so on (see arrows in table).

$\int x^3 \cos x \, dx = x^3 \sin x - 3x^2(-\cos x) + 6x(-\sin x) - 6(\cos x) + C$

$\qquad\qquad\quad = x^3 \sin x + 3x^2 \cos x - 6x \sin x - 6 \cos x + C$

Type 2

Find $\int e^x \sin x \, dx$. $\qquad\qquad$ Let $u = e^x$ and $v' = \sin x$.

Sign	Derivatives of u		Integrals of v
+	e^x		$\sin x$
−	e^x		$-\cos x$
+	e^x		$-\sin x$
−	e^x		$\cos x$

You will notice that neither of the functions goes to 0. In the second type we can stop at any multiplication as in Type 1 and finish with the integral of the product of the last 2 functions.

$\int e^x \sin x \, dx = e^x(-\cos x) - e^x(-\sin x) + \int e^x(-\sin x) \, dx + C$

It is preferable to stop when the product being integrated is the same as the original question, so we can bring it to the LHS.

$2 \int e^x \sin x \, dx = -e^x \cos x + e^x \sin x + C$

Hence:

$\int e^x \sin x \, dx = \dfrac{1}{2} e^x(\sin x - \cos x) + C$

Exercise 6.04 Integration by parts

1 Find each integral using integration by parts.

a $\int \ln (x+1) \, dx$

b $\int \ln x^2 \, dx$

c $\int x \cos x \, dx$

d $\int xe^{-x} \, dx$

e $\int x \sin 2x \, dx$

f $\int xe^{2x} \, dx$

g $\int x \ln x \, dx$

2 Find each integral using integration by parts. Some will need to be integrated twice.

a $\int e^x \sin x \, dx$

b $\int e^x \cos x \, dx$

c $\int x \, e^{x^2} \, dx$

d $\int x^2 \ln x \, dx$

e $\int x^2 \sin x \, dx$

f $\int x \tan^{-1} x \, dx$

g $\int x^2 e^{4x} \, dx$

3 Evaluate each each definite integral using integration by parts.

a $\int_1^2 \ln x \, dx$

b $\int_0^1 e^{\sqrt{x}} \, dx$, put $u = \sqrt{x}$

c $\int_1^e \ln x^2 \, dx$

d $\int_0^{0.5} \cos^{-1} 2x \, dx$

e $\int_0^{\pi} x \cos x \, dx$

f $\int_0^2 \ln (x^2 + 1) \, dx$

g $\int_{-\pi}^{\pi} x \sin x \, dx$

6.05 Recurrence relations

Sometimes when integrating a function with a high power, such as $(\cos x)^6$, we find the integral of the same function with a smaller power in the answer, such as $(\cos x)^5$.

A **recurrence relation** is a **recursive formula** that expresses an integral in terms of a similar integral with a smaller power.

Two examples of a recurrence relation for integrating a difficult function are:

$$\int x^n e^x \, dx = x^n e^x - n \int x^{n-1} e^x \, dx$$

$$\int \tan^n x \, dx = \frac{(\tan x)^{n-1}}{n-1} - \int (\tan x)^{n-2} \, dx$$

With repeated application of the formula, we can eventually reduce the power of the integral to 1 or 0, when it can be easily found.

EXAMPLE 23

If $I_n = \int \cos^n x \, dx$, prove that $I_n = \dfrac{1}{n}[\sin x \,(\cos x)^{n-1} + (n-1)I_{n-2}]$ for $n > 0$.

Solution

Let $u = (\cos x)^{n-1}$ and $v' = \cos x$

So $u' = (n-1)(\cos x)^{n-2}(-\sin x)$ and $v = \sin x$

$\quad = -(n-1)\sin x\,(\cos x)^{n-2}$

Using integration by parts:

$\int uv' \, dx = uv - \int vu' \, dx$

$I_n = \int \cos^n x \, dx$

$\quad = (\cos x)^{n-1}\sin x - \int -(n-1)\sin x\,(\cos x)^{n-2}\sin x \, dx$

$\quad = \sin x\,(\cos x)^{n-1} + (n-1)\int \sin^2 x\,(\cos x)^{n-2} \, dx$

$\quad = \sin x\,(\cos x)^{n-1} + (n-1)\int (1 - \cos^2 x)(\cos x)^{n-2} \, dx$

$\quad = \sin x\,(\cos x)^{n-1} + (n-1)\int (\cos x)^{n-2} - \cos^n x \, dx$

$\quad = \sin x\,(\cos x)^{n-1} + (n-1)\left[\int (\cos x)^{n-2} \, dx - \int \cos^n x \, dx\right]$

This last term is the same integral as in the question.

$\quad = \sin x\,(\cos x)^{n-1} + (n-1)\,[I_{n-2} - I_n]$

$\quad = \sin x\,(\cos x)^{n-1} + (n-1)I_{n-2} - (n-1)I_n$

Move $-(n-1)I_n$ to the LHS:

$I_n + (n-1)I_n = \sin x\,(\cos x)^{n-1} + (n-1)I_{n-2}$

$(1 + n - 1)I_n = \sin x\,(\cos x)^{n-1} + (n-1)I_{n-2}$

$nI_n = \sin x\,(\cos x)^{n-1} + (n-1)I_{n-2}$

$I_n = \dfrac{1}{n}[\sin x\,(\cos x)^{n-1} + (n-1)I_{n-2}]$ as required.

EXAMPLE 24

$I_n = \int x^n e^x\, dx$ for $n \geq 0$.

a Show that $I_n = x^n e^x - nI_{n-1}$

b Hence evaluate I_3.

Solution

a Let $u = x^n$ and $v' = e^x$

So $u' = nx^{n-1}$ and $v = e^x$

Using integration by parts:

$\int uv'\, dx = uv - \int vu'\, dx$

$$I_n = \int x^n e^x\, dx$$
$$= x^n e^x - \int e^x nx^{n-1}\, dx$$
$$= x^n e^x - n \int x^{n-1} e^x\, dx$$
$$= x^n e^x - nI_{n-1} \text{ as required.}$$

b From **a**, $I_3 = x^3 e^x - 3I_2$,

but $I_2 = x^2 e^x - 2I_1$

and $I_1 = x^1 e^x - 1I_0$

$$I_0 = \int x^0 e^x\, dx = e^x + C$$

Hence:

$$I_3 = x^3 e^x - 3I_2$$
$$= x^3 e^x - 3(x^2 e^x - 2I_1)$$
$$= x^3 e^x - 3[x^2 e^x - 2(x^1 e^x - 1I_0)]$$
$$= x^3 e^x - 3[x^2 e^x - 2(xe^x - e^x)] + C$$
$$= x^3 e^x - 3(x^2 e^x - 2xe^x + 2e^x) + C$$
$$= e^x(x^3 - 3x^2 + 6x - 6) + C$$

Exercise 6.05 Recurrence relations

1 Prove each recurrence relation.

   a $\int x^n \cos x \, dx = x^n \sin x - n \int x^{n-1} \sin x \, dx$

   b $\int (\ln x)^n \, dx = x(\ln x)^n - n \int (\ln x)^{n-1} \, dx$

   c $\int x^n e^{2x} \, dx = \dfrac{1}{2} x^n e^{2x} - \dfrac{n}{2} \int x^{n-1} e^{2x} \, dx$

   d $\int \tan^n x \, dx = \dfrac{(\tan x)^{n-1}}{n-1} - \int (\tan x)^{n-2} \, dx$ Hint: $\tan^n x = (\tan x)^{n-2} \tan^2 x$

2 Let $I_n = \int_0^1 (1-x^2)^{\frac{n}{2}} \, dx$, where $n \geq 0$ and is an integer.

   a   Show that $I_n = \dfrac{n}{n+1} I_{n-2}$, for $n \geq 2$.

   b   Evaluate I_5.

3 For every integer $n \geq 0$, $I_n = \int_0^1 x^n (x^2-1)^5 \, dx$.

   Prove that for $n \geq 2$, $I_n = \dfrac{n-1}{n+11} I_{n-2}$.

4 a Differentiate $\sin^{n-1} \theta \cos \theta$, and express your result in terms of $\sin \theta$.

   b   Hence, deduce that

$$\int_0^{\frac{\pi}{2}} \sin^n \theta \, d\theta = \frac{n-1}{n} \int_0^{\frac{\pi}{2}} \sin^{n-2} \theta \, d\theta, \text{ for } n \geq 1.$$

   c   Evaluate $\int_0^{\frac{\pi}{2}} \sin^4 \theta \, d\theta$.

5 For integers $n \geq 0$, $I_n = \int_1^{e^2} (\log_e x)^n \, dx$.

   Show that for $n \geq 1$, $I_n = e^2 2^n - nI_{n-1}$.

6 Let $I_n = \int_0^1 \dfrac{x^{2n}}{x^2+1} \, dx$, where n is an integer $n \geq 0$.

   a   Show that $I_0 = \dfrac{\pi}{4}$.

   b   Show that $I_n + I_{n-1} = \dfrac{1}{2n-1}$.

   c   Hence, find $\int_0^1 \dfrac{x^4}{x^2+1} \, dx$.

7 Let $I_n = \int_0^{\frac{\pi}{4}} \sec^n x \, dx$, where n is an integer.

   a   Show that $I_n = \dfrac{1}{n-1} \left[(\sqrt{2})^{n-2} + (n-2)I_{n-2} \right]$ for $n > 1$.

   b   Hence evaluate I_4.

1 Find $\int x^3 e^{6x^4+1}\, dx$.

2 Evaluate $\int_{\sqrt{2}}^{2} \dfrac{dx}{x\sqrt{x^2-1}}$, using the substitution $x = \sec\theta$.

3 Using an appropriate substitution, find $\int \sin\theta \sec^3\theta\, d\theta$.

4 Show that $\int_{0}^{\frac{\pi}{2}} \dfrac{d\theta}{2+\cos\theta} = \dfrac{\pi\sqrt{3}}{9}$.

5 Evaluate $\int_{0}^{3} \dfrac{dx}{(3x+1)^2}$.

6 Find $\int \dfrac{dx}{\sqrt{9-(3-x)^2}}$.

7 Evaluate $\int_{-\frac{\pi}{2}}^{0} \dfrac{2}{x^2+4}\, dx$.

8 Find $\int \dfrac{dx}{x^2-4x+7}$.

9 Evaluate $\int_{0}^{0.5} \dfrac{x^2}{x^2-1}\, dx$.

10 Find real numbers A and B such that $\dfrac{1}{x(x-1)} = \dfrac{A}{x} + \dfrac{B}{x-1}$. Hence, find $\int \dfrac{1}{x(x-1)}\, dx$.

11 Find real numbers A and B such that $\dfrac{1}{(x+1)(x-1)^2} = \dfrac{A}{x+1} + \dfrac{B}{x-1} + \dfrac{C}{(x-1)^2}$.

Hence, find $\int \dfrac{1}{(x+1)(x-1)^2}\, dx$.

12 Using partial fractions, find $\int \dfrac{2x-3}{x^2+3x+2}\, dx$.

13 Find $\int \dfrac{2x^2+3x-1}{x^3-x^2+x-1}\, dx$.

14 Find $\int_{0}^{\frac{1}{2}} \dfrac{x^2+3}{(x^2+1)(x^2+2)}\, dx$ to 2 decimal places.

15 Use integration by parts to evaluate $\int_{1}^{e} \dfrac{\ln x}{x^2}\, dx$.

16 Use integration by parts to show that $\int (\ln x)^2\, dx = x\,[\ln|x|]^2 - 2x\ln|x| + 2x + C$.

17 Use integration by parts to evaluate $\int_{1}^{e} \dfrac{\ln x}{\sqrt{x}}\, dx$.

18 Evaluate $\int_{0}^{\ln 2} xe^x\, dx$.

19 If $I_n = \int_{0}^{1}(x^2-1)^n\, dx$ for $n \geq 0$, show that $I_n = -\dfrac{2n}{2n+1}\,I_{n-1}$ for $n \geq 1$.

7.

MECHANICS

Mechanics, sometimes referred to as classical mechanics or Newtonian mechanics, is the study of the effects of forces on objects. In the early days of mechanics, mathematics was applied to understand the motion of objects. For example, the Greek astronomers and philosophers tried to understand the motion of the planets and other celestial bodies.

Mechanics can be classified into 3 main domains: statics, kinematics and dynamics.

Statics deals with the study of objects at rest, kinematics deals with motion without taking into account the causes of the motion (forces), and dynamics deals with motion but takes into account the forces responsible for the motion.

CHAPTER OUTLINE

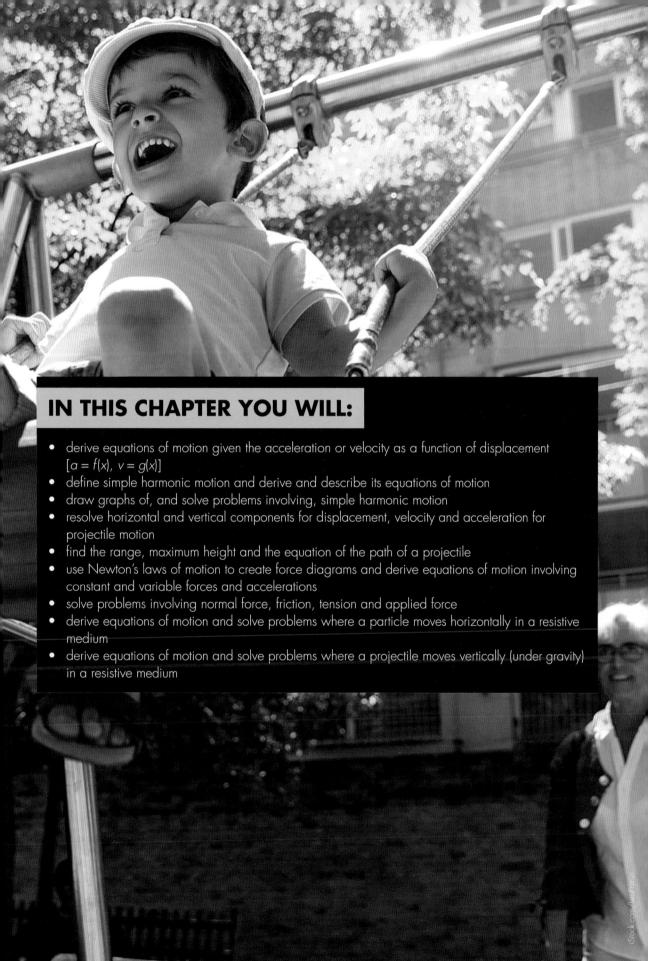

IN THIS CHAPTER YOU WILL:

- derive equations of motion given the acceleration or velocity as a function of displacement [$a = f(x)$, $v = g(x)$]
- define simple harmonic motion and derive and describe its equations of motion
- draw graphs of, and solve problems involving, simple harmonic motion
- resolve horizontal and vertical components for displacement, velocity and acceleration for projectile motion
- find the range, maximum height and the equation of the path of a projectile
- use Newton's laws of motion to create force diagrams and derive equations of motion involving constant and variable forces and accelerations
- solve problems involving normal force, friction, tension and applied force
- derive equations of motion and solve problems where a particle moves horizontally in a resistive medium
- derive equations of motion and solve problems where a projectile moves vertically (under gravity) in a resistive medium

TERMINOLOGY

acceleration: The rate of change of velocity with respect to time.

applied force: A force that is applied to an object by a person or another object; it can be a push or a pull action.

at rest: Stationary, at zero velocity.

displacement: A change in position relative to original position.

frictional force: The force exerted on an object by the surface it sits or moves on.

initially: At the start, when time is zero.

normal force: For an object on a surface, the reactive force of the surface on the object that is equal in size and acting perpendicular to the surface.

phase shift: A horizontal translation of a trigonometric function.

projectile: An object that is thrown or projected upwards.

resisted motion: Motion that encounters resisting forces, for example friction and air resistance.

simple harmonic motion: Motion in which an object's acceleration is proportional to its displacement.

spring force: A restoring force, always acting to restore the spring toward equilibrium.

tension force: The pulling force transmitted by means of a string, cable, chain, or similar.

terminal velocity: The constant velocity that a free-falling object will eventually reach when the resistance of the medium through which it is falling prevents further acceleration.

velocity: The rate of change of displacement with respect to time.

weight: The amount of gravitational force acting on matter.

Motion with variable forces

7.01 Velocity and acceleration in terms of x

Velocity ($\dot{x}$ or v) and acceleration ($\ddot{x}$ or a) are by definition functions of time, that is,

$v = \dfrac{dx}{dt}$ and $a = \dfrac{dv}{dt}$. However, if acceleration is determined by displacement, x, then we need to rewrite these expressions as derivatives with respect to x.

Acceleration

$$a = \frac{d}{dx}\left(\frac{1}{2}v^2\right)$$

Proof

$$a = \frac{d}{dx}\left(\frac{1}{2}v^2\right)$$

$$= \frac{d}{dv}\left(\frac{1}{2}v^2\right) \cdot \frac{dv}{dx} \qquad \text{by the chain rule}$$

$$= v\frac{dv}{dx}$$

$$= \frac{dx}{dt} \cdot \frac{dv}{dx}$$

$$= \frac{dv}{dt} \qquad \text{by the chain rule again}$$

Hence, if acceleration depends on time (t), write acceleration as $\dfrac{dv}{dt}$, but if the acceleration depends on position (x), we write acceleration as $\dfrac{d}{dx}\left(\dfrac{1}{2}v^2\right)$.

EXAMPLE 1

A particle moves in a straight line such that its velocity v cm s^{-1} when it is x cm from the origin is given by $v = 1 - 2e^{-x}$.

Find the acceleration of the particle at the origin.

Solution

$$a = \frac{d}{dx}\left(\frac{1}{2}v^2\right)$$

$$= \frac{d}{dx}\left(\frac{1}{2}(1 - 2e^{-x})^2\right)$$

$$= \frac{1}{2}\frac{d}{dx}(1 - 4e^{-x} + 4e^{-2x})$$

$$= \frac{1}{2}(4e^{-x} - 8e^{-2x})$$

$$= 2e^{-x} - 4e^{-2x}$$

When $x = 0$,

$$a = 2 - 4$$

$$= -2 \text{ cm s}^{-2}$$

EXAMPLE 2

The acceleration of a particle is given by $\dfrac{d^2x}{dt^2} = 12x^2 - 2x + 3$ where x is the displacement.

Find the velocity when the particle is 5 m from the origin if initially the particle is at the origin and has velocity -2 m s^{-1}.

Solution

$$\frac{d^2x}{dt^2} = 12x^2 - 2x + 3$$

$$\frac{d}{dx}\left(\frac{1}{2}v^2\right) = 12x^2 - 2x + 3$$

$$\frac{1}{2}v^2 = 4x^3 - x^2 + 3x + C$$

When $t = 0$, $x = 0$, $v = -2$

$\therefore C = 2$

So $v^2 = 8x^3 - 2x^2 + 6x + 4$

When $x = 5$:

$$v^2 = 984$$

$$v = \pm\sqrt{984}$$

$$= \pm 31.4 \text{ m s}^{-1}$$

Exercise 7.01 Velocity and acceleration in terms of x

1 Find an expression for the acceleration of a particle whose velocity is given by $v = \sqrt{x^2 + 2}$.

2 A particle moves on a line so that when it is x m from the origin its acceleration is $-3x$ m s^{-2}. It is released from rest at $x = 5$ m.

    a In which direction will it first move?

    b If its velocity is v m s^{-1}, express v^2 as a function of x.

    c Where will the particle next come to rest?

    d What is the direction of the acceleration at this point?

    e Describe the motion, including the greatest speed.

3 A rocket is fired vertically from the ground with an initial velocity of 60 m s^{-1}.
It is subject to a force that gives it a constant acceleration of -10 m s^{-2}.
After t seconds it is x m above the ground with velocity v m s^{-1}.

    a Express v^2 in terms of x.

    b Express v in terms of t.

    c What is the greatest height reached by the rocket?

    d When does it reach this height?

    e Where is the rocket 3 seconds after being fired?

    f When is the rocket 105 m above the ground?

4 A particle moves on a straight line so that, when x m from the origin, its velocity is v m s^{-1}, where $v^2 = 4(7 + 6x - x^2)$.

a Prove that its acceleration is $-4(x - 3)$ m s^{-2}.

b Find the positions at which it is at rest and its acceleration at these positions.

c Since $v^2 > 0$, then $4(7 + 6x - x^2) > 0$. Determine the interval of the line on which the particle is constrained to move.

d What is the particle's greatest speed?

5 When a particle is x m from the origin, its acceleration is $\dfrac{1}{\sqrt{4x + 9}}$ m s^{-2}. It is released from rest at the origin.

a In which direction will it first move?

b If its velocity at a position x is v m s^{-1}, express v^2 as a function of x.

c Prove that the particle does not change direction.

d Find its velocity at $x = 4$.

6 A particle is initially at the origin where it is given an initial velocity of 5 m s^{-1}. When x m from the origin, its acceleration is $-50e^{-4x}$ m s^{-2}.

a Determine its velocity as a function of x.

b Determine x as a function of time.

7 The velocity of a particle is given by $v = \dfrac{dx}{dt} = 3x$ cm s^{-1}. The particle is initially at rest 1 cm to the right of the origin.

a Find the particle's displacement after 3 seconds.

b Calculate the velocity after 3 seconds.

c Show that the particle is never at the origin.

d Show that the acceleration is $9x$ cm s^{-2}.

8 The acceleration of a particle is given by $\dfrac{1}{16 + x^2}$ m s^{-2}. The particle is initially at rest at $x = 0$ m.

a Find an expression for v^2 in terms of x.

b What is the limiting maximum speed of the particle?

9 The velocity of a particle is given by $v = \sqrt{4x + 6}$. If the particle is initially at the origin, find an expression for the displacement x at time t.

10 The acceleration of a particle is given by $-8e^{-x}$ cm s^{-2}. If initially the particle is at the origin, moving with a velocity of 4 cm s^{-1}, find an expression for the displacement x at time t.

11 A particle moves with a constant acceleration of 8 m s^{-2}. It starts from the origin with velocity of 4 m s^{-1}.

a Find the velocity in terms of displacement.

b Find v when $x = 3$ m.

c Find the displacement in terms of time.

12 A rocket launched vertically from Earth's surface has a downwards force producing a motion given by $a = -\dfrac{160\,000}{x^2}$ km s^{-2}, where x is the distance in kilometres from the centre of Earth. Given the radius of Earth is 6400 km and the initial velocity is 4 km s^{-1}, find at what distance above Earth's surface the rocket will have $v = 0$. Answer to the nearest km.

13 A mass has acceleration a m s^{-2} given by $a = v^2 - 8$ m s^{-2}, where v m s^{-1} is the velocity of the mass when it has a displacement of x metres from the origin. Find v in terms of x, given that $v = -3$ m s^{-1} where $x = 1$.

14 A particle moves in a straight line so that at time t seconds, it has acceleration a m s^{-2}, velocity v m s^{-1} and position x m from the origin. The velocity and position of the particle at any time t seconds are related by $v = -2x^2$ m s^{-1}. Initially $x = 4$ m. Find the initial acceleration of the particle and express x in terms of t.

7.02 Simple harmonic motion

In mechanics, **simple harmonic motion** is a type of periodic or oscillating motion in which the restoring force is directly proportional to the displacement and acts in the direction opposite to that of the displacement.

That is, the more you pull it one way, the more it wants to return to the centre of the motion. The classic example is a mass on a spring because the more the mass stretches it, the more it wants to return towards the centre of the motion. Under simple harmonic motion, an object moves back and forth about a central position in a cyclic way.

Newton's second law of motion states that force equals mass times acceleration ($F = ma$): the force acting on an object is proportional to, and in the same direction as, the acceleration of the object. If you imagine pulling a mass on a spring and then letting it go, it will bounce back and forth around an equilibrium position. With simple harmonic motion, the velocity is greatest in the centre of the motion, whereas the restoring force (hence acceleration) is greatest at the extremes of the motion or where displacement is a maximum. Other examples of simple harmonic motion are: a pendulum, though only if it swings at small angles as in a pendulum clock, the rise and fall of the tide in a river, and other wave motions.

Simple harmonic motion (about x = 0)

Simple harmonic motion is defined by

$$\ddot{x} = -n^2 x.$$

We can show that one solution to this differential equation is the displacement function:

$$x = A \cos(nt + \alpha).$$

If a particle is undergoing simple harmonic motion about the origin, then

$$x = A \cos(nt + \alpha), \text{ where } A, n \text{ and } \alpha \text{ are constants and } A > 0 \text{ and } n > 0.$$

Using our knowledge of trigonometric functions:

- the amplitude (A) is the maximum value of x
- the phase shift $\dfrac{\alpha}{n}$ is dependent on the initial conditions
- the period of the motion (T) is the time for the particle to complete one full oscillation (cycle), and $T = \dfrac{2\pi}{n}$
- the frequency (f) is the number of complete oscillations per second, $f = \dfrac{1}{T} = \dfrac{n}{2\pi}$.

Proof

Given $x = A \cos(nt + \alpha)$, then on differentiating we get:

$$\dot{x} = -An \sin(nt + \alpha)$$
$$\ddot{x} = -An^2 \cos(nt + \alpha)$$
$$= -n^2 A \cos(nt + \alpha)$$

Hence, $\ddot{x} = -n^2 x$

Note that the sine function $x = A \sin(nt + \alpha)$ can also be used to represent simple harmonic motion. It is better to use this result when the particle starts its motion at the centre.

Similarly for $x = A \sin(nt + \alpha)$ we can show that $\ddot{x} = -n^2 x$.

EXAMPLE 3

A particle is moving in simple harmonic motion with a period of $\dfrac{\pi}{2}$ seconds and amplitude of 5 cm. Find its displacement as a function of time given that it starts at $x = -5$.

Solution

Since $T = \dfrac{2\pi}{n} = \dfrac{\pi}{2}$

$4\pi = n\pi$

$n = 4$.

Given $A = 5$ and $x = A \cos(nt + \alpha)$, we get

$x = 5 \cos(4t + \alpha)$.

When $t = 0$, $x = -5$,

$-5 = 5 \cos \alpha$

$\alpha = \pi$.

Displacement is $x = 5 \cos(4t + \pi) = -5 \cos 4t$, as $\cos(\pi + \theta) = -\cos \theta$.

Velocity in simple harmonic motion

$$\dot{x}^2 = n^2(A^2 - x^2)$$

Proof

In simple harmonic motion the position of the object is proportional to the force acting on it. This force acts in the opposite direction to the displacement.

$F = -kx$ ⟵ where $k > 0$

$m\ddot{x} = -kx$ ⟵ $F = ma = m\ddot{x}$

$\ddot{x} = -\dfrac{k}{m}x$

We generally write this as

$\ddot{x} = -n^2 x$, where $n = \sqrt{\dfrac{k}{m}}$

$\ddot{x} = \dfrac{d}{dx}\left(\dfrac{1}{2}\dot{x}^2\right) = -n^2 x$.

Integrate both sides with respect to x,

$\dfrac{1}{2}\dot{x}^2 = -\dfrac{n^2 x^2}{2} + C$.

We know that $\dot{x} = 0$ when $x = \pm A$, which gives $C = \dfrac{n^2 A^2}{2}$.

Hence, $\dfrac{1}{2}\dot{x}^2 = -\dfrac{n^2 x^2}{2} + \dfrac{n^2 A^2}{2}$

$x^2 = -n^2 x^2 + n^2 A^2$

$\dot{x}^2 = n^2(A^2 - x^2)$

Simple harmonic motion about $x = c$

If a particle is undergoing simple harmonic motion about $x = c$, then the equations of motion are:

$\ddot{x} = -n^2(x - c)$

$\dot{x}^2 = n^2[A^2 - (x - c)^2]$

$x - c = A \cos(nt + \alpha)$ or $x - c = A \sin(nt + \alpha)$

These last 2 equations are generally written $x = A \cos(nt + \alpha) + c$ and $x = A \sin(nt + \alpha) + c$.

EXAMPLE 4

A ship requires 9 metres of water to enter the harbour safely. At low tide, the harbour is 8 metres deep and at high tide the harbour is 11 metres deep. Low tide is at 11 a.m. and high tide is at 5 p.m. The tidal motion is simple harmonic.

a State the amplitude and period of the tidal motion.

b Between what times of day is it possible for the ship to enter the harbour and how long is this period?

Solution

a $2A = 11 - 8 = 3$, so $A = \dfrac{1}{2} \times 3 = 1.5$ metres

Time between low and high tides = 5 p.m. – 11 a.m. = 6 h

$T = 6\,h \times 2 = 12$ hours

b Since $T = \dfrac{2\pi}{n} = 12$ then $n = \dfrac{\pi}{6}$, and the centre of motion is $\dfrac{8 + 11}{2} = 9.5$ metres

Let x be the depth of the water in metres at t hours after 11 a.m.

Equation describing tidal motion is $x = 1.5 \cos\left(\dfrac{\pi}{6}t + \alpha\right) + 9.5$

Using the fact that $x = 8$ when $t = 0$, we get:

$8 = 1.5 \cos\left(\dfrac{\pi}{6} \times 0 + \alpha\right) + 9.5$

$-1.5 = 1.5 \cos \alpha$

$$\cos \alpha = -1$$

$$\alpha = \pi$$

$$\therefore \ x = 1.5 \cos\left(\frac{\pi}{6}t + \pi\right) + 9.5$$

But $\cos\left(\dfrac{\pi t}{6} + \pi\right) = -\cos\dfrac{\pi t}{6}$ as $\cos(\pi + \theta) = -\cos \theta$

$$x = -1.5 \cos\left(\frac{\pi t}{6}\right) + 9.5$$

The earliest time the ship can enter the harbour is when $x = 9$.

$$9 = -1.5 \cos\left(\frac{\pi t}{6}\right) + 9.5$$

$$-0.5 = -1.5 \cos\left(\frac{\pi t}{6}\right)$$

$$\cos\left(\frac{\pi t}{6}\right) = \frac{1}{3}$$

$$\frac{\pi t}{6} = \cos^{-1}\left(\frac{1}{3}\right) \text{ or } 2\pi - \cos^{-1}\left(\frac{1}{3}\right) \quad \text{1st, 4th quadrants}$$

Make sure your calculator is in radian mode.

$$= 1.2309 \ldots \text{ or } 5.0522 \ldots$$

$$t = \frac{6}{\pi}(1.2309 \ldots) \text{ or } \frac{6}{\pi}(5.0522 \ldots)$$

$$= 2.3508 \ldots \text{ or } 9.6490 \ldots$$

$t = 2.35$ hours (after 11 a.m.) = 1.21 p.m., and

$t = 9.65$ hours (after 11 a.m.) = 8.39 p.m.

So the harbour is navigable for this ship between 1:21 p.m. and 8:39 p.m., a period of 7 hours and 18 minutes.

Hooke's law

When an elastic object, such as a spring, is stretched, the increased length is called its **extension**. The extension of an elastic object is directly proportional to the force applied to it.

This is **Hooke's law**, an example of simple harmonic motion. Its formula is $F = kx$, where k is called the **spring constant** and has units in $N\,m^{-1}$. The greater the value of k, the stiffer the spring.

x is the displacement or extension of the spring.

Objects cannot be stretched indefinitely. If the elastic object is stretched beyond its limit, known as the **elastic limit**, it may not retain its elasticity. The equation works as long as the elastic limit is not exceeded.

Provided the elastic limit is not exceeded, the graph of force versus displacement (x) produces a straight line that passes through the origin. The gradient of this line is the spring constant (k).

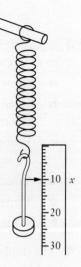

Properties of simple harmonic motion

For simple harmonic motion oscillating about the origin

$$\ddot{x} = -n^2 x \qquad \dot{x}^2 = n^2(A^2 - x^2) \qquad x = A\cos(nt + \alpha)$$

At $x = 0$, $\dot{x} = \pm nA$. The velocity of the particle is a maximum at the centre of motion with the sign indicating the direction of motion.

At $x = \pm A$, $\dot{x} = 0$. the velocity of the particle is zero at the extremities of the motion.

At $x = 0$, $\ddot{x} = 0$. The acceleration of the particle is zero at the centre of the motion.

At $x = \pm A$, $\ddot{x} = \mp n^2 A$. The acceleration is a maximum at the extremities of the motion and is always directed towards the centre of the motion.

Exercise 7.02 Simple harmonic motion

1 A particle is moving in simple harmonic motion. The graph of its displacement as a function of time is drawn below.

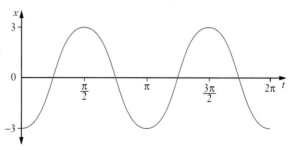

  a If $x = A\cos(nt + \alpha)$ is the displacement of the particle at any time t, find A, n and α.

  b Draw a graph of the velocity versus time for $0 \le t \le 2\pi$.

  c Draw a graph of acceleration versus time for $0 \le t \le 2\pi$.

2 A particle is x metres from the origin after t seconds where $x = 2 + \sqrt{3} \sin t$.

    a   Prove that its acceleration is $-(x - 2)$ m s^{-2}.

    b   Draw a graph of $x = 2 + \sqrt{3} \sin t$ for $0 \le t \le 2\pi$ and state the amplitude and period of the motion.

3 A particle is oscillating about a central point so that its displacement in metres at any time t seconds is given by $x = 3 \cos 2t$.

    a   Draw the graph of this function for displacement.

    b   What are the first 3 times when the particle has maximum displacement and what is this maximum displacement?

    c   Write an expression in terms of t for the velocity of the particle and draw a graph for velocity.

    d   What is the velocity when the particle is at its maximum displacement?

    e   Write an expression for the acceleration and draw its graph.

    f   What is the acceleration when the particle is at the origin?

4 The displacement for a particle is given by $x = 5 \cos 3t - 12 \sin 3t$, where x is in metres and t is in seconds.

    a   By deriving the equation for its acceleration, prove that the particle is moving in simple harmonic motion.

    b   Find the period of the motion.

    c   Find the maximum speed.

5 A particle is moving in simple harmonic motion with acceleration given by $\ddot{x} = -16x$, where x metres is the displacement of the particle from the centre of motion (origin). Initially the particle is at the origin, moving to the right at 6 m s^{-1}. Find the displacement as a function of time.

6 A mass is oscillating at the end of a spring with squared velocity given by $\dot{x}^2 = 225 - 625x^2$ where x cm is the displacement from the centre of motion.

    a   Find the acceleration of the mass as a function of x.

    b   Find the amplitude, period of the motion and maximum speed of the mass.

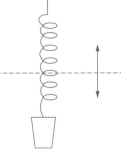

7 The squared velocity of a particle moving in simple harmonic motion in a straight line is given by $\dot{x}^2 = 6x - x^2$ m s^{-1}, where x is the displacement in metres.

    a   Find the 2 points between which the particle is oscillating.

    b   Find the centre of the motion.

    c   Find the maximum speed of the particle.

    d   Find the acceleration of the particle in terms of x.

    e   Find the period.

8 The squared velocity of a particle moving in simple harmonic motion is given by

$\dot{x}^2 = 60 - 8x - 4x^2$ m s^{-1}, where x is the displacement from the centre of motion in metres.

a Find the acceleration of the particle in terms of x.

b Write down the centre of the motion.

c Find the amplitude, period and frequency of the motion. $\boxed{\text{Frequency } f = \frac{1}{T}}$

d If initially the particle is at $x = -1$ and $\dot{x} > 0$, find an expression for the displacement in the form $x = A \cos(nt + \alpha) + c$.

9 A particle oscillating in simple harmonic motion is x metres from an origin after t seconds where $x = 3 \cos(2t + \alpha)$.

a State the amplitude, period and greatest speed of the particle.

b If its velocity is $\dot{x}$ m s^{-1}, prove that $\dot{x}^2 = 4(9 - x^2)$.

c Prove that its acceleration is $-4x$ m s^{-2}.

d If initially the particle was at $x = 1.5$ m with velocity $3\sqrt{3}$ m s^{-1}, find a suitable value for α.

10 At x metres from the origin, the squared velocity of a particle in m s^{-1} is $\dot{x}^2 = 28 - 24x - 4x^2$.

a Prove that its acceleration is $-4(x + 3)$ m s^{-2}.

b Explain how this shows the motion is simple harmonic and find the period of oscillation.

c Find the positions at which the particle is at rest and hence state the amplitude.

d What is the greatest speed?

e Show that $\dot{x}^2$ can be written in the form $\dot{x}^2 = n^2[A^2 - (x + 3)^2]$.

11 The displacement of a particle from an origin is x metres after t seconds, where

$$x = \cos 5t - \sqrt{3} \sin 5t$$

a Rewrite this in the form $x = A \sin(5t - \alpha)$, with $A > 0$ and $0 < \alpha < \frac{\pi}{2}$, and hence describe the motion.

b Find the first time at which its velocity is 5 m s^{-1}.

12 A particle P is moving on the x-axis according to $x = 3 \sin 4t$, where x cm is the displacement of P from O at time t seconds.

a Prove that P moves in simple harmonic motion and state the period of the motion.

b Find the first time when the particle is 1.5 cm right of the origin and its velocity at this point.

c Find the greatest speed of P and the interval in which it moves.

13 A particle is moving in simple harmonic motion with acceleration given by $\ddot{x} = -4x$. Initially the particle is at the centre of the motion and has a velocity of 12 m s^{-1}.

    a Find the velocity as a function of x.

    b Use the fact that $\dfrac{dt}{dx} = \dfrac{1}{\frac{dx}{dt}}$ to show that $x = 6 \sin 2t$.

    c Find the period and frequency of the motion.

    d Sketch a graph of acceleration against time for 2 periods.

14 A particle is moving according to the formula $\dot{x}^2 = \pi^2(4 - x^2)$, where x metres is the displacement of the particle from the origin at time t seconds.

    a Find the period and amplitude of the motion.

    b If the particle is initially at $x = 2$, write down an expression for the displacement as a function of time.

    c For what percentage of the period is the particle within $\sqrt{2}$ m of the origin?

15 The table below shows the depth of water (in metres) at the end of a wharf as it varies with the tides at different times during the morning.

Time (t)	midnight	2 a.m.	4 a.m.	6 a.m.	8 a.m.	10 a.m.	midday
Depth (d)	8.50	12.74	14.50	12.74	8.50	4.26	2.50

    a Assuming that the tidal motion is simple harmonic, find the amplitude and period of the tidal motion.

    b Calculate the depth of water at the wharf at 3:00 p.m.

    c A ship requires a depth of 10 m to dock safely at the wharf. For how long can the ship dock safely at the wharf on this morning?

7.03 Projectile motion

A **projectile** is a body in free fall that is subject only to the force of gravity.

We have studied **projectile motion** in Mathematics Extension 1, Chapter 10, *Further vectors*. The acceleration due to gravity is represented by g and near Earth's surface g is approximately 9.8 m s^{-2}, often rounded to 10 m s^{-2}.

An object must be dropped from a height, thrown vertically upwards or thrown at an angle to be considered a projectile. The path followed by a projectile is known as a **trajectory**. The existence of gravity forces the projectile to travel along a trajectory that has the shape of a parabola, and gravity pulls the object downwards.

For simplicity, we assume:

- the influence of air resistance is zero

- there is a single force of gravity acting downwards

Assume $g = 10$ m s^{-2} unless otherwise specified.

The factors that affect the trajectory are:

- the angle of projection, θ
- the initial speed, V
- the height of projection.

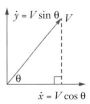

From the Mathematics Extension 1 course, Chapter 10, *Further vectors*, we know that the initial velocity has 2 components, $\dot{x} = V \cos \theta$ and $\dot{y} = V \sin \theta$, which are velocity components in the x and y directions (or horizontal and vertical directions) respectively.

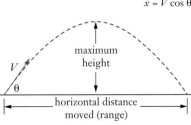

The horizontal component remains constant because there are no forces acting on the particle in this direction and so is independent of t, while the vertical component changes with t due to gravity.

Projection from the origin

Horizontal

$\ddot{x} = 0$

Integrating with respect to t:

$\dot{x} = C$, where C is constant

When $t = 0$, $C = V \cos \theta$

$\therefore \dot{x} = V \cos \theta$

Integrating with respect to t:

$x = Vt \cos \theta + D$, where D is constant

When $t = 0$, $x = 0$, so $D = 0$

$x = Vt \cos \theta$

Vertical

$\ddot{y} = -g$

Integrating with respect to t:

$\dot{y} = -gt + E$, where E is constant

When $t = 0$, $E = V \sin \theta$

$\therefore \dot{y} = -gt + V \sin \theta$

Integrating with respect to t:

$y = -\dfrac{1}{2}gt^2 + Vt \sin \theta + F$, where F is constant

When $t = 0$, $y = 0$, so $F = 0$

$y = -\dfrac{1}{2}gt^2 + Vt \sin \theta$

> The formulas on these pages do not need to be memorised, but they must be proved for each problem.

Maximum height

At maximum height, $\dot{y} = 0$

Hence $0 = -gt + V \sin \theta$

$t = \dfrac{V \sin \theta}{g}$

> Time to reach maximum height

Substitute into y, vertical displacement:

$$y = -\frac{1}{2}g\left(\frac{V \sin \theta}{g}\right)^2 + V\left(\frac{V \sin \theta}{g}\right) \sin \theta$$

$$= -\frac{V^2 \sin^2 \theta}{2g} + \frac{V^2 \sin^2 \theta}{g}$$

$$\therefore y_{max} = \frac{V^2 \sin^2 \theta}{2g}$$

Time of flight

Projectile returns to ground when $y = 0$.

$0 = -\dfrac{1}{2}gt^2 + Vt\sin\theta$

$0 = -gt^2 + 2Vt\sin\theta$

$gt^2 = 2Vt\sin\theta$

$gt = 2V\sin\theta \qquad\qquad t \neq 0$

$t = \dfrac{2V\sin\theta}{g}$

Note that the time of flight is double the time to reach the maximum height $t = \dfrac{V\sin\theta}{g}$.

Range

Substitute into x, horizontal displacement, for the range of flight:

$\therefore x = V\left(\dfrac{2V\sin\theta}{g}\right)\cos\theta$

$\quad = \dfrac{2V^2\sin\theta\cos\theta}{g}$

$\therefore x = \dfrac{V^2\sin2\theta}{g}$

Maximum range

For maximum range, $\sin2\theta = 1$.

$\therefore 2\theta = 90°$

$\theta = 45°$

So $x_{max} = \dfrac{V^2\sin(90°)}{g}$

$\therefore x_{max} = \dfrac{V^2}{g}$

Equation of the trajectory (path of the projectile)

Since $x = Vt\cos\theta$ and $y = -\dfrac{1}{2}gt^2 + Vt\sin\theta$

From the first equation $t = \dfrac{x}{V\cos\theta}$; substitute into the second equation:

$y = -\dfrac{1}{2}g\left(\dfrac{x}{V\cos\theta}\right)^2 + V\left(\dfrac{x}{V\cos\theta}\right)\sin\theta$

$\therefore y = -\dfrac{gx^2}{2V^2\cos^2\theta} + x\tan\theta$

This is the equation of a parabola of the form $y = ax^2 + bx$, concave down ($a < 0$).

MATHS IN FOCUS 12. Mathematics Extension 2 ISBN 9780170413435

Modelling projectile motion

Many sporting events use projectiles, for example games such as cricket, golf, football and basketball that use balls. Other pursuits that can be analysed using projectiles include shooting, archery, snowboarding, javelin, discus and shot put.

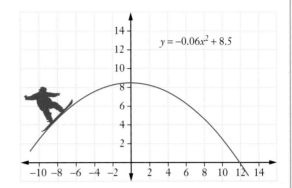

$y = -0.06x^2 + 8.5$

Many athletes and sporting coaches study projectiles. Analysing results and execution, through the use of technology, can assist in improving performance.

The parabolic graph shows the trajectory of a snowboarder excelling in the sport.

Using technology, we can determine the velocity at various points along the trajectory, the length of the trajectory and the maximum height of the event.

1 Using the Internet, obtain some images of an activity that involves a projectile. Using technology, determine the equation of the trajectory, the range and maximum height of the projectile.

2 After checking to ensure that it is safe, drop an object from a multilevel building. Take a video of the falling object from another location so that the whole drop can be observed. From the video, determine the height at a number of times during the flight. Using the data collected, determine the total time of flight, the maximum velocity and the gravitational constant.

3 Throw a basketball from the free throw line and use technology to determine the maximum height of the throw and the initial velocity if the basketball is to go into the hoop. Investigate the use of different angles for the free throw.

4 If a ball is dropped vertically from rest at the same time and height that a second ball is launched horizontally, which ball will strike the ground first? Use technology to investigate this problem.

5 Use technology to determine the range of a projectile on an inclined plane. What is the relationship between the range, the angle of projection and the angle of the inclined plane?

Trebuchets

A trebuchet is a type of catapult, and was a common type of siege weapon that used a swinging arm to throw a projectile. These weapons appeared in ancient China in the 4th century BCE as a traction trebuchet, using human power to swing the arm.

By the 12th century CE trebuchets appeared in the Mediterranean region, employing a counterweight to swing the arm.

The trebuchet was phased out as a weapon of war as the availability of gunpowder increased.

EXAMPLE 5

An object is projected vertically upward with initial velocity of 50 m s^{-1}.

a What is the time for the object to reach its maximum height?

b Find the maximum height reached.

c How long does the object take to return to its starting point?

Solution

a Since the object is projected vertically, $\theta = 90°$, $V = 50$ m s^{-1}. $g \approx 10$ m s^{-2}.

Using $y = -\dfrac{1}{2}gt^2 + Vt \sin \theta$ gives $y = -5t^2 + 50t$ ⬅ The formulas in these examples must be derived (proved) using the given values, not just quoted or memorised. They are quoted here just to save time.

Maximum height is attained when $\dot{y} = 0$:

$$\dot{y} = -10t + 50 = 0$$

$$-10t = -50$$

$$t = 5 \text{ s}$$

b Therefore, maximum height is:

$$y_{max} = -5t^2 + 50t$$

$$= -5(5)^2 + 50(5)$$

$$= 125 \text{ m}$$

c For the object to return to $y = 0$:

$$0 = -5t^2 + 50t$$

$$= -5t(t - 10)$$

$$\therefore t = 0 \text{ or } t = 10$$

The object takes 10 seconds to return to its starting point.

A rock is thrown with an initial vertical velocity of 30 m s^{-1} and an initial horizontal velocity of 15 m s^{-1}. Take gravity as 9.8 m s^{-2}.

a What will these 2 velocity components be when the rock reaches its highest point?

b How long will the rock be in the air?

c Find the range for the rock.

Solution

a $V \cos \theta = 15$ and $V \sin \theta = 30$.

Components of velocity at the highest point:

$\dot{x} = 15$ m s^{-1} (this is constant for entire motion because it is independent of t)

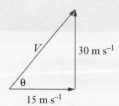

and

$\dot{y} = 0$ m s^{-1} (always 0 at the highest point)

b Time of flight is $t = \dfrac{2V \sin \theta}{g}$

$$= \frac{2 \times 30}{9.8}$$

$$\approx 6.12 \text{ s}$$

c Range $= Vt \cos \theta$

$$= 15 \times 6.12$$

$$\approx 92 \text{ m}$$

Projection from a height not at the origin

In the case of a projectile launched from a height h, such as shooting an arrow or cannonball from a fortress at the top of a hill, the horizontal components remain the same but the vertical displacement is different.

$\ddot{y} = -g$

$\dot{y} = -gt + V \sin \theta$

From here the equation for vertical displacement changes:

$y = -\dfrac{1}{2}gt^2 + Vt \sin \theta + D$, where D is constant

When $t = 0, y = h$ so $D = h$

$y = -\dfrac{1}{2}gt^2 + Vt \sin \theta + h$ ← This is just the standard vertical displacement function translated h units upwards.

Shutterstock.com/Gabbiere

EXAMPLE 7

A horizontal drain releases waste water 5 metres above a collection pond. The water comes out horizontally and enters the pond 3 metres out from the end of the pipe. Find the velocity at which the water escapes the drain.

Solution

Horizontally, $x = Vt \cos 0°$

$$= Vt$$

Vertically, $y = -5t^2 + Vt \sin 0° + 5$

$$= -5t^2 + 5$$

When $x = 3$, $y = 0$

$3 = Vt$

$t = \dfrac{3}{V}$

Substitute into y:

$0 = -5\left(\dfrac{3}{V}\right)^2 + 5$

$1 = \dfrac{9}{V^2}$

$V^2 = 9$

$V = 3 \quad (V > 0)$

$V = 3 \text{ m s}^{-1}.$

EXAMPLE 8

A cannon is fired from a cliff of height 25 metres. The cannonball has initial velocity of 60 m s^{-1} at an angle of 48° to the horizontal.

a How long will the cannonball take to land?

b Find the range of the cannon.

c What is the maximum height obtained by the cannonball?

d How much would the range change if the angle of projection is 38°?

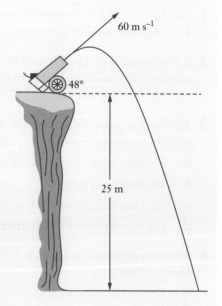

Solution

a Time of flight is t when $y = 0$

$$y = \frac{1}{2}gt^2 + Vt\sin\theta + h$$

$$0 = -5t^2 + 60t\sin 48° + 25$$

$$t = \frac{-60\sin 48° \pm \sqrt{(60\sin 48°)^2 + 500}}{-10}$$

Since $t > 0$, $t \approx 9.45$ s

b Range: $x = Vt\cos\theta$

$$= 60 \times 9.45 \times \cos 48°$$

$$\approx 379 \text{ m}$$

c Maximum height is y when $\dot{y} = 0$

That is, $t = \dfrac{V\sin\theta}{g} = \dfrac{60 \times \sin 48°}{10} \approx 4.46$ s

$$y = -5t^2 + 60t\sin 48° + 25$$

Hence, maximum height is given by

$$y_{max} = -5(4.46)^2 + 60(4.46)\sin 48° + 25$$

$$\approx 124.4 \text{ m}$$

d Range for angle of projection of 38°

Time of flight (as in part **a**):

$$t = \frac{-60\sin 38° \pm \sqrt{(60\sin 38°)^2 + 500}}{-10}$$

Since $t > 0$, $t \approx 8.01$ s

Range: $x = 60 \times 8.01 \times \cos 38°$

$$\approx 379 \text{ m}$$

Comparing with part **b**, there is no change in the range.

Exercise 7.03 Projectile motion

1 An object is projected vertically upward with initial velocity of 35 m s^{-1}.
Find the maximum height reached. Use $g = 10$ m s^{-2} and answer to 2 decimal places.

2 A small mass is dropped from a helicopter at 980 metres. Neglecting air resistance, calculate the time that it takes to fall to Earth. Use $g = 9.8$ m s^{-2} and give your answer in exact form.

3 An object is projected up into the air with a vertical velocity of 45 m s^{-1}. Neglect air resistance and use $g = 9.8$ m s^{-2}. Find correct to one decimal place:

  a the time of rise to the top of the motion

  b the maximum height

  c the time that the object is in the air

  d the velocity of the object after 8 seconds

4 A mass is projected at an angle of 45° to the horizontal with velocity 20 m s^{-1}.
Use $g = 10$ m s^{-2}.

  a Derive the horizontal and vertical equations of the motion.

  b What is the time taken to hit the ground? Give your answer in exact form.

  c Derive the Cartesian equation of the motion.

5 An object is projected at an angle of 30° to the horizontal. The target is 60 m from the point of projection. At what initial speed must it be projected to hit the target?

6 An object is projected at an angle of α to the horizontal from a 50 m cliff. The target is 500 m from the base of the cliff. If the projectile has initial speed of 100 m s^{-1}, at what angle must it be projected to hit the target?

7 A golf ball is hit at 50 m s^{-1}. At what angle should it leave the club in order to travel 250 m horizontally?

8 Show by example that 2 projectiles can be launched with the same speed but at different angles and still have the same range.

9 A ski jumper comes off the end of the jump horizontally and falls 90 m vertically before contacting the slope a record 180 m horizontally from the end of the jump. What was the initial speed of this jumper as she left the jump? Answer in exact form.

10 A projectile is fired horizontally with speed V m s^{-1} from a point h metres above horizontal ground.

  a Prove that the projectile will reach the ground after $\sqrt{2hg^{-1}}$ seconds.

  b If the projectile hits the ground at an angle of 60° to the horizontal with a speed of $2V$ m s^{-1}, show that $3V^2 = 2gh$.

11 A ball is thrown upwards from the top of a building at an angle of 22° above the horizontal with an initial velocity of 16 m s^{-1}. If the ball is in the air for 3 s, how tall is the building, to 2 decimal places? Use $g = 9.81$ m s^{-2}.

12 A stunt man jumps a canyon that is 10 metres wide. He rides his motorcycle up an incline of 12°. What minimum speed is required for him to cross the canyon successfully? Use $g = 9.81$ m s^{-2} and answer to 2 decimal places.

13 Amy is a 2-metre tall basketball player who is aiming at a basket that is 3 m above the court 10 m away. If she shoots at an angle of 45°, at what initial speed must she throw the basketball so that it goes into the basket without hitting the backboard? Answer in exact form.

14 A football kicked at 15 m s^{-1} just passes over the 4 m cross-bar from a distance of 15 m away. Show that if α is the angle of projection then $5 \tan^2 \alpha - 15 \tan \alpha + 9 = 0$.

15 A ball is kicked from ground level over 2 walls of height 6 m and distant 6 m and 12 m from the point of projection, as shown in the diagram.

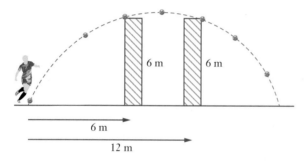

a Prove that if α is the angle of projection, then $\tan \alpha = \dfrac{3}{2}$.

b Prove that if the walls are h metres high and distant x_1 and x_2 metres from the point of projection, then $\tan \alpha = \dfrac{h(x_1 + x_2)}{x_1 x_2}$.

16 A golf ball is struck with initial speed V m s^{-1} at an angle of elevation of α.

a Prove that, when it is y metres above the point of projection, the vertical velocity of the ball is given by $\dot{y}^2 = V^2 \sin^2 \alpha - 2gy$.

b If speed at this point is S m s^{-1}, prove that $S^2 = V^2 - 2gy$.

7.04 Forces and equations of motion

Over 300 years ago, Englishman Sir Isaac Newton worked in many areas of mathematics and physics. He developed the theories of gravitation in 1666 when he was only 23 years old. Twenty years later, he presented his 3 laws of motion in the *Principia Mathematica Philosophiae Naturalis*.

Written in Latin, the *Principia* is a collection of 3 books written by Newton, covering his laws of motion, his law of universal gravitation and a derivation of Kepler's laws of planetary motion.

The *Principia* is considered by many to be the most important work in the history of science.

Newton's laws of motion

Newton's laws of motion consist of 3 fundamental laws of classical physics.

1 An object remains in a state of rest or uniform motion in a straight line unless it is acted upon by an external force.

2 For a constant mass, force equals mass × acceleration.

3 For every action, there is an equal and opposite reaction.

Newton's **first law** defines inertia, and states that if there is no net force acting on an object then the object will remain at constant velocity (including zero velocity if it was not already moving and the object remains at rest).

Newton's **third law** of motion states that for every force there is an equal and opposite force. This law is useful in explaining how a wing on a plane generates lift and an engine produces thrust.

The force exerted by one object on another object is equal to the (reaction) force of the second object on the first, acting in the opposite direction. For a person standing, the force of the person's feet on the ground is balanced by the force of the ground on the person's feet.

Force is measured in newtons (N). One newton is the force required to produce an acceleration of 1 m s^{-2} in a mass of 1 kg.

Two areas of mechanics, kinematics and statics, involve the motion of a particle, which is a dimensionless object and generally represented by a point. The first of these, kinematics, applies when the object is moving and the second, as you might expect, when the object is at rest.

In each case, there will be forces acting on the object that result in motion or no motion. The study of each is based on Newton's laws of motion.

Force diagrams

A force diagram shows all the forces acting on an object, the force's direction and its magnitude. All the forces included exist for that object in the given situation. Therefore, to construct force diagrams, it is extremely important to know the various types of forces.

The contact forces acting on a body may include:

Normal force ($F_{norm} = N$)

Frictional force ($F_{fric} = F$)

Tension force ($F_{tens} = T$)

Applied force (F_{app})

Air resistance force ($F_{air} = R$)

Spring force (F_{spring})

Gravity ($F_{grav} = mg$)

Normal force

The **normal force** is the support force exerted upon an object that is in contact with another stable object. For example, when 2 surfaces are in contact (such as a book on a desk), they exert a normal force on each other, perpendicular to the contacting surfaces.

Friction

Friction is the resistance to motion of one object moving relative to another. If 2 surfaces can move over each other without any resistance, they have smooth contact, whereas if resistance is experienced, then they have rough contact.

The force of friction is that resistance encountered at the point of contact of 2 bodies sliding over each other. Friction opposes the direction of motion and the frictional force is equal in size to the force causing the motion.

Tension

Tension force on an object is transmitted through a string, rope or wire when it is pulled tight from opposite ends. The tension is directed along the length of the wire and pulls equally on the objects on the opposite ends of the wire.

Applied force

An applied force is exerted on an object by a person or another object. For example, when a person pushes or pulls an object, they are applying a force to the object.

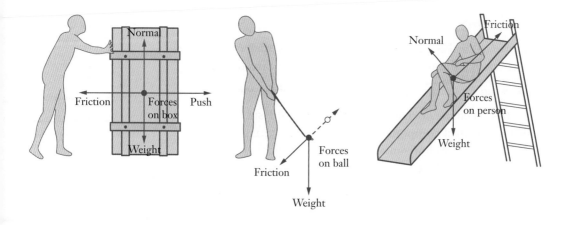

EXAMPLE 9

Draw a force diagram for each situation.

a   A book resting on a desk

b   A skydiver falling through the air

c   A box being dragged to the right

d   A pendant hanging around a neck

e   A stone dropped from a cliff (neglect air resistance)

f   A spring scale used to weigh a suitcase

Solution

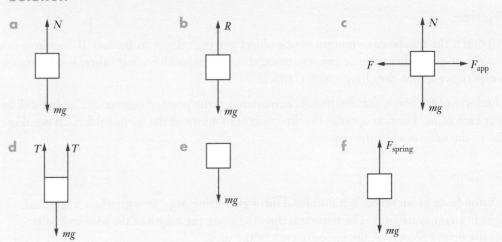

When all forces are equal and opposite, the object remains at rest.

If, however, the forces are not in balance then the object will move in the direction determined by the greater force.

Resolution of vector quantities

If a force (F) makes an angle of θ with the x-axis, then F can be written in terms of components in the x and y directions.

$F \cos \theta$ and $F \sin \theta$ are called the components of the force $\underset{\sim}{F}$.

$F \cos \theta$ is called the **horizontal component** and $F \sin \theta$ is called the **vertical component**. The process of writing a force in terms of its horizontal and vertical components is called **resolving a force**.

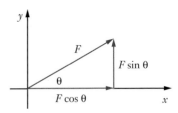

EXAMPLE 10

Resolve horizontally and vertically the force F that makes an angle of 60° to the x-axis.

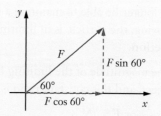

Solution

The horizontal component is $F \cos 60° = \dfrac{1}{2}F$

The vertical component is $F \sin 60° = \dfrac{\sqrt{3}}{2}F$

Equilibrium

If several forces act on an object and the body remains at rest, then the forces are in equilibrium.

The condition for equilibrium is $\Sigma F = 0$.

For example, if there are 3 forces in equilibrium, the forces represent the sides of a triangle.

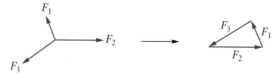

Since a closed triangle is formed then $\Sigma F = 0$, that is, $F_1 + F_2 + F_3 = 0$.

In this force diagram, the normal force N has 2 components: a horizontal component $N \sin \theta$ and a vertical component $N \cos \theta$.

In the next force diagram F is the frictional force. If the system is in equilibrium, then we can resolve horizontally and vertically.

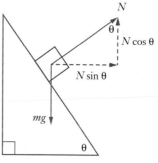

Horizontally:

$F \cos \theta = N \sin \theta$

Vertically:

$F \sin \theta + N \cos \theta = mg$

From the first equation, $F = N \tan \theta$. Substitute into the second equation: $N \tan \theta \sin \theta + N \cos \theta = mg$

$$N\left(\frac{\sin^2 \theta}{\cos \theta} + \frac{\cos^2 \theta}{\cos \theta}\right) = mg$$

Rearranging, $N = mg \cos \theta$ and $F = mg \sin \theta$

Coefficient of friction

There will be a point where, if the angle is increased sufficiently, the frictional force will no longer be able to maintain a system in equilibrium. If the object is just on the point of moving, the particle is in limiting equilibrium and the force of friction is called the **limiting friction.**

The magnitude of the limiting friction stays in constant ratio to the normal reaction on the object, and this constant ratio is called the coefficient of friction, μ (mu).

$$\mu = \frac{F}{N} \text{ or } F = \mu N.$$

The coefficient of friction depends on the material of the objects in contact. There are 2 coefficients of friction. The **static coefficient of friction** is applied when both objects are not moving. The **dynamic coefficient of friction** is applied when one or both objects are moving.

Remember, friction always opposes any motion or intended motion.

EXAMPLE 11

Weight is the force due to gravity.
Weight = mass × gravity

An object of weight 25 N, in rough contact with a plane inclined at θ to the horizontal, is just about to slide. If the static coefficient of friction between the plane and the object is $\sqrt{3}$, find the angle θ.

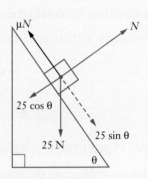

Solution

Resolving forces on the object horizontally and vertically for equilibrium:

Along the plane:

$$\mu N = 25 \sin \theta \qquad\qquad [1]$$

Perpendicular to the plane:

$$N = 25 \cos \theta \qquad\qquad [2]$$

Solving for θ: [1] ÷ [2]

$$\mu = \tan \theta$$

hence, $\tan \theta = \sqrt{3}$

$$\therefore \quad \theta = 60°$$

MATHS IN FOCUS 12. Mathematics Extension 2   ISBN 9780170413435

In this force diagram, the tensions T_1 and T_2 on particle P can be resolved into the horizontal and vertical components.

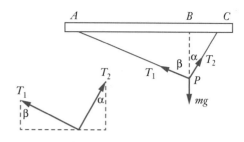

Horizontally:

$T_1 \sin \beta = T_2 \sin \alpha$

Vertically:

$T_1 \cos \beta + T_2 \cos \alpha = mg$

The angles α and β, or the lengths AB, BC and BP, allow you to determine the tensions T_1 and T_2.

EXAMPLE 12

Calculate the tension in a wire supporting a 60 kg tightrope walker where the weight of the tightrope walker at the centre of the wire causes it to sag by 8°. Let $g = 9.8$ m s^{-2}.

Solution

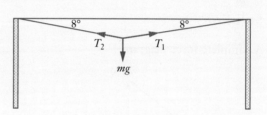

Horizontally;

$T_1 \cos 8° = T_2 \cos 8°$

Hence $T_1 = T_2$

Vertically:

$T_1 \sin 8° + T_2 \sin 8° = mg$

Solving simultaneously:

$2T \sin 8° = mg$ where $T = T_1 = T_2$

$$T = \frac{mg}{2 \sin 8°}$$

$$= \frac{60 \times 9.8}{2 \sin 8°}$$

$$\approx 2112 \text{ N}$$

The applied force in this force diagram can be resolved into the horizontal and vertical components.

Horizontally:

$F_{app} \cos \theta$ will provide the force to move the object in the horizontal direction.

Vertically:

$N + F_{app} \cos \theta = mg$

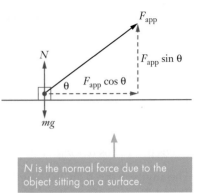

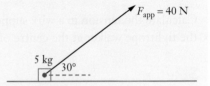

N is the normal force due to the object sitting on a surface.

EXAMPLE 13

A 40 N force is applied to a 5 kg box at an angle of 30° to the horizontal. If the dynamic coefficient of friction is 0.4, find the acceleration of the box.

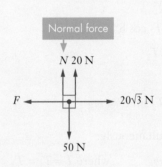

Solution

A complete force diagram:

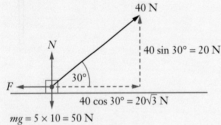

Normal force

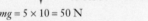

$mg = 5 \times 10 = 50$ N

Resolving forces vertically:

Normal force (N) = 50 N – 20 N = 30 N ◄──── newtons

Resolving forces horizontally:

$F = \mu N = 0.4 \times 30$, where μ is the dynamic coefficient of friction

$F = 12$ newtons

∴ Net force horizontally

$F_{net} = 20\sqrt{3} - 12 = 22.641...$ newtons

For the acceleration of the box, $F_{net} = m\ddot{x}$

$22.641... = 5\ddot{x}$

$\ddot{x} = 22.6 \div 5$

≈ 4.5 m s^{-2}

ISBN 9780170413435

Exercise 7.04 Forces and equations of motion

1 Calculate the magnitude of the horizontal and vertical components of a force of:

   a   16 N inclined at 60° to the horizontal

   b   20 N inclined at 30° to the horizontal

   c   24 N inclined at 20° to the vertical

   d   18 N inclined at 40° to the vertical

2 Express each force in the diagrams below as components along the plane and perpendicular to the plane.

a 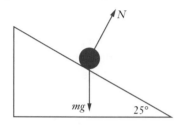   **b**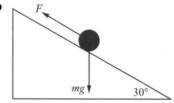

c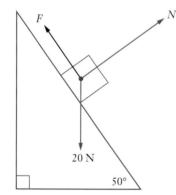

3 An object weighing 5 N is just about to slide up a plane where the inclination is 40° to the horizontal, under the action of a force inclined at an angle θ to the plane.

   a   If the static coefficient of friction is $\dfrac{1}{\sqrt{3}}$, find the magnitude of the applied force in terms of θ.

   b   What is the applied force if $\theta = 45°$?

4 An object of mass 0.5 kg is pulled along horizontally at constant velocity. If it takes 2 N of force to maintain this constant velocity, calculate the dynamic coefficient of friction to one decimal place using $g = 9.8$ m s^{-2}.

5 A 4.5 kg block is pulled up a 20° ramp at a constant velocity by an 8 kg counterbalance. Calculate the dynamic coefficient of friction using $g = 9.8$ m s^{-2}.

6 The dynamic coefficient of friction between a 170 kg box and a carpet floor is 0.85.

   a   How much force would it take to push the box at a constant velocity across the carpet floor? Use $g = 9.8$ m s^{-2}.

   b   How much force would be required to push the box and to accelerate it by 0.5 m s^{-2}?

7 A string of length 20 cm is attached to 2 points A and B at the same level and a distance of 10 cm apart. A ring of mass 5 g can slide on the string and a horizontal force F is applied so that the ring is in equilibrium vertically below B. Use $g = 9.8$ m s^{-2}.

   a   Find the tension in the string.

   b   Find F, the force applied to maintain equilibrium.

8 A particle of mass 15 kg is suspended by 2 strings 6 m and 8 m long, their other ends being fastened to a rod 10 m apart. If the rod is held at an angle such that the particle hangs directly below the middle of the rod, find the tensions of the strings. Use $g = 9.8$ m s^{-2}.

9 A box of mass 10 kg is being pulled along a smooth surface by a rope inclined at 45° to the horizontal. The tension in the rope is 12 N.

   a   Draw a force diagram for this situation.

   b   Find the acceleration for the box.

   c   Find the normal reaction between the box and the floor.

10 Two particles of masses M and m are connected by a light string. The first mass, M, is placed on a rough horizontal table with the string passing over a smooth pulley. The second mass, m, is hanging freely. The dynamic coefficient of friction between the mass M and the table is μ.

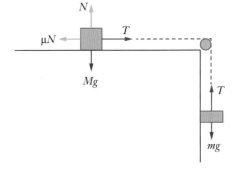

   a   Find the acceleration of the system.

   b   Find the tension in the string.

   c   Show that the system does not move if
   $$\mu \geq \frac{m}{M}.$$

11 A particle of mass M is just about to slide up a plane, of inclination α to the horizontal, under the action of an applied force F (pull) that is inclined at an angle θ to the plane. If the static coefficient of friction is μ, show that:

$$F = \left(\frac{\sin \alpha + \mu \cos \alpha}{\cos \theta + \mu \sin \theta} \right) Mg$$

THE PERIOD OF A PENDULUM

Italian astronomer and physicist Galileo Galilei (1564–1642) became interested in pendulums when he watched a chandelier swinging in a cathedral. He began to experiment with pendulums and discovered that the period of the pendulum is not affected by the amplitude.

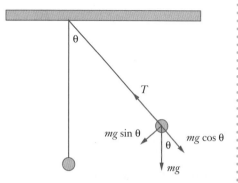

Devise an investigation to find an approximation for the acceleration due to gravity using a pendulum.

1 What effect does changing the mass on the end of the string make?

2 What effect does changing the length of the string have on a pendulum?

3 What difference does changing the angle of swing make?

Note: The periodic time for a swinging pendulum is constant only when amplitudes are small.

4 Plot a graph of periodic time, T, against length, l, getting a curve (a parabola). Try a few quick calculations to see whether the graph to produce a straight line is T, $\dfrac{1}{T}$, $\sqrt{T}$ or T^2 against l.

7.05 Resisted horizontal motion

When an object moves through a medium such as air, water or oil, the resistance (force), R, of the medium slows down the object's motion and varies as a function of the velocity. The higher the velocity of the object, the greater the resistance. The resistance causes the object to experience a **retardation**, which is a negative acceleration (or deceleration).

For horizontal motion, there are no other forces apart from the resistance due to the velocity or velocity squared, that is $R = -kv$ or $R = -kv^2$.

If $R = -kv$, or linear resistance, the object is moving at low speeds. In this case the drag is due to viscosity (thickness or 'stickiness') of the medium in which it is travelling.

$$R = -kv$$

Direction of motion

$$ma = -kv$$

$$a = -\frac{k}{m}v$$

If $R = -kv^2$, or quadratic resistance, the object is more likely at high speeds. In this case the drag is related to the momentum transfer between the moving object and the fluid in which it is travelling.

Direction of motion ←————————

$R = -kv^2$

It is conventional to take the direction of motion to be positive.

$$ma = -kv^2$$

$$a = -\frac{k}{m}v^2$$

EXAMPLE 14

A particle of mass m initially with speed v_0 moves horizontally against a resistance proportional to the square of the speed. Express its velocity in terms of the distance travelled and its displacement in terms of velocity.

Solution

$$ma = -kv^2$$

$$a = -\frac{k}{m}v^2$$

$$v\frac{dv}{dx} = -\frac{k}{m}v^2$$

On integrating,

$$\int \frac{1}{v}\,dv = -\frac{k}{m}\int dx$$

$$\ln v = -\frac{k}{m}x + C$$

$v > 0$ because it is speed

When $x = 0$, $v = v_0$ so $C = \ln v_0$

$$\ln v = -\frac{k}{m}x + \ln v_0$$

$$\ln v - \ln v_0 = -\frac{k}{m}x$$

$$\ln\left(\frac{v}{v_0}\right) = -\frac{k}{m}x \qquad [*]$$

$$\frac{v}{v_0} = e^{-\frac{k}{m}x}$$

$$v = v_0\, e^{-\frac{k}{m}x}$$

From [*]:

$$x = -\frac{m}{k}\ln\left(\frac{v}{v_0}\right)$$

EXAMPLE 15

A particle of mass 3 kg is propelled from the origin along the x-axis with initial velocity of u m s^{-1}. The only forces acting on the particle in the x direction are friction, which is a constant 18 N, and drag due to air resistance which equals v^2 N, where v is the velocity of the particle t seconds after leaving the origin.

a   Show that $\dfrac{dv}{dt} = -6 - \dfrac{1}{3}v^2$.

b   Show that $t = \dfrac{1}{\sqrt{2}}\tan^{-1}\left[\dfrac{3\sqrt{2}(u-v)}{18+uv} \right]$.

c   By using $\dfrac{dv}{dt} = v\dfrac{dv}{dx}$, find an expression for v in terms of x.

Solution

a $3a = -18 - v^2$

$a = -6 - \dfrac{1}{3}v^2$

Hence $\dfrac{dv}{dt} = -6 - \dfrac{1}{3}v^2$

$\xleftarrow{\hspace{2cm}} \overset{18+v^2}{\underset{\text{direction of}}{\circ}} \xrightarrow{\hspace{2cm}}$

direction of
motion

b   Rearranging the expression from part a, $\dfrac{dv}{18+v^2} = -\dfrac{1}{3}dt$

On integrating:

$$\int \frac{dv}{18+v^2} = -\frac{1}{3}\int dt$$

$$\frac{1}{3\sqrt{2}}\left[\tan^{-1}\left(\frac{v}{3\sqrt{2}}\right) - \tan^{-1}\left(\frac{u}{3\sqrt{2}}\right) \right] = -\frac{1}{3}t$$

$$\therefore \sqrt{2}t = \left[\tan^{-1}\left(\frac{u}{3\sqrt{2}}\right) - \tan^{-1}\left(\frac{v}{3\sqrt{2}}\right) \right]$$

Taking tan of both sides:

$$\tan\left(\sqrt{2}t\right) = \tan\left[\tan^{-1}\left(\frac{u}{3\sqrt{2}}\right) - \tan^{-1}\left(\frac{v}{3\sqrt{2}}\right) \right]$$

Simplify using $\tan(A-B) = \dfrac{\tan A - \tan B}{1+\tan A \tan B}$

where $A = \tan^{-1}\left(\dfrac{u}{3\sqrt{2}}\right)$ and $B = \tan^{-1}\left(\dfrac{v}{3\sqrt{2}}\right)$

$$\tan\left(\sqrt{2}t\right) = \frac{\dfrac{u}{3\sqrt{2}} - \dfrac{v}{3\sqrt{2}}}{1 + \dfrac{u}{3\sqrt{2}} \times \dfrac{v}{3\sqrt{2}}}$$

$$= \frac{u - v}{3\sqrt{2} + \dfrac{uv}{3\sqrt{2}}} \qquad \text{multiplying by } \frac{3\sqrt{2}}{3\sqrt{2}}$$

$$= \frac{3\sqrt{2}\left(u - v\right)}{\left(3\sqrt{2}\right)^2 + uv} \qquad \text{multiplying by } \frac{3\sqrt{2}}{3\sqrt{2}} \text{ again}$$

$$= \frac{3\sqrt{2}\left(u - v\right)}{18 + uv} \qquad \text{as required}$$

c
$$\frac{dv}{dt} = v\frac{dv}{dx} = -6 - \frac{1}{3}v^2$$

$$\frac{v}{-6 - \dfrac{1}{3}v^2}\,dv = dx$$

$$\frac{v}{18 + v^2}\,dv = -\frac{1}{3}\,dx$$

$$\frac{1}{2}\ln\left(18 + v^2\right) = -\frac{1}{3}x + C$$

$$\ln\left(18 + v^2\right) = -\frac{2}{3}x + D$$

$$18 + v^2 = e^{-\frac{2}{3}x + D} = Ae^{-\frac{2}{3}x}$$

When $x = 0$, $v = u$:

$$18 + u^2 = A$$

$$\therefore 18 + v^2 = \left(18 + u^2\right)e^{-\frac{2}{3}x}$$

$$v^2 = (18 + u^2)e^{-\frac{2}{3}x} - 18$$

Formulas for acceleration, a

- Use $a = \dfrac{dv}{dt}$ when $x(t)$ or $v(t)$ is required

- Use $a = v\dfrac{dv}{dx}$ when $v(x)$ is required

ISBN 9780170413435

Exercise 7.05 Resisted horizontal motion

1 A particle of mass m initially with speed u moves horizontally against a resistance proportional to its speed (v). Express its velocity in terms of the distance travelled and its displacement in terms of velocity.

2 A particle of mass m initially with speed u moves horizontally against a resistance proportional to the square of its speed (v^2). Express its velocity in terms of the distance travelled and its displacement in terms of velocity.

3 A particle of unit mass moves in a straight line against a resistance equal to $v + v^3$, where v is the velocity of this particle. Initially, the particle is at the origin and is travelling with speed $V > 0$. Show that v is related to the displacement x by the expression

> Unit mass means 1 kg.

$$x = \tan^{-1}\left(\frac{V - v}{1 + Vv}\right)$$

4 A high speed train of mass M starts from rest and moves along a straight track. At time t hours, the distance travelled by the train from its starting point is x km, and its velocity is v km h^{-1}.

The train is driven by a constant force F and has a resistive force of kv^2 in the opposite direction, where k is a positive constant. The resultant force is zero when the train is travelling at 430 km h^{-1}.

a Show that the equation of motion for the train is $Ma = F\left[1 - \left(\dfrac{v}{430}\right)^2\right]$.

b Find, in terms of F and M, the time taken for the train to reach a velocity of 400 km h^{-1}.

5 A particle of mass 1 kg moving in a straight line from the origin is subject to a resisting force of kv^3, where v is the speed at time t and k is a constant.
Initial speed is v_0 and x is the displacement of the particle.

a Show that $v = \dfrac{v_0}{kv_0 x + 1}$.

b Deduce that $t = \dfrac{1}{2}kx^2 + \dfrac{x}{v_0}$.

c A bullet is fired horizontally at a target 2400 m away. The bullet is observed to travel the first 800 m in 0.8 s and the next 800 m in 1 s. Assuming that air resistance is proportional to v^3 and gravity can be neglected, calculate the time taken to travel the last 800 m.

7. Mechanics

6 The acceleration of a particle moving in a straight line is given by $a = k(1 - v^2)$, $k > 0$ where v is its velocity at any time t. Initially the particle is at the origin and at rest.

    a   Find an expression for the velocity in terms of t and hence the velocity as $t \to \infty$.

    b   Find an expression for the position of the particle in terms of velocity.

7 A particle of mass m moves in a horizontal straight line. The particle is resisted by a constant force mk and a variable force mv^2, where k is a constant ($k > 0$) and v is the speed. Initially its speed is u at the origin.

    a   Show that the distance travelled is $\dfrac{1}{2} \ln \left(\dfrac{k + u^2}{k + v^2} \right)$.

    b   Show that the time taken for the particle to come to rest is $t = \dfrac{1}{\sqrt{k}} \tan^{-1} \dfrac{u}{\sqrt{k}}$.

8 The only force acting on a particle moving horizontally in a straight line is a resistance of $mk(c + v)$ acting in that line, where m is the mass of the particle, v is the velocity and k and c are positive constants. Initially the particle moves with speed $U > 0$, and it comes to rest at time T. At time $\dfrac{1}{2}T$ its velocity is $\dfrac{1}{8}U$.

    a   Show that the acceleration is given by $a = -k(c + v)$.

    b   Show that $c = \dfrac{1}{48}U$.

    c   Show that at time t, $\dfrac{48v}{U} = 49e^{-kt} - 1$.

7.06 Resisted vertical motion

When a particle is moving vertically (either upwards or downwards) the acceleration due to gravity is always towards Earth. There may also be a resistance, R, to the particle whose direction is always opposing the direction of travel. Again, $R = -kv$ or $R = -kv^2$.

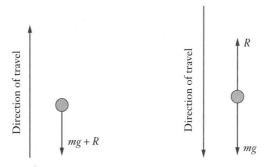

Sometimes the motion of the particle is composed of an upwards journey followed by its downwards journey. In these cases, we are required to treat each part of the journey separately.

Terminal velocity

As an object falls, air resistance may become so great that its magnitude is equal to that of the force due to gravity. This means there is zero net force acting and the object is no longer accelerating. It cannot fall any faster, and its velocity has reached what is called **terminal velocity**, v_T.

v_T occurs when $a = 0$ as $t \to \infty$.

EXAMPLE 16   Retardation is negative acceleration.

An object falls from rest and the retardation due to the air resistance is $0.2\,v^2$ m s^{-2}.

a   Use $g = 9.8$ m s^{-2} to show that $v = 7\left(\dfrac{e^{2.8t} - 1}{e^{2.8t} + 1}\right)$.

b   Show that $x = 5 \ln\left(e^{2.8t} + 1\right) - 7t - 5 \ln 2$.

c   Find the terminal velocity of the object.

Solution

a   Take the positive direction to be downwards.

$$ma = mg - 0.2mv^2$$

$$a = g - 0.2v^2 = 9.8 - 0.2v^2$$

$$\frac{dv}{dt} = \frac{98 - 2v^2}{10} = \frac{49 - v^2}{5}$$

Integrating,

$$\int \frac{dv}{49 - v^2} = \frac{1}{5}\int dt$$

$0.2\, mv^2$

mg

Let $\dfrac{1}{49-v^2} = \dfrac{1}{(7+v)(7-v)} = \dfrac{A}{7+v} + \dfrac{B}{7-v}$

$$1 = A(7-v) + B(7+v)$$
$$1 = 7A - Av + 7B + Bv$$
$$1 = 7A + 7B + v(B-A)$$

$$\therefore 7A + 7B = 1$$
$$B - A = 0$$
$$B = A$$
$$\therefore 7A + 7A = 1$$
$$14A = 1$$
$$A = \dfrac{1}{14}, B = \dfrac{1}{14}$$

$$\dfrac{1}{49-v^2} = \dfrac{1}{14}\left(\dfrac{1}{7+v} + \dfrac{1}{7-v}\right)$$

$$\int \dfrac{dv}{49-v^2} = \dfrac{1}{14}\int\left(\dfrac{1}{7+v} + \dfrac{1}{7-v}\right) dv$$

$$= \dfrac{1}{14}\left[\ln|7+v| - \ln|7-v|\right]$$

$$= \dfrac{1}{14}\ln\left|\dfrac{7+v}{7-v}\right|$$

So $\displaystyle\int \dfrac{dv}{49-v^2} = \dfrac{1}{5}\int dt$

$$\dfrac{1}{14}\ln\left|\dfrac{7+v}{7-v}\right| = \dfrac{1}{5}t + C$$

When $t = 0, v = 0$

$$\dfrac{1}{14}\ln\left|\dfrac{7+0}{7-0}\right| = \dfrac{1}{5}(0) + C$$

$$\dfrac{1}{14}\ln 1 = C$$

$$C = 0$$

$$\dfrac{1}{14}\ln\left|\dfrac{7+v}{7-v}\right| = \dfrac{1}{5}t$$

$$t = \dfrac{5}{14}\ln\left|\dfrac{7+v}{7-v}\right|$$

ISBN 9780170413435

Hence, $e^{\frac{14}{5}t} = \dfrac{7+v}{7-v}$

$(7-v)e^{\frac{14}{5}t} = 7+v$

$7\left(e^{2.8t}-1\right) = v\left(e^{2.8t}+1\right)$

$$v = 7\left(\dfrac{e^{2.8t}-1}{e^{2.8t}+1}\right)$$

b   Taking the result from part **a**,

$\dfrac{dx}{dt} = 7\left(\dfrac{e^{2.8t}-1}{e^{2.8t}+1}\right)$

$\quad = -7\left(\dfrac{1-e^{2.8t}}{e^{2.8t}+1}\right)$

$\quad = -7\left(\dfrac{1+e^{2.8t}-2e^{2.8t}}{e^{2.8t}+1}\right)$

$\quad = -7\left(1-\dfrac{2e^{2.8t}}{e^{2.8t}+1}\right)$

Integrating,

$x = -7\left[t-\dfrac{2}{2.8}\ln\left(e^{2.8t}+1\right)\right]+C$

$\quad = -7t+\dfrac{14}{2.8}\ln\left(e^{2.8t}+1\right)+C$

$\quad = -7t+5\ln\left(e^{2.8t}+1\right)+C$

When $t=0$, $x=0$.

$0 = 0+5\ln 2+C$

$C = -5\ln 2$

$x = -7t+5\ln\left(e^{2.8t}+1\right)-5\ln 2$

$x = 5\ln\left(e^{2.8t}+1\right)-7t-5\ln 2$

c The terminal velocity occurs as $t \to \infty$.

Since $v = 7\left(\dfrac{e^{2.8t}-1}{e^{2.8t}+1}\right)$ (from part **a**), we can divide numerator and denominator by

$e^{2.8t}$ to get $v = 7\left(\dfrac{1-e^{-2.8t}}{1+e^{-2.8t}}\right)$. Hence as $t \to \infty$, $v \to 7\left(\dfrac{1-0}{1+0}\right) = 7$ m s^{-1}.

Alternatively, we could simply make acceleration equal to 0.

Thus $\dfrac{49-v^2}{5} = 0$

$v^2 = 49$

$v = 7$ m s^{-1}

Many problems involving an object projected vertically through a resistive medium require the motion to be analysed during its upward motion *and* then separately during its downward motion. In the upward and downward motions the positive direction is taken to be in the direction of the motion. These problems are usually about velocity, displacement or time to reach significant points.

EXAMPLE 17

An object of mass m is projected vertically upwards with initial velocity v_0 in a medium where the resistance is $R = mkv$.

a Prove that the maximum height is:

$$x = \frac{v_0}{k} + \frac{g}{k^2}\ln\left(\frac{g}{g+kv_0}\right)$$

b Prove that the time it takes to reach the maximum height is:

$$t = \frac{1}{k}\ln\left(\frac{g+kv_0}{g}\right)$$

c Given that $v_0 = \dfrac{v_T}{2}$ (where v_T is the terminal velocity), show that the velocity v, of the object on returning to its original launch point satisfies the equation:

$$\frac{kv}{g} + \ln\left(\frac{g}{g-kv}\right) + \frac{1}{2} + \ln\frac{2}{3} = 0$$

 ISBN 9780170413435

Solution

a For the upward motion (positive direction upwards):

$ma = -mg - mkv$

$a = -g - kv$

For maximum height, we want x when $v = 0$.

$v\dfrac{dv}{dx} = -g - kv$

$\dfrac{dv}{dx} = \dfrac{-g - kv}{v} = \dfrac{-(g + kv)}{v}$

$\dfrac{dx}{dv} = -\dfrac{v}{g + kv}$

$= -\dfrac{1}{k}\left(\dfrac{g + kv - g}{g + kv}\right)$

$= -\dfrac{1}{k}\left(1 - \dfrac{g}{g + kv}\right)$

Integrating wrt v,

$x = -\dfrac{1}{k}\left(v - \dfrac{g}{k}\ln\left[g + kv\right]\right) + C$ $\qquad\qquad (g, k > 0, \ v \geq 0)$

When $x = 0$, $v = v_0$:

$0 = -\dfrac{1}{k}\left(v_0 - \dfrac{g}{k}\ln\left[g + kv_0\right]\right) + C$

$C = \dfrac{1}{k}\left(v_0 - \dfrac{g}{k}\ln\left[g + kv_0\right]\right)$

Hence

$x = -\dfrac{1}{k}\left(v - \dfrac{g}{k}\ln\left[g + kv\right]\right) + \dfrac{g}{k}\left(v_0 - \dfrac{1}{k}\ln\left[g + kv_0\right]\right)$

$= \dfrac{1}{k}\left[v_0 - \dfrac{g}{k}\ln\left[g + kv_0\right] - \left(v - \dfrac{g}{k}\ln\left[g + kv\right]\right)\right]$

$= \dfrac{1}{k}\left[v_0 - v + \dfrac{g}{k}\ln\left(\dfrac{g + kv}{g + kv_0}\right)\right]$

Maximum height when $v = 0$:

$$x = \frac{1}{k}\left[v_0 - 0 + \frac{g}{k}\ln\left(\frac{g + k(0)}{g + kv_0}\right)\right]$$

$$= \frac{1}{k}\left[v_0 + \frac{g}{k}\ln\left(\frac{g}{g + kv_0}\right)\right]$$

$$= \frac{v_0}{k} + \frac{g}{k^2}\ln\left(\frac{g}{g + kv_0}\right)$$

b For time to reach maximum height, we want t when $v = 0$.

$$\frac{dv}{dt} = -g - kv$$

$$\frac{dt}{dv} = \frac{-1}{g + kv}$$

Integrating,

$$t = -\frac{1}{k}\ln(g + kv) + D$$

When $t = 0$, $v = v_0$:

$$0 = -\frac{1}{k}\ln(g + kv_0) + D$$

$$D = \frac{1}{k}\ln(g + kv_0)$$

Hence $t = -\frac{1}{k}\ln(g + kv) + \frac{1}{k}\ln(g + kv_0)$

$$= \frac{1}{k}\ln\left(\frac{g + kv_0}{g + kv}\right)$$

Maximum height when $v = 0$:

$$t = \frac{1}{k}\ln\left(\frac{g + kv_0}{g + k(0)}\right)$$

$$= \frac{1}{k}\ln\left(\frac{g + kv_0}{g}\right)$$

c For the downward motion (positive direction downwards):

$$ma = mg - mkv$$

$$a = g - kv$$

ISBN 9780170413435

Terminal velocity v_T when $a = 0$

$$0 = g - kv_T$$

$$kv_T = g$$

$$v_T = \frac{g}{k}$$

$$\therefore v_0 = \frac{v_T}{2} = \frac{g}{2k}$$

Find velocity in terms of displacement:

$$a = v\frac{dv}{dx} = g - kv$$

$$\frac{dv}{dx} = \frac{g - kv}{v}$$

$$\frac{dx}{dv} = \frac{v}{g - kv}$$

$$= -\frac{1}{k}\left(\frac{g - kv - g}{g - kv}\right)$$

$$= -\frac{1}{k}\left(1 - \frac{g}{g - kv}\right)$$

Integrating wrt v,

$$x = -\frac{1}{k}\left(v - \frac{g}{k}\ln\left[g - kv\right]\right) + E \qquad \left(g - kv = a > 0\right)$$

When $x = 0$, $v = 0$:

$$0 = -\frac{1}{k}\left(0 - \frac{g}{k}\ln\left|g + k(0)\right|\right) + E$$

$$E = \frac{1}{k}\left(-\frac{g}{k}\ln g\right) = -\frac{g}{k^2}\ln g$$

Hence

$$x = -\frac{1}{k}\left(v - \frac{g}{k}\ln\left[g - kv\right]\right) - \frac{g}{k^2}\ln g$$

$$= -\frac{1}{k}\left[v - \frac{g}{k}\ln\left[g - kv\right] + \frac{g}{k}\ln g\right]$$

$$= -\frac{1}{k}\left[v + \frac{g}{k}\ln\left(\frac{g}{g - kv}\right)\right]$$

Now, distance up equals distance down.

$$x = \frac{v_0}{k} + \frac{g}{k^2} \ln\left(\frac{g}{g+kv_0}\right) = -\frac{1}{k}\left[v + \frac{g}{k}\ln\left(\frac{g}{g-kv}\right)\right]$$

Substitute $v_0 = \dfrac{g}{2k}$:

$$\frac{\frac{g}{2k}}{k} + \frac{g}{k^2}\ln\left[\frac{g}{g+k\left(\frac{g}{2k}\right)}\right] = -\frac{1}{k}\left[v + \frac{g}{k}\ln\left(\frac{g}{g-kv}\right)\right]$$

$$\frac{g}{2k^2} + \frac{g}{k^2}\ln\left[\frac{g}{\left(\frac{3g}{2}\right)}\right] = -\frac{v}{k} - \frac{g}{k^2}\ln\left(\frac{g}{g-kv}\right)$$

$$\frac{g}{2k^2} + \frac{g}{k^2}\ln\frac{2}{3} = -\frac{v}{k} - \frac{g}{k^2}\ln\left(\frac{g}{g-kv}\right)$$

$$\frac{g}{2} + g\ln\frac{2}{3} = -kv - g\ln\left(\frac{g}{g-kv}\right)$$

$$\frac{1}{2} + \ln\frac{2}{3} = -\frac{kv}{g} - \ln\left(\frac{g}{g-kv}\right)$$

$$\frac{kv}{g} + \ln\left(\frac{g}{g-kv}\right) + \frac{1}{2} + \ln\frac{2}{3} = 0$$

Exercise 7.06 Resisted vertical motion

1 An object of mass 1 kg is dropped from a height of h m above the ground. It experiences air resistance proportional to the square of its velocity.

   a Show that the motion of the object is given by $a = g - kv^2$.

   b Find v^2 in terms of x.

   c Find the velocity of the object as it hits the ground in terms of g, k and h.

2 A particle of unit mass is projected vertically upwards from the ground with initial speed U m s^{-1}. It experiences an air resistance that is proportional to the square of its speed.

   a Find the time to reach the maximum height.   Unit mass means 1 kg.

   b Find the maximum height.

3 A body with mass 1 kg is projected vertically from a point on level ground with velocity of 40 m s^{-1}. The forces acting on the body are gravity and air resistance of $\dfrac{v}{4}$ newtons, where v is the velocity of the body.

a Show that the equation of motion of the body is $a = -\dfrac{v+40}{4}$.

b Find the maximum height reached by the body.

c Find the time taken for the body to reach its maximum height.

4 An object of mass m kg is dropped from rest in a medium where the resistance has a magnitude of $0.1\,mv^2$ newtons, where the velocity of the object is v m s^{-1}. After t seconds, the object has fallen a distance of x metres, and has velocity v m s^{-1} and acceleration of a m s^{-2}. The object hits the ground in $\ln(1+\sqrt{2})$ seconds after it is dropped.

a Draw a diagram showing the forces acting on the object.

b Show that $a = 0.1(100 - v^2)$ m s^{-2}.

c Express v as a function of t.

d Hence, show that the speed at which the object hits the ground is $5\sqrt{2}$ m s^{-1}.

e Show that the distance the object falls is $5\ln 2$ metres.

5 Zoe drops a stone of unit mass into an empty well to determine its depth. When the stone is dropped it experiences air resistance proportional to its velocity, v m s^{-1}.

a Explain why the acceleration of the stone is given by $a = g - kv$, where k is a constant and g is the acceleration due to gravity.

b Show that the velocity of the stone is given by $v = \dfrac{g}{k}(1 - e^{-kt})$ m s^{-1}.

c Hence, find the terminal velocity of the stone.

d Derive an expression for x, the distance travelled, as a function of the velocity, v, at any time.

e Given that $k = 0.2$, $g = 10$ and Zoe counted 3 seconds before the stone hit the bottom of the well, how deep was the empty well?

6 A particle of mass m is projected vertically upwards and experiences a resistance of magnitude mkv^2 newtons. During its downward motion, the terminal velocity of the particle is V m s^{-1}. Find the position of the particle below its maximum height when it reaches 50% of its terminal velocity.

7 A stone of mass 2 kg is launched vertically upward into the air from the ground with initial speed of 15 m s^{-1}. The stone experiences a resistive force of $\dfrac{1}{3}v^2$ newtons in the opposite direction to its velocity. The acceleration of this stone until it reaches its maximum height is $a = -\dfrac{60+v^2}{6}$.

a Find the time taken by the stone to reach its maximum height.

b Show that $v^2 = 285e^{-\frac{x}{3}} - 60$.

c Find H, the maximum height reached by the stone.

8 Maximus fires an arrow, of mass m, vertically upwards with initial speed of V m s^{-1}. The arrow experiences a resistance equal to mkv^2, $k > 0$. If the initial speed is equal to its terminal speed, show that the final speed when the arrow returns to its initial position is $\dfrac{V\sqrt{2}}{2}$ m s^{-1}.

9 An object of unit mass is dropped from an altitude of 1200 m in a medium whose resistance is proportional to v^2.

    a Show that $v^2 = \dfrac{g}{k}(1 - e^{-2kx})$, where x metres is the distance fallen.

    b If $k = 0.003$, find the speed at which the object hits the ground.

10 A particle is projected vertically upward under gravity with initial velocity of v_0. Air resistance is proportional to the square of the velocity.

    a Show that the greatest height reached is $\dfrac{1}{2k}\ln\left(\dfrac{g + kv_0^2}{g}\right)$.

    b The particle then falls from its greatest height. Find the terminal velocity.

    c The particle then returns to its point of projection with speed V. Show that $(g + kv_0^2)(g - kV^2) = g^2$.

7.07 Resisted projectile motion

Consider an object launched with a velocity V at an angle of θ to the horizontal (ground). Without air resistance, the trajectory followed by this projectile is known to be a parabola.

The extent to which air resistance affects various projectiles is determined by the speed, shape, size and surface texture of the projectile. Air density can also be a factor.

Linear drag model

In the case of a resistive force that grows linearly with velocity ($R = -kv$), we can still separate the motion into horizontal and vertical components.

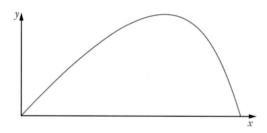

The components for acceleration in the horizontal and vertical directions can be found using Newton's laws of motion.

$m\ddot{x} = -k\dot{x}$ and $m\ddot{y} = -mg - k\dot{y}$

so we get $\ddot{x} = -\dfrac{k}{m}\dot{x}$ and $\ddot{y} = -g - \dfrac{k}{m}\dot{y}$

Horizontal

$$\frac{dv_x}{dt} = -\frac{k}{m}v_x \text{ where } v_x = \dot{x}$$

$$\frac{dv_x}{v_x} = -\frac{k}{m}dt$$

Integrating:

$$\ln v_x = -\frac{k}{m}t + C$$

When $t = 0$, $v_x = V\cos\theta$

so $C = \ln(V\cos\theta)$

On substituting and rearranging:

$$\ln v_x = -\frac{k}{m}t + \ln(V\cos\theta)$$

$$\therefore \ln\left(\frac{v_x}{V\cos\theta}\right) = -\frac{k}{m}t$$

so $\dot{x} = v_x = (V\cos\theta)\,e^{-\frac{k}{m}t}$

Integrating again:

$$x = V\cos\theta\int e^{-\frac{k}{m}t}\,dt$$

$$= -\frac{m}{k}V\cos\theta\, e^{-\frac{k}{m}t} + D$$

When $t = 0$, $x = 0$

so $D = \dfrac{m}{k}V\cos\theta$

On substituting and rearranging,

$$x = -\frac{m}{k}(V\cos\theta)\,e^{-\frac{k}{m}t} + \frac{m}{k}V\cos\theta$$

$$= \frac{mV\cos\theta}{k}\left(1 - e^{-\frac{k}{m}t}\right)$$

Vertical

$$\frac{dv_y}{dt} = -g - \frac{k}{m}v_y \text{ where } v_y = \dot{y}$$

$$\frac{dv_y}{g + \dfrac{k}{m}v_y} = -dt$$

Integrating:

$$\frac{m}{k}\ln\left(g + \frac{k}{m}v_y\right) = -t + E$$

When $t = 0$, $v_y = V\sin\theta$

so $E = \dfrac{m}{k}\ln\left(g + \dfrac{k}{m}V\sin\theta\right)$

On substituting and rearranging:

$$\frac{m}{k}\ln\left(g + \frac{k}{m}v_y\right) = -t + \frac{m}{k}\ln\left(g + \frac{k}{m}V\sin\theta\right)$$

$$t = \frac{m}{k}\ln\left(g + \frac{k}{m}V\sin\theta\right) - \frac{m}{k}\ln\left(g + \frac{k}{m}v_y\right)$$

$$t = \frac{m}{k}\ln\left(\frac{g + \dfrac{k}{m}V\sin\theta}{g + \dfrac{k}{m}v_y}\right)$$

$$e^{\frac{k}{m}t} = \frac{mg + kV\sin\theta}{mg + kv_y}$$

Making v_y the subject:

$$mg + kv_y = \frac{mg + kV\sin\theta}{e^{\frac{k}{m}t}}$$

$$kv_y = (mg + kV\sin\theta)\,e^{-\frac{k}{m}t} - mg$$

$$\dot{y} = v_y = \left(\frac{m}{k}g + V\sin\theta\right)e^{-\frac{k}{m}t} - \frac{m}{k}g$$

Vertical (continued)

Integrating again:

$$y = \int \left[\left(\frac{m}{k} g + V \sin \theta \right) e^{-\frac{k}{m}t} - \frac{m}{k} g \right] dt$$

$$= -\frac{m}{k} \left(\frac{mg}{k} + V \sin \theta \right) e^{-\frac{k}{m}t} - \frac{m}{k} gt + F$$

When $t = 0$, $y = 0$

so $F = \dfrac{m}{k} \left(\dfrac{mg}{k} + V \sin \theta \right)$

On substituting and rearranging,

$$y = -\frac{m}{k} \left(\frac{mg}{k} + V \sin \theta \right) e^{-\frac{k}{m}t} - \frac{m}{k} gt + \frac{m}{k} \left(\frac{mg}{k} + V \sin \theta \right)$$

$$= \frac{m}{k} \left(\frac{mg}{k} + V \sin \theta \right) - \frac{m}{k} \left(\frac{mg}{k} + V \sin \theta \right) e^{-\frac{k}{m}t} - \frac{mgt}{k}$$

$$= \frac{m}{k} \left(\frac{mg}{k} + V \sin \theta \right) \left[1 - e^{-\frac{k}{m}t} \right] - \frac{mgt}{k}$$

Trajectory of a projectile in a resistive medium

To find the equation of the path of the projectile, eliminate t from the 2 displacement equations above.

$$x = \frac{mV \cos \theta}{k} \left(1 - e^{-\frac{k}{m}t} \right) \tag{1}$$

$$\frac{kx}{mV \cos \theta} = 1 - e^{-\frac{k}{m}t} \tag{*}$$

$$\therefore \qquad t = -\frac{m}{k} \ln \left(1 - \frac{kx}{mV \cos \theta} \right)$$

Also

$$y = \frac{m}{k} \left(\frac{mg}{k} + V \sin \theta \right) \left(1 - e^{-\frac{k}{m}t} \right) - \frac{mgt}{k} \tag{2}$$

MATHS IN FOCUS 12. Mathematics Extension 2 ISBN 9780170413435

Substituting in [2] and rearranging to eliminate t:

$$y = \frac{m}{k}\left(g\frac{m}{k} + V\sin\theta\right)\left(\frac{kx}{mV\cos\theta}\right) - \frac{mg}{k}\left[-\frac{m}{k}\ln\left(1 - \frac{kx}{mV\cos\theta}\right)\right] \text{ from [*]}$$

$$y = \frac{m}{k}\left(\frac{mg}{k} + V\sin\theta\right)\left(\frac{kx}{mV\cos\theta}\right) - \frac{mg}{k}\left[-\frac{m}{k}\ln\left(1 - \frac{kx}{mV\cos\theta}\right)\right]$$

$$= \left(\frac{mg}{k} + V\sin\theta\right)\left(\frac{x}{V\cos\theta}\right) + \frac{m^2g}{k^2}\ln\left(1 - \frac{kx}{mV\cos\theta}\right)$$

$$= \left(\frac{mg}{kV\cos\theta} + \tan\theta\right)x + \frac{m^2g}{k^2}\ln\left(1 - \frac{kx}{mV\cos\theta}\right)$$

The path is not a parabola because the function is more complicated and of the form $y = ax + b\ln(1 - cx)$.

Time for maximum height

At maximum height $\dot{y} = 0$

$$\left(\frac{m}{k}g + V\sin\theta\right)e^{-\frac{k}{m}t} - \frac{m}{k}g = 0$$

on rearranging,

$$t = \frac{m}{k}\ln\left(1 + \frac{kV}{mg}\sin\theta\right)$$

Speed at any point

The speed at any point on the trajectory depends on both the vertical and horizontal velocities.

$$S = \sqrt{\dot{x}^2 + \dot{y}^2}$$

Terminal velocity

Terminal velocity occurs when the projectile is descending and $\ddot{y} = 0$.

$$-g + \frac{k}{m}\dot{y} = 0.$$

$$v_T = \dot{y} = \frac{m}{k}g$$

7. Mechanics

A ball is kicked at an angle of 30° to the horizontal at 12 m s^{-1} and experiences air resistance proportional to the velocity in both the x and y directions. Find the equations of motion, the equation of the path of the ball and an expression for the maximum height reached.

Solution

Resolving forces, $m\ddot{x} = -mk\dot{x}$ and $m\ddot{y} = -mg - mk\dot{y}$

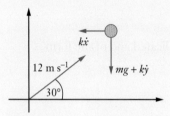

So we get

$$\frac{dv_x}{dt} = -kv_x \text{ and } \frac{dv_y}{dt} = -10 - kv_y$$

Horizontal

In the x direction we get a differential equation like those we have seen in exponential growth and decay problems.

So, $v_x = Ae^{-kt}$

When $t = 0$, $v_x = 12 \cos 30°$, so $A = 6\sqrt{3}$, and we get

$\dot{x} = v_x = 6\sqrt{3}\,e^{-kt}$.

Integrating, we get

$$x = -\frac{6\sqrt{3}}{k}e^{-kt} + C$$

When $t = 0$, $x = 0$ and so $C = \dfrac{6\sqrt{3}}{k}$

$$\therefore\ x = -\frac{6\sqrt{3}}{k}e^{-kt} + \frac{6\sqrt{3}}{k}$$

$$= \frac{6\sqrt{3}}{k}(1 - e^{-kt}) \qquad\qquad\qquad [1]$$

Note that $1 - e^{-kt} = \dfrac{kx}{6\sqrt{3}}$, which we will use on page 308. [2]

Vertical

In the y direction the equation is now more complicated.

$$\frac{dv_y}{dt} = -10 - kv_y$$

$$\frac{dt}{dv_y} = \frac{1}{-(10 + kv_y)}$$

$$-dt = \frac{dv_y}{10 + kv_y}$$

Integrating, $-t = \frac{1}{k} \ln (10 + kv_y) + D$

When $t = 0$, $v_y = 12 \sin 30° = 6$; hence, $D = -\frac{1}{k} \ln (10 + 6k)$

So, $-kt = \ln (10 + kv_y) - \ln (10 + 6k)$

$$e^{-kt} = \frac{10 + kv_y}{10 + 6k}$$

which gives

$$\dot{y} = v_y = \frac{1}{k}[(10 + 6k)e^{-kt} - 10]$$

Integrating, we get

$$y = \frac{1}{k}\left[\frac{10 + 6k}{-k} e^{-kt} - 10t \right] + E$$

When $t = 0$, $y = 0$ and so $E = \frac{10 + 6k}{k^2}$

$$\therefore y = \frac{1}{k}\left(\frac{10 + 6k}{-k} e^{-kt} - 10t \right) + \frac{10 + 6k}{k^2}$$

$$= \frac{10 + 6k}{k^2}(1 - e^{-kt}) - \frac{10}{k}t \qquad [3]$$

Path of projectile

The equation of motion is found by first making t the subject of [1]:

$$\frac{kx}{6\sqrt{3}} = 1 - e^{-kt}$$

$$e^{-kt} = 1 - \frac{kx}{6\sqrt{3}}$$

$$-kt = \ln\left(1 - \frac{kx}{6\sqrt{3}} \right)$$

$$t = -\frac{1}{k}\ln\left(1 - \frac{kx}{6\sqrt{3}}\right)$$

Substitute t and [2] into [3]: $y = \frac{10+6k}{k^2}\cdot\frac{kx}{6\sqrt{3}} - \frac{10}{k}\left[-\frac{1}{k}\ln\left(1 - \frac{kx}{6\sqrt{3}}\right)\right]$

$$= \frac{10+6k}{6k^2\sqrt{3}}kx + \frac{10}{k^2}\ln\left(1 - \frac{kx}{6\sqrt{3}}\right)$$

$$= \frac{5+3k}{3k\sqrt{3}}x + \frac{10}{k^2}\ln\left(1 - \frac{kx}{6\sqrt{3}}\right)$$

Maximum height

At the maximum height, $\dot{y} = 0$

$$0 = \frac{1}{k}\left[(10+6k)e^{-kt} - 10\right]$$

$$(10+6k)e^{-kt} = 10$$

$$e^{-kt} = \frac{10}{10+6k} \qquad [4]$$

$$e^{kt} = \frac{5+3k}{5} = 1 + 0.6k$$

$$kt = \ln(1 + 0.6k)$$

$$t = \frac{1}{k}\ln(1 + 0.6k)$$

Substituting into the equation for vertical displacement [3]:

$$y = \frac{10+6k}{k^2}\left(1 - e^{-kt}\right) - \frac{10}{k}t$$

$$= \frac{10+6k}{k^2}\left(1 - \frac{10}{10+6k}\right) - \frac{10}{k}\left[\frac{1}{k}\ln(1 + 0.6k)\right] \quad \text{from}[4]$$

$$= \frac{10+6k}{k^2}\left(\frac{10+6k-10}{10+6k}\right) - \frac{10}{k^2}\ln(1 + 0.6k)$$

$$= \frac{6k}{k^2} - \frac{10}{k^2}\ln(1 + 0.6k)$$

$$= \frac{6}{k} - \frac{10}{k^2}\ln(1 + 0.6k)$$

Note: If you know the value of k, which is the coefficient of drag, then the height can be calculated.

Using geometry software

If we use a slider and let $k = 1$ in this example, it is possible to identify the time to reach maximum height as $t = \ln 1.6 \approx 0.47$s. The maximum height is $y = 1.32$ m at $x \approx 3.89$ m.

As well, it is possible to show that the range ≈ 6.67 m.

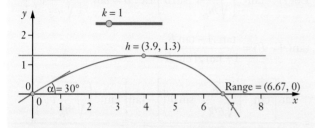

Quadratic drag model

In the case of a resistive force that grows with the square of the velocity (kv^2), we still separate the motion into horizontal and vertical components.

The components for acceleration in the horizontal and vertical can be found using Newton's laws of motion:

$m\ddot{x} = -k\dot{x}^2$ and $m\ddot{y} = -mg - k\dot{y}^2$

so we get $\ddot{x} = -\dfrac{k}{m}\dot{x}^2$ and $\ddot{y} = -g - \dfrac{k}{m}\dot{y}^2$

Horizontal

$\dfrac{dv_x}{dt} = -\dfrac{k}{m}v_x^2$

$\dfrac{dv_x}{v_x^2} = -\dfrac{k}{m}dt$

Integrating:

$-\dfrac{1}{v_x} = -\dfrac{k}{m}t + C$

When $t = 0$, $v_x = V\cos\theta$

$\therefore C = -\dfrac{1}{V\cos\theta}$

Hence:

$\dfrac{k}{m}t = \dfrac{1}{v_x} - \dfrac{1}{V\cos\theta}$

so $v_x = \dfrac{mV\cos\theta}{m + kV(\cos\theta)t}$

$\dot{x} = v_x = \dfrac{V\cos\theta}{1 + \dfrac{k}{m}V(\cos\theta)t}$

Integrating:

$x = \dfrac{m}{k}\ln\left(1 + \dfrac{k}{m}V(\cos\theta)t\right) + D$

When $t = 0$, $x = 0$

$\therefore D = 0$

Hence

$x = \dfrac{m}{k}\ln\left(1 + \dfrac{k}{m}V(\cos\theta)t\right)$

ISBN 9780170413435

Vertical

$$\frac{dv_y}{dt} = -g - \frac{k}{m}v_y^2$$

$$\frac{dv_y}{g + \frac{k}{m}v_y^2} = -dt$$

Integrating:

$$\sqrt{\frac{m}{gk}}\tan^{-1}\left(\sqrt{\frac{k}{mg}}v_y\right) = -t + E$$

When $t = 0$, $v_y = V\sin\theta$

$$\therefore E = \sqrt{\frac{m}{gk}}\tan^{-1}\left(\sqrt{\frac{k}{mg}}V\sin\theta\right)$$

Hence:

$$t = \sqrt{\frac{m}{gk}}\left\{\tan^{-1}\left(\sqrt{\frac{k}{mg}}V\sin\theta\right) - \tan^{-1}\left(\sqrt{\frac{k}{mg}}v_y\right)\right\}$$

Let $A = \tan^{-1}\left(\sqrt{\frac{k}{mg}}V\sin\theta\right)$ Let $B = \tan^{-1}\left(\sqrt{\frac{k}{mg}}v_y\right)$.

$$\tan(A - B) = \frac{\tan A - \tan B}{1 + \tan A \tan B}$$

$$= \frac{\tan\left[\tan^{-1}\left(\sqrt{\frac{k}{mg}}V\sin\theta\right)\right] - \tan\left[\tan^{-1}\left(\sqrt{\frac{k}{mg}}v_y\right)\right]}{1 + \tan\left[\tan^{-1}\left(\sqrt{\frac{k}{mg}}V\sin\theta\right)\right]\tan\left[\tan^{-1}\left(\sqrt{\frac{k}{mg}}v_y\right)\right]}$$

$$= \frac{\sqrt{\frac{k}{mg}}V\sin\theta - \sqrt{\frac{k}{mg}}v_y}{1 + \sqrt{\frac{k}{mg}}V\sin\theta\sqrt{\frac{k}{mg}}v_y}$$

$$= \frac{\sqrt{\frac{k}{mg}}\left(V\sin\theta - v_y\right)}{1 + \frac{k}{mg}Vv_y\sin\theta}$$

so $A - B = \tan^{-1}\left\{\dfrac{\sqrt{\frac{k}{mg}}\left(V\sin\theta - v_y\right)}{1 + \frac{k}{mg}Vv_y\sin\theta}\right\}$ and

$$t = \sqrt{\frac{m}{gk}}\tan^{-1}\left\{\frac{\sqrt{\frac{k}{mg}}\left(V\sin\theta - v_y\right)}{1 + \frac{k}{mg}Vv_y\sin\theta}\right\}$$

Rearranging (after considerable algebra):

$$\dot{y} = v_y = \frac{\sqrt{mgk}\,V\sin\theta - mg\,\tan\left(\sqrt{\frac{gk}{m}}\,t\right)}{Vk\sin\theta\,\tan\left(\sqrt{\frac{gk}{m}}\,t\right) + \sqrt{mgk}}$$

This last step is left to the student as an exercise.

It is not possible to integrate this function to find y, the vertical displacement.

MATHS IN FOCUS 12. Mathematics Extension 2

ISBN 9780170413435

Terminal velocity

In the case of a falling object, when the magnitude of the resistive force equals the object's weight, the object reaches its **terminal velocity**. The magnitude of the resistive force depends on the size and shape of the object and on the properties of the medium through which the object is moving.

A skydiver weighing 75 kg has a terminal velocity of 216 km h^{-1}, a golf ball 158 km h^{-1} and a raindrop 32 km h^{-1}.

EXAMPLE 19

A 5 kg cannonball is launched from the origin $(0, 0)$ with velocity of 400 m s^{-1} at an angle of 30° to the horizontal. A drag force of magnitude kv^2 is experienced with a proportion constant of $k = 0.0001$.

a Write an expression for the range of the cannonball and find how long it takes it to reach 1000 m.

b Find how long it takes the cannonball to reach its maximum height and its speed at that point. Use $g = 9.8$ m s^{-2}.

Solution

a $m\ddot{x} = -kv_x^2$

$\quad 5\ddot{x} = -0.0001v_x^2$

$\quad\quad \ddot{x} = -0.00002v_x^2$

$\quad \dfrac{dv_x}{dt} = -0.00002v_x^2$

$\quad \dfrac{dt}{dv_x} = -\dfrac{1}{0.00002v_x^2} = -\dfrac{50000}{v_x^2}$

Integrating:

$t = \dfrac{50000}{v_x} + C$

When $t = 0$, $v_x = V\cos\theta = 400\cos 30° = 400\left(\dfrac{\sqrt{3}}{2}\right) = 200\sqrt{3}$

$0 = \dfrac{50000}{200\sqrt{3}} + C$

$ = \dfrac{250}{\sqrt{3}} + C$

$$C = -\frac{250}{\sqrt{3}}$$

Hence $t = \dfrac{50\,000}{v_x} - \dfrac{250}{\sqrt{3}}$

$$t + \frac{250}{\sqrt{3}} = \frac{50\,000}{v_x}$$

$$v_x = \frac{50\,000}{t + \dfrac{250}{\sqrt{3}}}$$

$$v_x = \frac{50\,000\sqrt{3}}{t\sqrt{3} + 250}$$

Integrating:

$$x = 50\,000\sqrt{3}\left[\frac{1}{\sqrt{3}}\ln\left(t\sqrt{3} + 250\right)\right] + D$$

$$x = 50\,000\ln\left(t\sqrt{3} + 250\right) + D$$

When $t = 0$, $x = 0$

$0 = 50\,000\ln(0 + 250) + D$

$D = -50\,000\ln 250$

Hence,

$$x = 50\,000\ln\left(t\sqrt{3} + 250\right) - 50\,000\ln 250$$

$$x = 50\,000\ln\left(\frac{t\sqrt{3} + 250}{250}\right)$$

For $x = 1000$

$$1000 = 50\,000\ln\left(\frac{t\sqrt{3} + 250}{250}\right)$$

$$0.02 = \ln\left(\frac{t\sqrt{3} + 250}{250}\right)$$

$$e^{0.02} = \frac{t\sqrt{3} + 250}{250}$$

$$250e^{0.02} = t\sqrt{3} + 250$$

$$t\sqrt{3} = 250e^{0.02} - 250$$

$$t = \frac{250\left(e^{0.02} - 1\right)}{\sqrt{3}}$$

$$\approx 2.9 \text{ s}$$

ISBN 9780170413435

b   Maximum height is when $v_y = 0$

That is, when $\sqrt{mgk}\, V \sin\theta - mg \tan\left(\sqrt{\dfrac{gk}{m}}\, t\right) = 0$

$$\tan\left(\sqrt{\dfrac{gk}{m}}t\right) = \sqrt{mgk}\,(V\sin\theta)$$

$$\tan\left(\sqrt{\dfrac{gk}{m}}t\right) = \dfrac{\sqrt{mgk}}{mg}\,(V\sin\theta)$$

$$\tan\left(\sqrt{\dfrac{9.8\times0.001}{5}}t\right) = \dfrac{\sqrt{5\times9.8\times0.0001}}{5\times9.8}\,(400\sin30°)$$

$$\tan(0.014t) = \dfrac{\sqrt{5\times9.8\times0.0001}}{5\times9.8}(200)$$

$\tan(0.014t) = 0.2857...$

$0.014t = 0.2782...$

$t = 19.8785...$

$t \approx 19.9$ s

Now, for speed we know $S = \sqrt{v_x^2 + v_y^2}$, but $v_y = 0$.

$\therefore S = v_x$ at $t = 19.9$ s

and we know from the quadratic drag model that:

$$v_x = \dfrac{V\cos\theta}{1 + \dfrac{k}{m}V(\cos\theta)t}$$

$$= \dfrac{400\times\cos30°}{1 + \dfrac{0.0001}{5}\times400\times\cos30°\times19.9}$$

$$= 304.4370...$$

$$\approx 304$$

Hence the cannonball has speed 304 m s^{-1} at the maximum height of the trajectory.

Dimples on golf balls

Why do golf balls have dimples? Because their aerodynamic design causes them to travel higher and further than balls with smooth surfaces. The dimples make the air flow more smoothly around the ball, resulting in less drag. The dimples also help the air to move faster *above* the ball, which decreases the pressure there and, similar to the airflow over the wings of a plane, causes the ball to lift. Together, less drag and more lift cause dimpled golf balls to travel further.

Shutterstock.com/cloki

Exercise 7.07 Resisted projectile motion

1   A projectile of mass 5 kg is fired from the origin with velocity of 100 m s^{-1} at an angle of elevation of 15°. In addition to gravity, assume that air resistance provides a force that is proportional to the velocity and that opposes the motion.

   a   Show that the vertical component of acceleration of the projectile is $-10 - \dfrac{k}{5}v$.

   b   Given the constant of proportionality is 2.5, find the terminal velocity.

   c   What is the maximum height reached by the projectile?

2   A bullet of mass 7.5 g is fired at 380 m s^{-1} at an angle of 5° to the horizontal. The resistance due to drag is mkv, where v is the velocity, $k = 0.3$ is the coefficient of drag and the acceleration due to gravity is 9.8 m s^{-2}.

   a   Find correct to 2 significant figures the horizontal displacement of the bullet at 2 seconds.

   b   How long does it take to double the distance travelled in part **a**?

   c   How long does it take to halve the distance travelled in part **a**?

3   A baseball of mass 145 g is thrown at an angle of 10° to the horizontal at a speed of 30 m s^{-1} and reaches the batter 18 m away in exactly 1 s. If the resistive force is proportional to the velocity of the baseball, show that the coefficient of drag is 0.158.

4   A cricket ball with mass 160 g is hit at 50 m s^{-1} at an angle of 5° to the horizontal. The drag force is proportional to the speed and the coefficient of drag is 0.09.

   a   How long does it take the ball to reach its maximum height? Use $g = 9.8$ m s^{-2}

   b   What is the horizontal distance travelled when it is at its maximum point?

   c   What is the speed of the ball after 1 second?

5 A golfer drives a 46 g golf ball a distance of 155 m. The drag force is proportional to the square of the velocity (v^2) and the terminal velocity is 44 m s^{-1}.

   a Calculate the drag coefficient (k) for the golf ball. Use $g = 9.8$ m s^{-2}.

   b If the initial speed is 60 m s^{-1} and the time of flight is 4.53 seconds, calculate the angle of projection.

7. TEST YOURSELF

1 A particle moves in a straight line such that its velocity v cm s^{-1} when it is x cm from the origin is given by $v = 2 - e^{-2x}$. Find the acceleration of the particle at the origin.

2 A particle is initially at the origin where it is given an initial velocity of 10 m s^{-1}. When x metres from the origin, its acceleration is $a = -40e^{-8x}$ m s^{-2}. Determine its velocity $v(x)$.

3 A particle is moving in simple harmonic motion with acceleration given by $\ddot{x} = -36x$, where x metres is the displacement of the particle from the centre of motion. Initially, the particle is at the origin, moving to the left at 5 m s^{-1}. Find the displacement as a function of time.

4 When x metres from an origin, the velocity of a particle is v m s^{-1}, where

$$v^2 = 40 - 8x - 4x^2$$

  a Prove that its acceleration is $a = -4(x + 1)$.

  b Find the positions where the particle is at rest.

  c What will be the greatest speed?

  d Write v^2 in the form $v^2 = n^2[A^2 - (x - c)^2]$

5 An object is projected vertically upwards with initial velocity of 25 m s^{-1}. Find the maximum height reached by the object and the total time taken to reach the ground again, assuming no air resistance. Assume $g = 9.8$ m s^{-2} and answer correct to one decimal place.

6 A cricket ball is hit for a six, and just clears the 1 m high fence on the boundary at 104.2 m away. If the ball leaves the bat at 75.4 m s^{-1}, at what angles to the horizontal must it leave the bat? Use $g = 9.8$ m s^{-2} and answer correct to 1 decimal place.

7 An object of weight 20 N, in rough contact with a plane inclined at θ to the horizontal, is just about to slide. If the coefficient of friction between the plane and the object is $\dfrac{1}{\sqrt{3}}$, find the angle θ.

8 A 50 N force is applied to a 10 kg box at an angle of 30° to the horizontal. If the coefficient of friction is 0.3, find the acceleration of the box in terms of g.

9 Two particles of masses 10 kg and 5 kg are connected by a light inelastic string. The first mass is placed on a rough horizontal table with the string passing over a smooth pulley on the edge of the table and the second mass is hanging freely. The coefficient of dynamic friction between the first mass and the table is 0.6. Assuming $g = 9.8$ m s^{-2}, find, correct to one decimal place: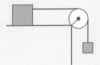

  a the acceleration of the system

  b the tension in the string

MATHS IN FOCUS 12. Mathematics Extension 2   ISBN 9780170413435

10 A car of mass m initially at the origin with speed v_0 moves horizontally against a resistance proportional to v^3. Express its displacement in terms of velocity and its velocity in terms of displacement.

11 A particle of unit mass is moving in a straight line with an initial velocity of 30 m s^{-1} in a medium which causes resistance equivalent to $\dfrac{v}{40}$ m s^{-1} when the velocity is v. Find the velocity and distance travelled in 40 s.

12 A skydiver falling through the air at velocity v m s^{-1} experiences a resistance of $\dfrac{1}{6}v$ newtons.

   a   Find her terminal velocity in terms of v.

   b   If the effect of a parachute is to increase the resistance to $2v$, find an expression for the approximate speed at which she hits the ground.

13 A 1 kg stone is thrown vertically upwards from the ground with a speed of 13 m s^{-1}. Assuming that the air resistance at velocity v is $\dfrac{v}{10}$, find both the time taken to reach maximum height and the maximum height reached by the stone. Use $g = 9.8$ m s^{-2}.

14 A ball is thrown at an angle of 30° to the horizontal at 8 m s^{-1} and experiences air resistance proportional to the velocity in both the x and y directions. The terminal velocity is $2g$ m s^{-1}. Show that the equation of motion is $y = \left(\dfrac{g+2}{2\sqrt{3}}\right)x + 4g\,\ln\left(1 - \dfrac{x}{8\sqrt{3}}\right)$.

15 A particle is projected with speed V at an angle α to the horizontal, up a plane that is inclined at an angle of θ to the horizontal.

   a   Show that the range R of the particle along the plane is given by

$$R = \frac{2V^2 \cos\alpha\,\sin(\alpha - \theta)}{g\cos^2\theta} \text{ and that the maximum range is } R = \frac{V^2}{g(1 + \sin\theta)}.$$

   b   Hence, show that the particle can never reach points outside the parabola given by $x^2 + \dfrac{2V^2}{g}y - \dfrac{V^4}{g^2} = 0$.

Practice set 2

In Questions **1** to **12**, select the correct answer **A**, **B**, **C** or **D**.

1 In the proof by mathematical induction that $1 + 2 + 3 + \ldots + n = \dfrac{n(n+1)}{2}$, for all positive integers n, the inductive hypothesis would assume that:

A $n = 1$

B $1 = \dfrac{1(1+1)}{2}$

C $1 + 2 + 3 + \ldots + k = \dfrac{k(k+1)}{2}$

D $1 + 2 + 3 + \ldots + n = \dfrac{n(n+1)}{2}$

2 For all positive integers n, $\displaystyle\sum_{k=1}^{n} k^3 = $:

A $\dfrac{n^2(n+1)^2}{4}$

B $\dfrac{n^2(n+1)^2}{2}$

C $\dfrac{n^3(n+1)^3}{2}$

D $\dfrac{n(n+1)}{2}$

3 Let $P(n)$ be the statement that $1^2 + 2^2 + 3^2 + \ldots + n^2 = \dfrac{n(n+1)(2n+1)}{6}$, $n > 0$. What is the statement $P(1)$?

A $1^2 + 2^2 + \ldots + n^2 = \dfrac{n(n+1)(2n+1)}{6}$

B $n = 1$

C $0^2 = \dfrac{0(0+1)(0+1)}{6}$

D $1^2 = \dfrac{1(1+1)(2+1)}{6}$

4 If $10^n + 3(4^{n+2}) + k$ is divisible by 9 for all $n > 0$, then the least positive integer value of k will be:

A 1

B 3

C 5

D 7

5 What would be a suitable substitution to determine $\displaystyle\int \dfrac{2}{x\sqrt{x^2 - 4}}\, dx$?

A $x = \cos \theta$

B $x = 2 \sin \theta$

C $u^2 = x^2 - 4$

D $x = 2 \sec \theta$

6 Find $\displaystyle\int \dfrac{2}{x^2 + 4x + 13}\, dx$.

A $\dfrac{1}{3} \tan^{-1}\left(\dfrac{x+2}{3}\right) + C$

B $\dfrac{2}{3} \tan^{-1}\left(\dfrac{x+2}{3}\right) + C$

C $\dfrac{1}{9} \tan^{-1}\left(\dfrac{x+2}{9}\right) + C$

D $\dfrac{2}{9} \tan^{-1}\left(\dfrac{x+2}{9}\right) + C$

7 Find $\int x \log_e x \, dx$.

A $\quad \dfrac{x^2}{2} \log_e x - \dfrac{x^2}{4} + C$

B $\quad \dfrac{x^2}{2} \log_e x - \dfrac{x}{2} + C$

C $\quad x \log_e x - \dfrac{x^2}{4} + C$

D $\quad x \log_e x - \dfrac{x}{2} + C$

8 Find $\int \dfrac{x}{(x-1)(x+4)} \, dx$.

A $\quad \dfrac{1}{4} \log_e |x+1| + \dfrac{5}{4} \log_e |x+4| + C$

B $\quad \dfrac{1}{5} \log_e |x+1| + \dfrac{4}{5} \log_e |x+4| + C$

C $\quad \dfrac{1}{4} \log_e |x-1| + \dfrac{5}{4} \log_e |x+4| + C$

D $\quad \dfrac{1}{5} \log_e |x-1| + \dfrac{4}{5} \log_e |x+4| + C$

9 A particle of mass m is moving in a straight line under the action of an applied force $F = \dfrac{m}{x^3} (8 + 10x)$. What is the equation for its velocity at any position, if the particle starts from rest at $x = 1$?

A $\quad v = \pm \dfrac{1}{x} \sqrt{7x^2 - 5x - 2}$

B $\quad v = \pm x \sqrt{7x^2 - 5x - 2}$

C $\quad v = \pm \dfrac{2}{x} \sqrt{7x^2 - 5x - 2}$

D $\quad v = \pm 2x \sqrt{7x^2 - 5x - 2}$

10 A particle moves in simple harmonic motion after starting at the origin with amplitude 4 m and period $\dfrac{\pi}{2}$ s. Which function could describe its motion?

A $\quad x = 4 \sin t$ **B** $\quad x = 4 \sin 4t$ **C** $\quad x = 4 \cos t$ **D** $\quad x = 4 \cos 4t$

11 A particle of unit mass falls from rest and the resistance is proportional to v^2, where v is its speed and k is a positive constant.
Which is the correct formula for v^2 where x is the distance fallen?

A $\quad v^2 = \dfrac{g}{k}(1 - e^{-2kx})$

B $\quad v^2 = \dfrac{g}{k}(1 + e^{-2x})$

C $\quad v^2 = \dfrac{g}{k}(1 - e^{2x})$

D $\quad v^2 = \dfrac{g}{k}(1 + e^{2x})$

12 A stone is thrown vertically upwards with a speed of 21 m s^{-1} from the edge of a cliff 20 m above the water. How long does the stone remain in the air before it hits the water?

A   4.2 s   **B**   5 s   **C**   8.4 s   **D**   10 s

13 Use mathematical induction to prove that 2 is a factor of $(n + 1)(n + 2)$, for all positive integers n.

14 Prove that $-2 - 4 - 6 - \ldots - 2n = -n(n + 1)$ for any positive integer n.

15 Show by using mathematical induction that $f^{(n)}(x) = a^n \sin\left(ax + \dfrac{n\pi}{2}\right)$ for all integers $n > 0$ if $f(x) = \sin ax$.

16 Prove that $\displaystyle\sum_{r=1}^{n} x^{r-1} = \dfrac{1 - x^n}{1 - x}$, for all integers $n \geq 1$.

17 Prove that $2^{3n} - 3^n$ is divisible by 5, for all integers $n \geq 1$.

18 Prove that $4^n > 3n + 7$ for all integers $n > 1$.

19 Prove by mathematical induction that $\dfrac{d}{dx}(x^n) = nx^{n-1}$ for all integers $n \geq 1$.

20 A sequence is given by the recurrence relation $a_1 = 3$, $a_n = a_{n-1} + 5$ for $n > 1$. Prove, using mathematical induction, that the general formula for the sequence is $a_n = 5n - 2$, $n > 1$.

21 Given $u_1 = 8$, $u_2 = 20$ and $u_n = 4u_{n-1} - 4u_{n-2}$ for $n \geq 3$, show by mathematical induction that $u_n = (n + 3)2^n$ for $n \geq 1$.

22 Simplify $\cos(x - y) - \cos(x + y)$. Hence, prove by induction that for all $n > 0$ that
$$\sin\theta + \sin 3\theta + \ldots + \sin(2n - 1)\theta = \dfrac{\sin^2 n\theta}{\sin\theta}$$

23 Find $\displaystyle\int \dfrac{dx}{\sqrt{4 + 4x - x^2}}$.

24 Evaluate $\displaystyle\int_0^{\frac{\pi}{2}} \dfrac{1}{\cos x + 1}\, dx$, using an appropriate substitution.

25 Find $\displaystyle\int \dfrac{dx}{x^2 - 2x + 10}$.

26 Find $\displaystyle\int \dfrac{1}{x(x - 2)}\, dx$.

27 Using partial fractions, find $\displaystyle\int \dfrac{2x - 1}{x^2 + 3x + 2}\, dx$.

28 Find $\displaystyle\int \dfrac{x}{(x + 2)(x + 3)}\, dx$.

29 Use integration by parts to evaluate $\displaystyle\int_1^e \dfrac{\ln x}{x}\, dx$.

30 Evaluate $\displaystyle\int_0^{\ln 3} x^2 e^x\, dx$.

31 Find $\displaystyle\int e^x \sin x\, dx$.

32 Find $\displaystyle\int \sin^{-1} x\, dx$.

33 A particle is moving in a straight line from a stationary position at the origin. Its velocity is v when its displacement from the origin is x. If the acceleration of the particle is given by $a = \dfrac{1}{(x + 3)^2}$, find v in terms of x.

ISBN 9780170413435

34 The depth of water in a harbour is assumed to rise and fall with time in simple harmonic motion. On a certain day the low tide had a height of 12 m at 12:30 p.m. and the following high tide had a height of 18 m at 6:30 p.m.
If a ship requires a depth of 16 m of water before it can leave the harbour, find the earliest time after 12 p.m. that the ship can leave the harbour.

35 A long jumper leaves the ground at an angle of 15° to the horizontal and at a speed of 12 m s^{-1}. How far does she jump on this occasion? Assume $g = 9.8$ m s^{-2}.

36 A projectile is fired at a target (T) so that the projectile leaves the gun at the same time the target is dropped from rest. Show that if the gun is initially aimed at the stationary target (and has sufficient range), the projectile hits the target.

37 A ball is thrown vertically upward and is caught 2 s later by the thrower.
Find the initial velocity of the ball and the maximum height the ball reaches.

38 A mass is oscillating at the end of a spring with a velocity given by $v^2 = 256 - 64x^2$ where x cm is the displacement from the centre of motion. Find:

   a   the acceleration of the mass as a function of x

   b   the maximum speed of the mass.

39 An object of mass m kilograms is dropped from the top of a cliff 40 m above a body of water and travels x metres. In the air, the resistance to its motion has magnitude $\dfrac{1}{10}mv$ when the object has speed v m s^{-1}. After the object enters the water, the resistance has magnitude $\dfrac{1}{10}mv^2$. Use $g = 10$ m s^{-2}.

   a   Write an expression for the acceleration a ms^{-2} in terms of v before the object enters the water.

   b   Show that $\dfrac{dv}{dx} = \dfrac{100 - v}{10v}$, and show that the speed of the object as it enters the water satisfies $\dfrac{v}{100} + \ln\left(1 - \dfrac{v}{100}\right) + 0.04 = 0$.

   c   Write an expression for a after the object enters the water.

   d   Given that the object slows on entry, find its terminal velocity in the water.

40 An object of weight 40 N, in rough contact with a plane inclined at θ to the horizontal, is just about to slide. If the coefficient of friction between the plane and the object is $\dfrac{1}{\sqrt{3}}$, find the angle θ.

41 A force of 820 N is applied to a 2 tonne vehicle at an angle of 10° to the horizontal. If the coefficient of friction is 0.04 and $g = 10$ m s^{-2}, find the acceleration of the vehicle.

42 A ball is kicked at 30° to the horizontal at 15 m s^{-1} and experiences air resistance proportional to its velocity in both the x and y directions. Given that its terminal velocity is 11 m s^{-1}, find the time for it to reach its maximum height. Use $g = 10$ m s^{-2}.

ANSWERS

Answers are based on full calculator values and only rounded at the end, even when different parts of a question require rounding. This gives more accurate answers. Answers based on reading graphs may not be accurate.

Chapter 1

Exercise 1.01

1 a $2i$ **b** $i\sqrt{7}$ **c** $\dfrac{i}{3}$ **d** $2i\sqrt{3}$

   e $\dfrac{i\sqrt{6}}{5}$ **f** $4i\sqrt{2}$ **g** $-i$ **h** i

   i $-i$ **j** $i-1$ **k** $-i$ **l** i

2 a $x=\pm 2i$ **b** $x=\pm 3i$

   c $z=\pm\dfrac{i}{6}$ **d** $z=\pm 2i\sqrt{5}$

3 a $x=-1\pm i\sqrt{2}$ **b** $x=\dfrac{1}{2}\pm\dfrac{i\sqrt{23}}{2}$

   c $z=\dfrac{-3\pm i\sqrt{3}}{2}$ **d** $z=\dfrac{5\pm i\sqrt{83}}{6}$

4 a $x=1\pm i\sqrt{2}$ **b** $x=2\pm i\sqrt{7}$

   c $z=-4\pm 2i$ **d** $z=1\pm i\sqrt{3}$

5 a $x=1\pm i$ **b** $v=3\pm i\sqrt{3}$

   c $w=-2\pm i\sqrt{6}$ **d** $z=-1\pm i\sqrt{6}$

   e $z=-\dfrac{1}{2}\pm\dfrac{i\sqrt{3}}{2}$ **f** $z=\dfrac{3}{2}\pm\dfrac{i\sqrt{7}}{2}$

6 a $\text{Re}(z)=\sqrt{3}$, $\text{Im}(z)=1$

   b $\text{Re}(z)=\dfrac{5}{2}$, $\text{Im}(z)=-\dfrac{\sqrt{2}}{2}$

   c $\text{Re}(z)=-3$, $\text{Im}(z)=6$

   d $\text{Re}(z)=x+3$, $\text{Im}(z)=-y+2$

   e $\text{Re}(z)=\dfrac{a}{a^2+4b^2}$, $\text{Im}(z)=\dfrac{2b}{a^2+4b^2}$

   f $\text{Re}(z)=\dfrac{x-4-6y}{x^2+y^2}$, $\text{Im}(z)=\dfrac{-1+x+y}{x^2+y^2}$

7 a $\sqrt{3}-i$ **b** $\dfrac{5+i\sqrt{2}}{2}$

   c $-6i-3$ **d** $x+3+iy-2i$

   e $\dfrac{a-2ib}{a^2+4b^2}$ **f** $\dfrac{x-4-6y+i-ix-yi}{x^2+y^2}$

8 a 13 **b** 3 **c** -41

   d 1 **e** $\dfrac{17}{9}$ **f** $\dfrac{5}{32}$

9 a 61 **b** 4 **c** 25

   d 1 **e** $\dfrac{1}{17}$ **f** 8

   g $4a^2+9b^2$ **h** $2x^2+2y^2$

10 Proof: See Worked solutions

11 Proof: See Worked solutions

12 a $x=2, y=-8$ **b** $x=3. y=-4$

   c $x=5, y=2$ **d** $x=4, y=2$

13 $x+y=6$

14 a $-4+4i$ **b** $4-5i$ **c** $-77-36i$

   d $23-7i$ **e** 21 **f** $-30+45i$

   g $(4+\sqrt{3})+i(\sqrt{2}-2\sqrt{6})$ **h** $-4xyi$

15 Proof: See Worked solutions

16 a Constant multiples of each answer below is also a solution.

     i $z^2-4z+5=0$ **ii** $z^2-2\sqrt{3}\,z+28=0$

     iii $z^2-z+1=0$ **iv** $z^2+8z+21=0$

   b A quadratic equation with complex conjugate roots will have *real* coefficients.

17 a $\dfrac{2+i}{5}$, $\text{Im}(z)=\dfrac{1}{5}$ **b** $\dfrac{-1+3i}{5}$, $\text{Im}(z)=\dfrac{3}{5}$

   c $\dfrac{-13-41i}{25}$, $\text{Im}(z)=-\dfrac{41}{25}$

   d $\dfrac{1-2\sqrt{6}\,i}{5}$, $\text{Im}(z)=-\dfrac{2\sqrt{6}}{5}$

18 Proof: See Worked solutions

19 a $\dfrac{-3-4i}{25}$ **b** $\dfrac{-i}{3}$

20 a $z^2-(5+8i)z+(15+25i)=0$, $a=1$, $b=-5-8i$, $c=15+25i$

   b A quadratic equation with complex non-conjugate roots will have some coefficients that are not *real*.

21 a $x = -\dfrac{6}{25}, y = -\dfrac{17}{25}$ **b** $x = -\dfrac{3}{26}, y = \dfrac{11}{26}$

22 a $-13 + 11i$ **b** $8 - 3i$

c $\dfrac{-17 + i}{10}$ **d** $13 - 14i$

Exercise 1.02

1 Taking $a > 0$:

a $2 + i$ **b** $3 - 2i$

c $3 + i$ **d** $\sqrt{2} + i\sqrt{2}$

2 a $4 - i, -4 + i$ **b** $1 - 2i, -1 + 2i$

c $5 - 2i, -5 + 2i$ **d** $1 + 5i, -1 - 5i$

e $\dfrac{3 - 3i}{\sqrt{2}}, \dfrac{-3 + 3i}{\sqrt{2}}$ **f** $\dfrac{-1 - i}{\sqrt{2}}, \dfrac{1 + i}{\sqrt{2}}$

3 a $5 + 4i, -5 - 4i$ **b** $3 + 4i, -3 - 4i$

c $4 - 2i, -4 + 2i$

4 a $1 + 2i$ **b** $1 - i, 2 + i$

5 a $3 + 2i, -1 + i$ **b** $2i, 3$

c $-2i, i$

Exercise 1.03

1 a

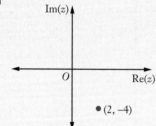

b

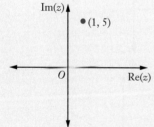

c

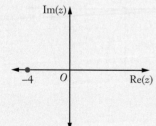

d

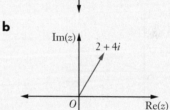

2 $A = -3 + i, B = 4 + 2i, C = 5 - 3i, D = -4 - 5i,$
$E = 2, F = -3i$

3 a

b

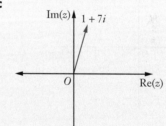

c

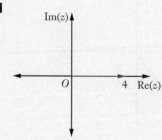

d

e

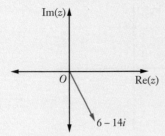

f

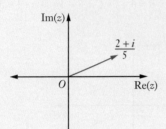

g

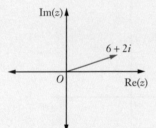

4 a

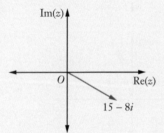

b

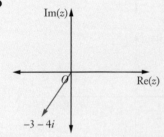

c

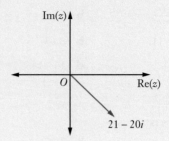

d

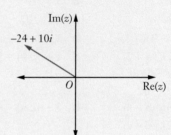

e

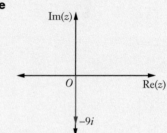

f

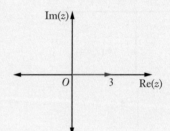

5 a

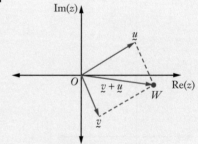

b

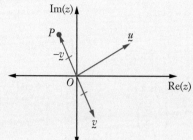

c

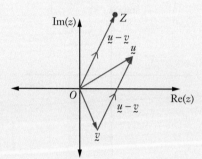

6 a

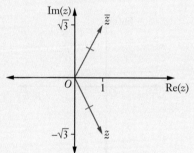

b

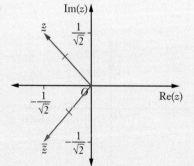

c

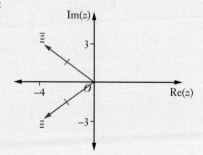

7 a $z = 4 + i, v = -2 + 2i, k = -5 - 2i, w = 2 - 3i$

b $\overline{z} = 4 - i, \overline{v} = -2 - 2i, \overline{k} = -5 + 2i, \overline{w} = 2 + 3i$

c

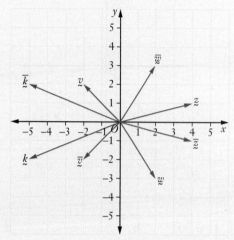

8 a

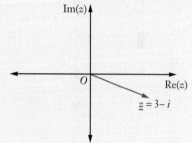

b

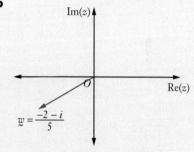

c

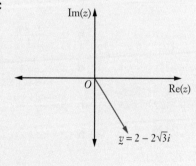

ISBN 9780170413435

9 a $w = -3 + 2i, v = -2 - 3i, u = 3 - 2i$

b

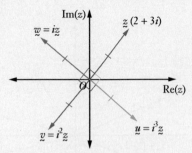

c Multiplying a complex number by i rotates the vector by 90° anticlockwise.

Exercise 1.04

1 a $\sqrt{2}$ **b** $2\sqrt{5}$ **c** $\sqrt{53}$ **d** $\sqrt{5}$

   e $\dfrac{\sqrt{37}}{7}$ **f** $3\sqrt{3}$ **g** 1

2 a $\dfrac{\pi}{4}$ **b** $\dfrac{\pi}{6}$ **c** $-\dfrac{\pi}{4}$ **d** $\dfrac{5\pi}{6}$

   e $-\dfrac{3\pi}{4}$ **f** 0 **g** $\dfrac{\pi}{2}$

3 a $\sqrt{2}\,\mathrm{cis}\left(-\dfrac{\pi}{4}\right)$ **b** $2\,\mathrm{cis}\dfrac{2\pi}{3}$

   c $\dfrac{2\sqrt{2}}{3}\,\mathrm{cis}\left(-\dfrac{3\pi}{4}\right)$ **d** $\mathrm{cis}\dfrac{\pi}{3}$

   e $\dfrac{2}{7}\,\mathrm{cis}\dfrac{\pi}{4}$ **f** $4\,\mathrm{cis}\left(-\dfrac{5\pi}{6}\right)$

   g $\sqrt{6}\,\mathrm{cis}\,\pi$

4 a $z = 1 + i\sqrt{3}$ **b** $z = \dfrac{\sqrt{3}}{4} + \dfrac{i}{4}$

   c $z = \dfrac{3 + 3i}{\sqrt{2}}$ **d** $z = \dfrac{1}{2} - \dfrac{i\sqrt{3}}{2}$

   e $z = \dfrac{-\sqrt{2} - i\sqrt{6}}{2}$ **f** $z = -\sqrt{3} - i$

5 a $z = \dfrac{1}{\sqrt{6}} - \dfrac{i}{\sqrt{6}}$ **b** $z = \dfrac{-3 - i\sqrt{3}}{2}$

   c $z = -2i$ **d** $z = -\dfrac{1}{\sqrt{2}} + \dfrac{i}{\sqrt{2}}$

   e $z = -3\sqrt{2}$ **f** $z = \dfrac{-\sqrt{3}}{2} + \dfrac{i}{2}$

   g $z = \dfrac{3 + 3i}{2}$ **h** $z = -\sqrt{2} - i\sqrt{2}$

6 a $z = 3\,\mathrm{cis}\dfrac{\pi}{3}$ **b** $w = 5\,\mathrm{cis}\left(-\dfrac{\pi}{4}\right)$

   c $u = \sqrt{3}\,\mathrm{cis}\dfrac{5\pi}{6}$ **d** $v = 2\,\mathrm{cis}\left(-\dfrac{3\pi}{5}\right)$

   e $z = 3\,\mathrm{cis}\,\pi$ **f** $w = 6\,\mathrm{cis}\left(-\dfrac{\pi}{2}\right)$

7 a $2\,\mathrm{cis}\left(-\dfrac{\pi}{2}\right)$

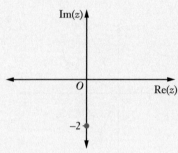

b $\sqrt{3}\,\mathrm{cis}\,\pi$

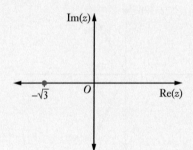

c $\dfrac{1}{3}\,\mathrm{cis}\dfrac{\pi}{2}$

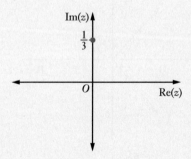

d $\frac{1}{2} \text{cis}\left(-\frac{\pi}{6}\right)$

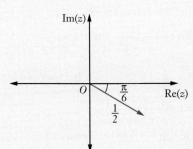

e $\text{cis}\left(-\frac{\pi}{4}\right)$

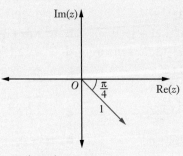

f $4 \text{cis}\left(-\frac{2\pi}{3}\right)$

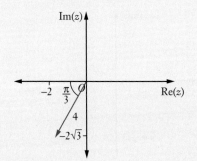

g $2 \text{cis } 0$

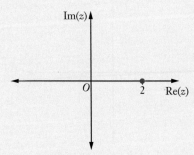

8 a $\sqrt{5} \text{ cis } 0.464$ **b** $\sqrt{74} \text{ cis } 2.19$

c $\frac{\sqrt{17}}{5} \text{ cis } (-0.245)$ **d** $4 \text{cis}\left(-\frac{5\pi}{6}\right)$

e $2 \text{cis } \frac{\pi}{4}$ **f** $\sqrt{15} \text{ cis } (-2.19)$

g $\text{cis}\left(-\frac{\pi}{3}\right)$ **h** $\sqrt{2} \text{ cis } \frac{\pi}{6}$

i $\frac{1}{2} \text{cis}\left(-\frac{\pi}{6}\right)$ **j** $\text{cis}\left(-\frac{3\pi}{4}\right)$

Exercise 1.05

1, 2 Proof: See Worked solutions

3 $z_1 = \sqrt{2} \text{ cis}\left(-\frac{\pi}{4}\right), z_2 = 2 \text{ cis } \frac{\pi}{6}$

   a $-\frac{\pi}{12}$ **b** $-\frac{5\pi}{12}$

   c $\frac{5\pi}{6}$ **d** $\frac{\pi}{2}$

4 a $2\sqrt{3} \text{ cis } \frac{8\pi}{15}$ **b** $\frac{\sqrt{3}}{2} \text{ cis } \frac{2\pi}{15}$

   c $\frac{1}{2} \text{cis}\left(-\frac{\pi}{5}\right)$ **d** $9\sqrt{3} \text{ cis}\left(-\frac{\pi}{3}\right)$

5 a $\text{cis } 3\alpha$ **b** $\text{cis } (-3\beta - 2\lambda)$ **c** $\text{cis } 4$

6–8 Proof: See Worked solutions

9 Yes, except undefined for $z = 0$.

10 a $2 \cos \theta$ **b** $2 \cos 2\theta$ **c** $2 \cos 3\theta$

   d $2i \sin 2\theta$ **e** $2i \sin 3\theta$

11 a $5\sqrt{26}$ **b** 8 **c** $\frac{1}{\sqrt{2}}$

   d $\frac{1}{5}$ **e** $\frac{1}{8}$

12 a $-\frac{\pi}{12}$ **b** $-\frac{11\pi}{12}$ **c** $\frac{3\pi}{4}$

   d $-\frac{5\pi}{12}$ **e** $\frac{\pi}{2}$

13 a $z = \frac{3\sqrt{3} + 3i}{2}, w = \sqrt{2} + i\sqrt{2}$

   b $\frac{w}{z} = \frac{2}{3} \text{ cis } \frac{\pi}{12}, \frac{w}{z} = \frac{\sqrt{6} + \sqrt{2} - i\sqrt{2} + i\sqrt{6}}{6}$

   c i $\cos \frac{\pi}{12} = \frac{\sqrt{6} + \sqrt{2}}{4}$ **ii** $\sin \frac{\pi}{12} = \frac{\sqrt{6} - \sqrt{2}}{4}$

14 Proof: See Worked solutions

Exercise 1.06

1 a i cis 0   **ii** 1

    b i $\sqrt{2}\operatorname{cis}\left(-\dfrac{\pi}{3}\right)$ **ii** $\dfrac{\sqrt{2}-i\sqrt{6}}{2}$

    c i 5 cis 3   **ii** $-4.95+0.706i$

    d i $\operatorname{cis}\left(-\dfrac{\pi}{2}\right)$ **ii** $-i$

2 a $\sqrt{3}\,e^{\frac{3i\pi}{4}}$ **b** $2e^{-\frac{i\pi}{3}}$ **c** $\dfrac{1}{2}e^{-\frac{i\pi}{5}}$

    d $\sqrt{3}e^{-\frac{5i\pi}{6}}$ **e** $6e^{i}$ **f** $4\sqrt{2}e^{-\frac{i\pi}{4}}$

    g $2e^{\frac{5i\pi}{6}}$ **h** $e^{\frac{2i\pi}{3}}$ **i** $e^{\frac{i\pi}{2}}$

    j $\dfrac{1}{2}e^{2i\pi}$ or $\dfrac{1}{2}e^{0i}$

Exercise 1.07

1 a $e^{i\pi}$ or -1 **b** $3\sqrt{2}\,e^{\frac{7i\pi}{12}}$ **c** $e^{\frac{3i\pi}{8}}$

    d $\sqrt{5}\,e^{\frac{5i\pi}{6}}$

2 a $e^{i(\theta_1+\theta_2)}$ **b** $e^{i(\theta_1-\theta_2)}$ **c** $\dfrac{1}{4}e^{\frac{i\pi}{2}}$

    d $\dfrac{1}{\sqrt{2}}e^{-\frac{2i\pi}{5}}$ **e** $\sqrt{5}\,e^{-i(2\alpha+5\lambda)}$ **f** $4\sqrt{2}\,e^{\frac{5i\pi}{12}}$

    g $e^{\frac{i\pi}{3}}$ **h** $\dfrac{4}{5}e^{\frac{i\pi}{4}}$

3 Proof: See Worked solutions

4 a $e^{-i\theta}=\cos(-\theta)+i\sin(-\theta)=\cos\theta-i\sin\theta$

    b Proof: See Worked solutions

Test yourself 1

1 a $5i$ **b** $3i\sqrt{2}$ **c** $\dfrac{2i\sqrt{2}}{3}$

2 a i **b** $2i\sqrt{10}$ **c** -1

    d $-i$ **e** $-1\pm2i$

3 a $x=\pm7i$ **b** $x=-3\pm2i$

4 a $x=2\pm i\sqrt{5}$ **b** $x=-3\pm i\sqrt{6}$

    c $x=\dfrac{-3\pm3i\sqrt{7}}{4}$

5 a $x=1\pm i$ **b** $x=-4\pm2i$

    c $x=\dfrac{1\pm i\sqrt{11}}{2}$

6 a $x=-2\pm2i$ **b** $x=4\pm3i$

    c $x=-5\pm4i$

7 a $\operatorname{Re}(z)=\dfrac{\sqrt{3}}{4},\operatorname{Im}(z)=-\dfrac{1}{2}$

    b $\operatorname{Re}(z)=2,\operatorname{Im}(z)=-9$

    c $\operatorname{Re}(z)=\dfrac{5x}{x^2+y^2},\operatorname{Im}(z)=\dfrac{-2y}{x^2+y^2}$

8 a $\bar{z}=-6-11i$ **b** $\bar{w}=\dfrac{3+i\sqrt{2}}{2}$

    c $\bar{u}=\dfrac{-a-7b-2i-ai}{a^2+b^2}$

9 Proof: See Worked solutions

10 a $x=3,y=-2$ **b** $x=-1,y=5$

11 $\operatorname{Im}(V)=0,\therefore\ 3x+2y+7=0$

12 a $15+4i$ **b** $-4+33i$

    c $31-3i$ **d** $-46-20i$

13 a $x^2-2x+3=0$ **b** $x^2+6x+34=0$

    c $x^2-2\sqrt{7}x+16=0$ **d** $x^2+x+1=0$

14 a $\dfrac{1+2i}{5}$ **b** $-i\sqrt{2}$

    c $\dfrac{4\sqrt{3}-5-i}{26}$ **d** $\dfrac{-1+4i\sqrt{3}}{7}$

15 a $\pm(2-i)$ **b** $\pm(3+i)$

    c $\pm(4-i)$ **d** $\pm\left(\dfrac{5+i}{\sqrt{2}}\right)$

16 a

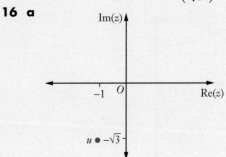

    b

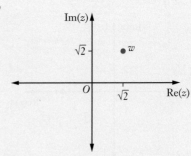

c

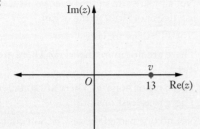

d

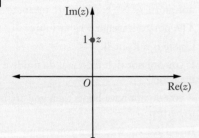

17

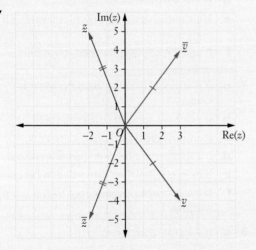

18

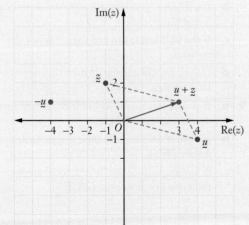

19

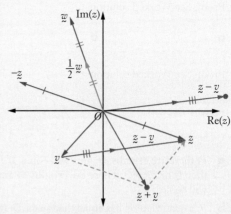

20 a $|z| = 2$, $\text{Arg}\, z = -\dfrac{\pi}{4}$

b $|z| = 2$, $\text{Arg}\, z = \dfrac{\pi}{6}$

c $|z| = 4$, $\text{Arg}\, z = \dfrac{2\pi}{3}$

d $|z| = \dfrac{1}{\sqrt{2}}$, $\text{Arg}\, z = -\dfrac{3\pi}{4}$

21 a $\text{cis}\,\dfrac{5\pi}{6}$ **b** $\sqrt{2}\,\text{cis}\left(-\dfrac{2\pi}{3}\right)$ **c** $\dfrac{1}{2}\,\text{cis}\,\dfrac{\pi}{6}$

22 a $z = \sqrt{2} + i\sqrt{2}$ **b** $\dfrac{1}{4} + \dfrac{i}{4}$

c $z = \dfrac{\sqrt{2} - i\sqrt{6}}{2}$

23 a $\sqrt{3}\,\text{cis}\,\dfrac{11\pi}{12}$ **b** $\dfrac{1}{3\sqrt{3}}\,\text{cis}\,\dfrac{7\pi}{12}$

c $\dfrac{1}{3}\,\text{cis}\left(-\dfrac{\pi}{6}\right)$ **d** $\dfrac{1}{243\sqrt{3}}\,\text{cis}\,\dfrac{\pi}{4}$

24 a $\arg z = \theta - 2\beta$ **b** $\arg z = 2\lambda - 3\alpha$

c $\arg z = \dfrac{\delta}{2} - \dfrac{\alpha}{4} + 2\phi$ **d** $\arg z = 3\varepsilon$

25 a $|z| = 5\sqrt{26}$ **b** $|z| = \dfrac{\sqrt{5}}{\sqrt{19}}$ **c** $|z| = \dfrac{1}{5}$

26 a $z = \sqrt{2} + i\sqrt{2},\, w = 2 + 2i\sqrt{3}$

b $zw = 8\,\text{cis}\,\dfrac{7\pi}{12},\, zw = 2\sqrt{2} - 2\sqrt{6} + i(2\sqrt{2} + 2\sqrt{6})$

c i $\cos\dfrac{7\pi}{12} = \dfrac{\sqrt{2} - \sqrt{6}}{4}$ **ii** $\sin\dfrac{7\pi}{12} = \dfrac{\sqrt{2} + \sqrt{6}}{4}$

27 a $e^{\frac{3i\pi}{5}}$ **b** $\sqrt{2}e^{-3i}$

28 a $\text{cis}\,\dfrac{\pi}{4}$ **b** $\dfrac{1}{2}\,\text{cis}\,(-2)$

29 Proof: See Worked solutions

30 a $e^{-\frac{i\pi}{3}}$ **b** 1

31 i

32 Proof: See Worked solutions

Chapter 2

Exercise 2.01

1 a P: there are crumbs, Q: ants will come. If P then Q or $P \Rightarrow Q$. There are crumbs $\Rightarrow$ ants will come.

b P: a quadrilateral has equal diagonals, Q: it is a square. If P then Q or $P \Rightarrow Q$. A quadrilateral has equal diagonals $\Rightarrow$ it is a square.

c P: people are unemployed, Q: they are bored. If P then Q or $P \Rightarrow Q$. People are unemployed $\Rightarrow$ they are bored.

2 a If you go skiing then you live in Cooma.

b If you have friends then you like maths.

c If you can debate then you are a politician.

d If an animal is a bird then it can fly.

3 a If you eat meat then you are a carnivore. FALSE. May be an omnivore.

b If you are on a boat then you are seasick. FALSE. May not get sea sick.

c If a shape has equal sides then it is a square. FALSE. May be many other shapes, such as a rhombus or pentagon.

d If an animal can sting then it is a honeybee. FALSE. May be many other types of animal, such as a wasp or jellyfish.

4 a If $x = 9$ then $x - 5 = 4$. TRUE. $\therefore\ x - 5 = 4$ iff $x = 9$ OR $\therefore\ x - 5 = 4 \Leftrightarrow x = 9$.

b If a quadrilateral has diagonals that are perpendicular then it is a rhombus . FALSE. Could be a kite.

c If $\dfrac{1}{a} < \dfrac{1}{b}$ then $a > b > 0$. FALSE. a could be negative and b could be positive.

d If you passed a driving test then you have a driver's licence. TRUE. You passed a driving test $\Leftrightarrow$ you have a driver's licence.

5 a It is not white.

b I do not know everything.

c Not all fish swim in the ocean, *or* At least one fish does not swim in the ocean.

d Not all babies are cute.

e There are not more than 5, *or* There are 5 or less.

f There is not none, *or* There are some, *or* There is at least one.

g Not no one passed the test, *or* Someone passed the test.

h Not some teachers are mean, *or* Teachers are not mean, *or* No teachers are mean.

i The potatoes weigh 3 kg or more.

j Cassie is not small.

6 a If you are not rich then you do not live in a mansion.

b If you do not have boots then you are not in the army.

c If you are not wise then you are not old.

d If $x^2 \neq 9$ then $x \neq 3$.

e If an animal does not have four legs then it is not a horse.

f If you are not superior then you are not a woman.

7 a If it is global warming then the water is rising.

b If you have accidents then you speed.

c If animals die then there is a drought.

d If a number is a fraction then it is rational.

e If he is not lazy then Sam will pass his exams.

f If a number has a square root then it is not negative.

8 a If $\dfrac{1}{n} < \dfrac{1}{n+1}$ then $n < 1$.

b If a line has gradient not equal to zero then it is not horizontal.

c If the bulldust is not red then it is not the outback.

d If they are not mammals then they are not blue whales.

9 a If your heart rate does not increase then you do not exercise. TRUE.

b If a plant does not die then it gets sufficient water. TRUE

c If a triangle does not have 2 equal angles, then it is not isosceles. TRUE

d If a number is not real then it is not an integer. TRUE

e If $x^2 \leq 4$ then $x \leq 2$. TRUE

10 a If an animal does not have a beak then it is not a bird. TRUE

b If a quadrilateral is not a rhombus then it does not have 2 pairs of opposite angles equal. TRUE

c If it does not have fins then it is not a fish. TRUE

d If $x^2 > 25$ then $x > 5$. FALSE (try $x = -6$)

e If a number is not prime then it is not odd. FALSE (try $x = 9$).

11 If a quadrilateral is not a square, then it does not have 4 equal angles. FALSE (rectangle)

12 D

13 a $\forall\, x \in \mathbb{N}, \exists\, y \in \mathbb{N}: y > x$

b If $x \in \mathbb{Q}$ then $\exists\, p, q \in \mathbb{Z}, q \neq 0: x = \dfrac{p}{q}$

c $\forall\, a \in \mathbb{Z}, a \neq 0, \exists\, b \in \mathbb{Q}: b = \dfrac{1}{a}$

d $\forall\, (x, y)$ and $(w, v), x, y, w, v \in \mathbb{R}, \exists\, (c, d): x < c < w, y < d < v$

e $\forall\, x \in \mathbb{R}, x \geq 0, \exists\, y \in \mathbb{R}, y \geq 0: y = \sqrt{x}$

14 a For all natural numbers m, there exists an integer n such that $n + m = 0$.

b For all integers a and b where b is non-zero, there exist rational numbers p and q such that $\dfrac{1}{a + b\sqrt{2}} = p + q\sqrt{2}$.

15 C

Exercise 2.02

Proof: See Worked solutions

Exercise 2.03

1 a $n = \dfrac{1}{2}$ **b** $n = -\dfrac{1}{2}$ **c** $n = 2$

d $n = 2$ **e** $x = -1$

2 a $n = -10$ **b** $x = 4$ **c** skew lines

d Many lizards lay eggs.

3 a False. rhombus or kite.

b False. $p = -1$

c False. $x = 2, y = -1$

d False. If $q = 0$ then r and p do not have to be equal.

e False. A rectangle 3 cm × 1 cm is not similar to a rectangle 4 cm × 1 cm.

4 a Yes

b No. The counterexample needs to find a dog that is not domesticated.

5 False. $a = 1, b = -2$

6 Yes, true

7 False. Undefined if $n = 1$.

8 All squares are rhombuses. Not all rhombuses are squares.

9 No. 3 points determine a unique circle. Choose a 4th point not on the circle.

10 Yes for $n \geq 3$.

11 No. Try arrowhead shape.

12 True

13 No. If $k < 0$ then $nk < mk$.

Exercise 2.04

1 Let $M = 2m - 1$ and $N = 2n - 1$ for some $m, n \in \mathbb{N}$.

Then $M \times N = (2m - 1) \times (2n - 1)$
$$= 4mn + -2m - 2n + 1$$
$$= 2(mn - m - n) + 1$$
$$= 2P + 1 \text{ where } P \in \mathbb{N}$$

Since $2P$ is even then $2P + 1$ is odd.

Therefore the product of two odd numbers is odd. QED.

2–8 Proof: See Worked solutions

9 $\dfrac{|x|}{x} = \begin{cases} 1 & \text{for } x > 0 \\ -1 & \text{for } x < 0 \end{cases}$

Exercise 2.05

1–11 Proof: See Worked solutions

10 f Equality holds when $a = b = c = d$.

Test yourself 2

1 a I get a lot of sleep $\Rightarrow$ I am healthy. $P =$ I get a lot of sleep, $Q =$ I am healthy

b A polygon has 5 sides $\Rightarrow$ it is a pentagon. $P =$ A polygon has 5 sides, $Q =$ it is a pentagon

c The teacher is nice $\Rightarrow$ I will learn. $P =$ The teacher is nice, $Q =$ I will learn.

2 a $B \Rightarrow A$ **b** $Q \Rightarrow \neg P$

c $\neg M \Rightarrow N$ **d** $\neg F \Rightarrow \neg B$

b If I can buy a car then I can save money.

c If I am bored then my computer is broken.

d If $a^3 = b^3$ then $a = b$.

3 $P \Rightarrow Q$ AND $Q \Rightarrow P$ together, that is, $P \Leftrightarrow Q$, an equivalence. For example, if a quadrilateral is a rhombus then the diagonals bisect each other at right angles.

4 a If a quadrilateral is a square then it has equal diagonals. Not an equivalence. Original statement false.

b If $\dfrac{1}{x} < 1$ then $x > 1$. Not an equivalence.

If $x < 0$ then $\dfrac{1}{x} < 1$ but $x \not> 1$.

c If I study hard then I pass my exams. Not necessarily true.

d If $a^2 = 9$ then $a = 3$. Not necessarily true: $a = -3$.

e If a triangle is isosceles then it has 2 equal angles. Equivalence. A triangle is isosceles $\Leftrightarrow$ it has two equal angles.

5 a It is not raining.

b The apple is ripe.

c Not all koalas are cute. At least one koala is not cute. Some koalas are not cute.

d No people are sexist.

e They are not all correct.

f $x > 4$ **g** $p \not\in \mathbb{N}$

6 No. Negation is there were less than or equal to 10.

7 a $\neg B \Rightarrow \neg A$ **b** $Q \Rightarrow \neg P$

c $M \Rightarrow \neg N$ **d** $F \Rightarrow B$

e If the boy does not have blue eyes then he does not have red hair.

f If the citizens don't have money then the country is not rich.

g If a quadrilateral does not have adjacent sides equal in length, then it is not a kite.

h If $x^2 \neq y^2$ then $x \neq y$

i If $a \not\in \mathbb{Z}$ then $a \not\in \mathbb{N}$

8 If $A \Rightarrow B$ and $B \Rightarrow A$ then we can write $A \Leftrightarrow B$ or A iff B. This is an equivalence. Now if $P \Rightarrow Q$ is true and $\neg Q \Rightarrow \neg P$ is also true then $P \Rightarrow Q$ implies $\neg Q \Rightarrow \neg P$; that is, $(P \Rightarrow Q) \Rightarrow (\neg Q \Rightarrow \neg P)$. Also $\neg Q \Rightarrow \neg P$ implies $P \Rightarrow Q$; that is, $(\neg Q \Rightarrow \neg P) \Rightarrow (P \Rightarrow Q)$. Then we can write $(P \Rightarrow Q) \Leftrightarrow (\neg Q \Rightarrow \neg P)$ so they are equivalences. This is also true if $P \Rightarrow Q$ is false.

9 a If $a = b$ then $a^2 = b^2$. True.

b If the car starts then the battery is not flat. True.

c If a number is not rational then it is not an integer. True

d If a quadrilateral is not a rhombus, then it does not have diagonals that bisect each other at right angles. True.

e If $ab \le b^2$ then $a \le b$. False. Try $b = -2$.

f If an animal is not a fish then it does not live in the water. False.

10 a $\forall\, x, y \in \mathbb{R}, x, y > 0, (x > y) \Rightarrow (x^2 > y^2)$

b If $a, b \in \mathbb{Z}, a < b, \exists\, c \in \mathbb{Q}, a < c < b\colon c = \dfrac{a+b}{2}$.

c Let $n \in \mathbb{Z}, n > 0\colon \forall\, n$,

$1 + 2 + 3 + \ldots + n = \dfrac{n(n+1)}{2}$.

11 a For all positive integers n and m, if

$n < m$ then $\dfrac{1}{n} > \dfrac{1}{m}$.

b For all real numbers a and b, $a^2 + b^2 \ge 2ab$.

c For all rational numbers p and q where $p < q$, there exists a real number r such that $p < r < q$.

12 Proof: See Worked solutions

13 a $x = 3, y = -3$ **b** $n = 1$

c A cicada sheds its skin.

d $a = -3, b = 4, c = 5$

e If $k = 17$ then $17(17 - 1) + 17 = 17 \times 17$ is not prime.

f $(c = -3) \Rightarrow (c^2 = 9)$

14 False; $a = 3, b = -1, c = -2, d = -3$

15 False; $x = -5, y = -6$

16 Proof: See Worked solutions

17 $k^6 - m^6 = (k^2 - m^2)(k^4 + k^2m^2 + m^4)$
$= (k - m)(k + m)(k^4 + k^2m^2 + m^4)$ and
$k^6 - m^6$
$= (k^3 - m^3)\,(k^3 + m^3)$
$= (k - m)(k^2 + km + m^2)(k + m)(k^2 - km + m^2)$.
Hence equate.

18–24 Proof: See Worked solutions

Chapter 3

Exercise 3.01

1 5

2 a … in the parallelogram of vectors
   b … the shorter diagonal in the parallelogram of vectors
   c … one vector is a scalar times the other
   d … the dot product of the vectors is 0

3 a $2, 90°$ **b** $5, 0°$
   c $5\sqrt{5}, \tan^{-1}(-0.5)$ **d** $4\sqrt{2}, 135°$
   e $4, 60°$

4 a $3\sqrt{2}\,\underset{\sim}{i} + 3\sqrt{2}\,\underset{\sim}{j}$ **b** $4\sqrt{3}\,\underset{\sim}{i} + 4\,\underset{\sim}{j}$
   c $-\sqrt{2}\,\underset{\sim}{i} + \sqrt{2}\,\underset{\sim}{j}$ **d** $5\underset{\sim}{i} - 5\sqrt{3}\,\underset{\sim}{j}$
   e $-3\sqrt{3}\underset{\sim}{i} - 3\underset{\sim}{j}$

5 a $\begin{pmatrix} 3\sqrt{2} \\ 3\sqrt{2} \end{pmatrix}$ **b** $\begin{pmatrix} 4\sqrt{3} \\ 4 \end{pmatrix}$ **c** $\begin{pmatrix} -\sqrt{2} \\ \sqrt{2} \end{pmatrix}$
   d $\begin{pmatrix} 5 \\ -5\sqrt{3} \end{pmatrix}$ **e** $\begin{pmatrix} -3\sqrt{3} \\ -3 \end{pmatrix}$

6 a 15   **b** 0   **c**   4
   d −3   **e** 10

7 a $0, 90°$ **b** $34, 0°$ **c** $-6, 180°$
   d $-20, 141°$ **e** $-8, 173°$

8 a $\underset{\sim}{v}$ and $\underset{\sim}{w}$ **b** $\underset{\sim}{u}$ and $\underset{\sim}{w}$

9 a $\underset{\sim}{v}$ and $\underset{\sim}{w}$ **b** $\underset{\sim}{u}$ and $\underset{\sim}{w}$

10 $\overrightarrow{PQ}: 2\underset{\sim}{i} + 4\underset{\sim}{j}$, $\overrightarrow{PR}: 4\underset{\sim}{i} + 8\underset{\sim}{j}$; hence $\overrightarrow{PR} = 2\overrightarrow{PQ}$ so P, Q and R are collinear.

Exercise 3.02

1 a $\sqrt{26}$ **b** $5\sqrt{5}$ **c** 7   **d** 5

2 a $\dfrac{\underset{\sim}{i} + \underset{\sim}{j} + \underset{\sim}{k}}{\sqrt{3}}$ **b** $\dfrac{2\underset{\sim}{i} - \underset{\sim}{j} + 2\underset{\sim}{k}}{3}$
   c $\dfrac{3\underset{\sim}{i} + 4\underset{\sim}{j} - 12\underset{\sim}{k}}{13}$ **d** $\dfrac{\underset{\sim}{i} + 3\underset{\sim}{j} + 2\underset{\sim}{k}}{\sqrt{14}}$

3 a $3\underset{\sim}{i} - 4\underset{\sim}{j}, 5$ **b** $-2\underset{\sim}{i}, 2$
   c $-\underset{\sim}{i} - \underset{\sim}{j} - \underset{\sim}{k}, \sqrt{3}$ **d** $5\underset{\sim}{i} + 6\underset{\sim}{j} + 8\underset{\sim}{k}, 5\sqrt{5}$

4 a $\sqrt{5}$ **b** 0   **c** $\sqrt{17}$

5 a $2\underset{\sim}{i}$ **b** $9\underset{\sim}{i} - 6\underset{\sim}{j} + 18\underset{\sim}{k}$
   c $3\underset{\sim}{i} - 4\underset{\sim}{j}$ **d** $2\underset{\sim}{i} - 2\underset{\sim}{j} + 2\underset{\sim}{k}$

6 a $\underset{\sim}{i} - 3\underset{\sim}{j} - \underset{\sim}{k}$ **b** $3\underset{\sim}{i} - 4\underset{\sim}{j} + 7\underset{\sim}{k}$
   c $2\underset{\sim}{i} - 3\underset{\sim}{j} + 4\underset{\sim}{k}$ **d** $-\underset{\sim}{j} - 2\underset{\sim}{k}$

7 a $\sqrt{17}$ **b** $\sqrt{29}$ **c** $2\sqrt{6}$ **d** $\sqrt{38}$

8 a 3   **b** 5
   c $\dfrac{2}{3}\underset{\sim}{i} + \dfrac{2}{3}\underset{\sim}{j} + \dfrac{1}{3}\underset{\sim}{k}$ **d** $\dfrac{3}{5}\underset{\sim}{i} - \dfrac{4}{5}\underset{\sim}{k}$
   e $5\underset{\sim}{i} + 2\underset{\sim}{j} - 3\underset{\sim}{k}$ **f** $-\underset{\sim}{i} + 2\underset{\sim}{j} + 5\underset{\sim}{k}$

9 a $\dfrac{3\sqrt{2}}{2}\underset{\sim}{i} - \sqrt{2}\underset{\sim}{j} + \dfrac{\sqrt{2}}{2}\underset{\sim}{k}$ **b** $\sqrt{3}\underset{\sim}{i} - \sqrt{3}\underset{\sim}{j} + \sqrt{3}\underset{\sim}{k}$
   c $4\underset{\sim}{i} - 4\underset{\sim}{j} + 2\underset{\sim}{k}$

10 $180°$; the vectors point in opposite directions.

Exercise 3.03

1 a 12   **b** 0   **c** 4   **d** 6
   e 0

2 A and C, B and C

3 a $83°$ **b** $55°$ **c** $25°$ **d** $71°$

4 $-\dfrac{1}{3}$

5 a Proof: See Worked solutions
   b $\underset{\sim}{u}$ and $\underset{\sim}{v} - \underset{\sim}{w}$ are perpendicular

6 scalar product = 0

7 $c(\underset{\sim}{i} + 3\underset{\sim}{j} + 5\underset{\sim}{k})$, c constant

8 a $\overrightarrow{AC} = 3\underset{\sim}{i} + 2\underset{\sim}{j}, \overrightarrow{BD} = -3\underset{\sim}{i} + 2\underset{\sim}{j}$
   b $112°36'$

9 Proof: See Worked solutions

Exercise 3.04

1

2–4 Proof: See Worked solutions

5 $x = 2, y = 3$

6–8 Proof: See Worked solutions

9 $\overrightarrow{OP} = \dfrac{n}{m+n}\underset{\sim}{a} + \dfrac{m}{m+n}\underset{\sim}{b}$

Exercise 3.05

1 a 3   **b** 13   **c** 15   **d** 33

  e 7

2 a $\dfrac{2\underset{\sim}{i} - 2\underset{\sim}{j} + \underset{\sim}{k}}{3}$ **b** $\dfrac{3\underset{\sim}{i} - 4\underset{\sim}{j} + 12\underset{\sim}{k}}{13}$

  c $\dfrac{2\underset{\sim}{i} + 5\underset{\sim}{j} + 14\underset{\sim}{k}}{15}$ **d** $\dfrac{4\underset{\sim}{i} + 7\underset{\sim}{j} - 32\underset{\sim}{k}}{33}$

  e $\dfrac{-3\underset{\sim}{i} - 2\underset{\sim}{j} + 6\underset{\sim}{k}}{7}$

3 a $\theta = 33.6°$ **b** $\theta = 40.2°$ **c** $\theta = 116.4°$

  d $\theta = 80.4°$ **e** $\theta = 65.9°$

4 a $y = \pm5$ **b** $z = \pm42$ **c** $x = \pm10$

5 $m = \pm4$

6 Any vector that satisfies $2a - 3b + 4c = 0$, where a, b and c are real, for example

$$\begin{pmatrix} 8 \\ 4 \\ -1 \end{pmatrix} \text{ or } \begin{pmatrix} 1 \\ 2 \\ 1 \end{pmatrix}.$$

7 $m = \pm\sqrt{6}$

Exercise 3.06

1 a

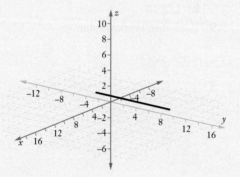

b

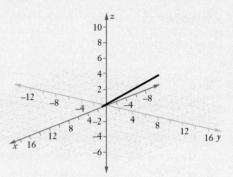

c

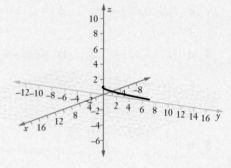

d

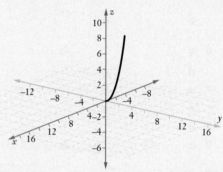

e

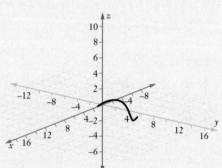

A helix (spiral) of radius 1 unit revolving around the y-axis, with endpoints $(-1, 0, 0)$ and $(-1, \pi, 0)$.

2 a

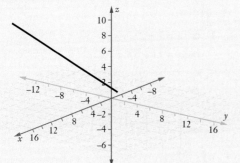

b

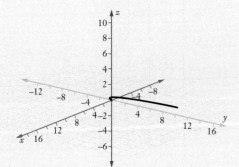

c

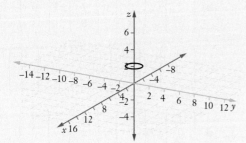

d

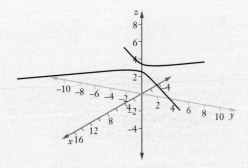

e

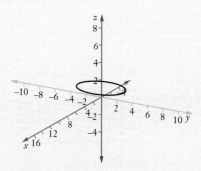

3 a vertical line going through $(1, 1, 0)$

b helix (spiral) starting at $(1, 0, 0)$ in both downward and anti-clockwise directions around the negative z-axis

c ellipse (oval) 1 unit in front of the y-z plane, centred on $(1, 0, 0)$, 4 units long on the y-axis, 2 units high on the z-axis

d helix in z direction starting at $(0, 0, 0)$ and increasing in radius and height

4 $x = 3 \cos t$, $y = 3 \sin t$ and $z = 2 - 3 \sin t$

5 x–y

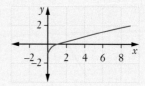

x–z

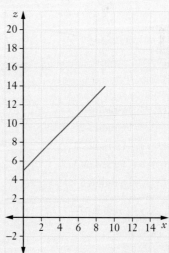

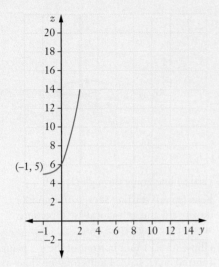

$$6 \quad \begin{pmatrix} t \\ t \\ 2t^2 \end{pmatrix} \text{ or } x = t, y = t, z = 2t^2$$

$$7 \quad x = t, y = \frac{1}{2}(t^2 - 1), z = \frac{1}{2}(t^2 + 1)$$

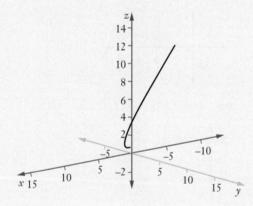

8 a centre $(1, -1, 1)$; radius 1
b centre $(-2, 3, 1)$; radius 2
c centre $(3, -1, -1)$; radius 3
d centre $(-1, -1, 1)$; radius 3
e centre $(2, 3, -1)$; radius 5

9 centre $\left(-2, -3, \frac{7}{4}\right)$, radius $\frac{5\sqrt{7}}{4}$

Exercise 3.07

1 a $x = 1 + 2\lambda, y = 1 - 3\lambda, z = 0$
b $x = 11 + 3\lambda, y = 2, z = 0$
c $x = 3 + 6\lambda, y = -9\lambda, z = -1 + \lambda$
d $x = 5 + 7\lambda, y = -2 - 4\lambda, z = 1 + 2\lambda$

2 a $\begin{pmatrix} 2 \\ -1 \\ 3 \end{pmatrix} + \lambda \begin{pmatrix} 1 \\ 2 \\ -1 \end{pmatrix}$

b $\begin{pmatrix} -2 \\ 1 \\ -3 \end{pmatrix} + \lambda \begin{pmatrix} -1 \\ 2 \\ 3 \end{pmatrix}$

c $\begin{pmatrix} 1 \\ -1 \\ 1 \end{pmatrix} + \lambda \begin{pmatrix} 2 \\ -2 \\ 1 \end{pmatrix}$

3 a $\begin{pmatrix} 3 \\ -5 \end{pmatrix} \pm \lambda \begin{pmatrix} 5 \\ 3 \end{pmatrix}$ or $\begin{pmatrix} -2 \\ -8 \end{pmatrix} \pm \lambda \begin{pmatrix} 5 \\ 3 \end{pmatrix}$

b $\begin{pmatrix} 6 \\ 2 \\ 5 \end{pmatrix} \pm \lambda \begin{pmatrix} 3 \\ 0 \\ 3 \end{pmatrix}$ or $\begin{pmatrix} 9 \\ 2 \\ 8 \end{pmatrix} \pm \lambda \begin{pmatrix} 3 \\ 0 \\ 3 \end{pmatrix}$

c $\begin{pmatrix} 1 \\ 1 \\ -3 \end{pmatrix} \pm \lambda \begin{pmatrix} 0 \\ 2 \\ 2 \end{pmatrix}$ or $\begin{pmatrix} 1 \\ -1 \\ -5 \end{pmatrix} \pm \lambda \begin{pmatrix} 0 \\ 2 \\ 2 \end{pmatrix}$

d $\begin{pmatrix} 1 \\ 0 \\ 3 \end{pmatrix} \pm \lambda \begin{pmatrix} 0 \\ 2 \\ 1 \end{pmatrix}$ or $\begin{pmatrix} 1 \\ 2 \\ 4 \end{pmatrix} \pm \lambda \begin{pmatrix} 0 \\ 2 \\ 1 \end{pmatrix}$

4 a neither; skew   **b** parallel
   c intersect at $(3, 6, 5)$ **d** same line

5 Vector equation: $\begin{pmatrix} x \\ y \\ z \end{pmatrix} = \begin{pmatrix} 3 \\ 2 \\ 6 \end{pmatrix} + \lambda \begin{pmatrix} 2 \\ 1 \\ 1 \end{pmatrix}$

or other variations for λ
Parametric equations: $x = 3 + 2\lambda, y = 2 + \lambda$,
$z = 6 + \lambda$
Cartesian form: $x - 2y = -1$ and $z - y = 4$
or $\dfrac{x - 3}{2} = y - 2 = z - 6$

6 $a_1 = -\dfrac{2}{5}$ and $a_3 = \dfrac{2}{5}$

7 Proof: See Worked solutions

8

$$\begin{pmatrix} x \\ y \\ z \end{pmatrix} = \begin{pmatrix} 2 \\ \dfrac{1}{3} \\ -\dfrac{1}{2} \end{pmatrix} + \lambda \begin{pmatrix} -7 \\ \dfrac{5}{3} \\ \dfrac{3}{2} \end{pmatrix}$$

9 -1

Exercise 3.08

1 See worked solutions: in each case $\lambda_2 = c\lambda_1$.

2 See worked solutions: where the vectors are of the form $(a_1, a_2, a_3) + \lambda(b_1, b_2, b_3)$ and $(c_1, c_2, c_3) + \lambda(d_1, d_2, d_3)$, then $b_1 d_1 + b_2 d_2 + b_3 d_3 = 0$.

3 a Any vector of the form
$(x_1, y_1, z_1) + \lambda(2, -1, 3)$, λ a constant

   b Any vector of the form
$(x_1, y_1, z_1) + \lambda(3, 1, -2)$, λ a constant

   c Any vector of the form
$(x_1, y_1, z_1) + \lambda(4, 3, -1)$, λ a constant

   d Any vector of the form
$(x_1, y_1, z_1) + \lambda(2, 2, -1)$, λ a constant

4 Vectors in the form $(x_1, y_1, z_1) + \lambda(x_2, y_2, z_2)$ where:

   a $2x_2 - y_2 + 3z_2 = 0$

   b $3x_2 + y_2 - 2z_2 = 0$

   c $4x_2 + 3y_2 - z_2 = 0$

   d $2x_2 + 2y_2 - z_2 = 0$

5 **a** and **d**, **b** and **d**, **c** and **d**

6 $\theta = 54.7°$

7 a See worked solutions:

AC: $(x, y, z) = (2, -3, 3) + \lambda_1(-4, 6, -4)$

BD: $(x, y, z) = (-3, 2, 1) + \lambda_2(6, -4, 0)$

Solving $\lambda_1 = \lambda_2 = \dfrac{1}{2}$

   b $36.2°$

8

$$\begin{pmatrix} x \\ y \\ z \end{pmatrix} = \begin{pmatrix} 2 \\ -1 \\ 3 \end{pmatrix} + \lambda \begin{pmatrix} 1 \\ 5 \\ 2 \end{pmatrix}$$

9

$$\begin{pmatrix} x \\ y \\ z \end{pmatrix} = \begin{pmatrix} 4 \\ 12 \\ 15 \end{pmatrix} + \lambda \begin{pmatrix} 8 \\ 5 \\ 2 \end{pmatrix}$$

10 Any 2 lines of the form $\begin{pmatrix} 2 \\ -1 \\ 3 \end{pmatrix} + \lambda \begin{pmatrix} a_1 \\ b_1 \\ c_1 \end{pmatrix}$ and

$\begin{pmatrix} 2 \\ -1 \\ 3 \end{pmatrix} + \lambda \begin{pmatrix} a_2 \\ b_2 \\ c_2 \end{pmatrix}$ where $a_1 a_2 + b_1 b_2 + c_1 c_2 = 0$, for

example $\begin{pmatrix} 2 \\ -1 \\ 3 \end{pmatrix} + \lambda \begin{pmatrix} 1 \\ 2 \\ -1 \end{pmatrix}$ and $\begin{pmatrix} 2 \\ -1 \\ 3 \end{pmatrix} + \lambda \begin{pmatrix} -1 \\ 1 \\ 1 \end{pmatrix}$.

Test yourself 3

1 $4, 36° 52'$

2 Perpendicular $(\underset{\sim}{u} \cdot \underset{\sim}{v} = 0)$

3 $4\underset{\sim}{i} + 4\sqrt{3}\underset{\sim}{j}, \dfrac{1}{8}(4\underset{\sim}{i} + 4\sqrt{3}\underset{\sim}{j})$

4 $\sqrt{29}, 111° 48'$

5 $\underset{\sim}{i} + \underset{\sim}{j} + 4\underset{\sim}{k}$

6 0; vectors are perpendicular

7 $\dfrac{1}{\sqrt{3}} \begin{pmatrix} 1 \\ 1 \\ -1 \end{pmatrix}$

8 $\underset{\sim}{a} + 3\underset{\sim}{b}$

9 a $\overrightarrow{AB} = \underset{\sim}{b} - \underset{\sim}{a}$

   b Proof: See Worked solutions

10 $|\underset{\sim}{u}| = 3, |\underset{\sim}{v}| = 3\sqrt{2}, \theta = 113.97°$

11 Any vector $a\underset{\sim}{i} + b\underset{\sim}{j} + c\underset{\sim}{k}$ where $c = a - 2b$,

for example $2\underset{\sim}{i} - \underset{\sim}{j} + 4\underset{\sim}{k}$

12

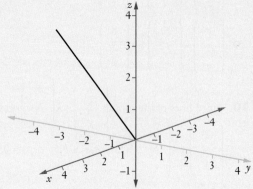

13

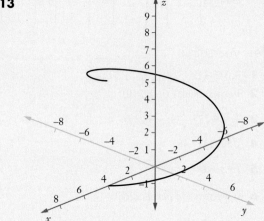

14 Radius $\sqrt{5}$, centre $(0, 4, -1)$

15 Vector $\begin{pmatrix} x \\ y \\ z \end{pmatrix} = \begin{pmatrix} -3 \\ 2 \\ -1 \end{pmatrix} + \lambda \begin{pmatrix} 2 \\ 1 \\ 3 \end{pmatrix};$

parametric $x = 2\lambda - 3, y = \lambda + 2, z = 3\lambda - 1$

16 $x = -7, z = 11\frac{1}{3}$

17, 18 Proof: See Worked solutions

19 $\theta = 78.5°$

20 $\begin{pmatrix} 1 \\ -3 \\ 2 \end{pmatrix} + \lambda \begin{pmatrix} 7 \\ 5 \\ 1 \end{pmatrix}$

Chapter 4

Exercise 4.01

1 a $\cos 5\theta + i \sin 5\theta$ **b** $\cos 3\theta - i \sin 3\theta$

c $\cos 7\theta - i \sin 7\theta$ **d** $\cos \dfrac{5\theta}{2} - i \sin \dfrac{5\theta}{2}$

e $\cos 3\theta + i \sin 3\theta$

2 a i $32\left(\cos \dfrac{5\pi}{6} + i \sin \dfrac{5\pi}{6}\right)$ **ii** $-16\sqrt{3} + 16i$

b i $9\sqrt{3}\left(\cos\left(\dfrac{-3\pi}{4}\right) + i \sin\left(\dfrac{-3\pi}{4}\right)\right)$

ii $-\dfrac{9\sqrt{6}(1+i)}{2}$

c i $\dfrac{1}{4\sqrt{2}}\left(\cos\left(-\dfrac{\pi}{3}\right) + i \sin\left(-\dfrac{\pi}{3}\right)\right)$

ii $\dfrac{1}{8\sqrt{2}} - \dfrac{i\sqrt{3}}{8\sqrt{2}}$

d i $243\left(\cos \dfrac{\pi}{2} + i \sin \dfrac{\pi}{2}\right)$ **ii** $243i$

3 $\cos\left(-\dfrac{2\pi}{5}\right) + i \sin\left(-\dfrac{2\pi}{5}\right)$

4 a $-2 - 2i$ **b** $-8 - 8\sqrt{3}i$

c $16\sqrt{2} - 16\sqrt{2}i$ **d** $\dfrac{-1}{\sqrt{2}} - \dfrac{i}{\sqrt{2}}$

e $\dfrac{\sqrt{3}}{2} + \dfrac{i}{2}$

5 a $\dfrac{1}{324}(\cos \pi + i \sin \pi)$

b $\dfrac{1}{512}\left[\cos\left(-\dfrac{\pi}{2}\right) + i \sin\left(-\dfrac{\pi}{2}\right)\right]$

6 Proof: See Worked solutions

7 a $\cos 4\theta - i \sin 4\theta$

b $\cos(4\alpha - 6\beta) + i \sin(4\alpha - 6\beta)$

c $\cos 18\delta + i \sin 18\delta$

d $\cos 2\beta + i \sin 2\beta$

e i

8 $\cos 2\alpha = \cos^2 \alpha - \sin^2 \alpha$, $\sin 2\alpha = 2 \sin \alpha \cos \alpha$,

$\tan 2\alpha = \dfrac{2 \tan \alpha}{1 - \tan^2 \alpha}$

9 a $2\dfrac{2}{3}$ **b** $\dfrac{5}{24}$

10 Proof: See Worked solutions

11 a $\tan 3\theta = \dfrac{\sin 3\theta}{\cos 3\theta}$

  b, c Proof: See Worked solutions

12 Proof: See Worked solutions

13 $\dfrac{-7\sqrt{3}}{32} + \dfrac{\pi}{8}$.

14 Proof: See Worked solutions

15 a $2i \sin \theta$ **b** $2i \sin 2\theta$ **c** $2i \sin n\theta$

16 a $2 \cos \dfrac{\pi}{12}$ **b** $i\sqrt{3}$ **c** -2

17,18 Proof: See Worked solutions

Exercise 4.02

1 a $x = 3i, -i$ **b** $x = -i, -5i$

  c $x = 2 + i, 1 + i$ **d** $x = 2i, \dfrac{-i}{3}$

2 a $z = \pm\left(\dfrac{1}{\sqrt{2}} + \dfrac{i}{\sqrt{2}}\right)$ **b** $z = \pm\left(\dfrac{3}{\sqrt{2}} - \dfrac{3i}{\sqrt{2}}\right)$

  c $z = \pm\left(\dfrac{\sqrt{3}}{2} + \dfrac{i}{2}\right)$ **d** $z = \pm\sqrt[4]{2} \operatorname{cis}\left(-\dfrac{\pi}{8}\right)$

  e $z = \pm i$ **f** $z = \pm 2\left(1 - i\sqrt{3}\right)$

  g $z = \pm(\cos 2 + i \sin 2)$

3 $1 + i$

4 a $x^2 - (4 + 6i)x + 10 + 20i = 0$

  b $x^2 - (3 + i)x + 20 - 12i = 0$

  c $3x^2 - (5 - 2i)x + 3 - i = 0$

5 $a = 2, p = -4, q = 2 + i$

6 a $3 + i$ $(a > 0)$ **b** $2 - i, -4 - 3i$

7 a $1, i$ **b** $2 + i, -i$

  c $2 - i, 1 - 2i$ **d** $4 + i, 2i$

  e $1 - i, 2$

Exercise 4.03

1 a i $z(z^2 + 1)$ **ii** $z(z + i)(z - i)$

  b i $z(z^2 - 6z + 10)$

    ii $z(z - 3 - i)(z - 3 + i)$

  c i $(z + 1)(z^2 - z + 1)$

    ii $(z+1)\left(z - \dfrac{1}{2} - \dfrac{\sqrt{3}}{2}i\right)\left(z - \dfrac{1}{2} + \dfrac{\sqrt{3}}{2}i\right)$

  d i $(z - 2)(z^2 + 2z + 4)$

    ii $(z-2)\left(z+1-i\sqrt{3}\right)\left(z+1+i\sqrt{3}\right)$

  e i $(z - 1)(z + 1)(z^2 + 4)$

    ii $(z - 1)(z + 1)(z + 2i)(z - 2i)$

  f i $(z^2 + 9)(z^2 + 1)$

    ii $(z - i)(z + i)(z - 3i)(z + 3i)$

  g i $(z + 1)(z^2 + 1)$

    ii $(z + 1)(z - i)(z + i)$

  h i $(z - 1)(z^2 + 2)$

    ii $(z - 1)\left(z - i\sqrt{2}\right)\left(z + i\sqrt{2}\right)$

2 a $z = -2, 2 + i, 2 - i$ **b** $z = i, -i, 3, -1$

  c $z = 1, e^{\frac{i\pi}{4}}, e^{-\frac{i\pi}{4}}$ **d** $z = -1, 5, 1 \pm i\sqrt{2}$

  e $z = -2, 2 \pm i\sqrt{5}$

3 $p = 25, q = -54, z = 1 \pm 3i, 2 \pm i\sqrt{3}$

4 a $z = \pm i, 3$ **b** $z = -2, 2, \dfrac{1}{2} \pm \dfrac{i\sqrt{3}}{2}$

5 a $z^3 - 7z^2 + 12z - 10 = 0$

  b $z^3 + 8 = 0$

  c $z^4 - 4z^3 + 19z^2 - 74z + 238 = 0$

  d $9z^3 - 42z^2 + 28z - 16 = 0$

  e $z^3 + 2z^2 - 2z + 3 = 0$

6 a 4 **b** 5

Exercise 4.04

1 a $|z_1||z_2| = 6$, $\arg z_1 z_2 = \dfrac{2\pi}{3}$

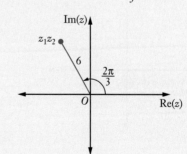

b $|z_1||z_2| = \sqrt{2}, \arg z_1 z_2 = \pi$

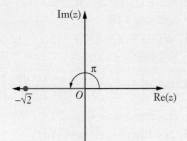

c $|z_1||z_2| = 2\sqrt{5}, \arg z_1 z_2 = \dfrac{\pi}{6}$

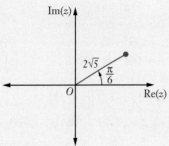

d $|z_1||z_2| = 1, \arg z_1 z_2 = \dfrac{3\pi}{4}$

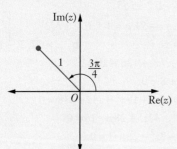

2 a $\dfrac{|z_1|}{|z_2|} = 1, \arg \dfrac{z_1}{z_2} = \dfrac{\pi}{2}$

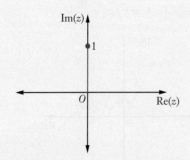

b $\dfrac{|z_1|}{|z_2|} = 2, \arg \dfrac{z_1}{z_2} = \dfrac{\pi}{4}$

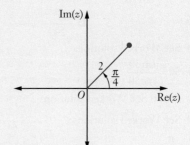

c $\dfrac{|z_1|}{|z_2|} = \sqrt{2}, \arg \dfrac{z_1}{z_2} = -\dfrac{\pi}{2}$

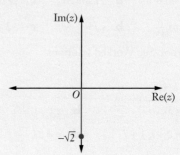

d $\dfrac{|z_1|}{|z_2|} = \dfrac{1}{2}, \arg \dfrac{z_1}{z_2} = -\dfrac{\pi}{3}$

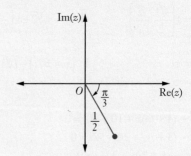

3 $z_1 z_2 = 2\sqrt{2} \operatorname{cis}\left(-\dfrac{5\pi}{6}\right)$

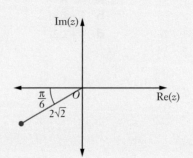

4 $\dfrac{z_1}{z_2} = 3\left(\cos\dfrac{\pi}{4} + i\sin\dfrac{\pi}{4}\right)$

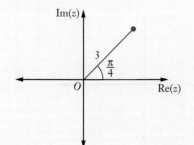

5 $\dfrac{z_1 z_2}{z_3} = \sqrt{2}\left(\cos\left(-\dfrac{\pi}{6}\right) + i\sin\left(-\dfrac{\pi}{6}\right)\right)$

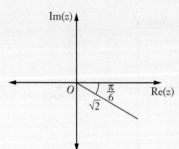

6

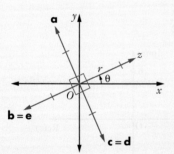

7

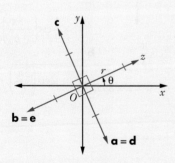

8

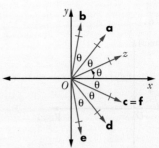

$\bar{z} = \cos\theta - i\sin\theta$

$\quad = \cos(-\theta) + i\sin(-\theta)$

$\quad = (\cos\theta + i\sin\theta)^{-1}$

$\quad = z^{-1}$

$\quad = \dfrac{1}{z}$

Only true if $|z| = 1$.

9 a $v = \dfrac{u}{i} = -iu$ **b** $v = -u = i^2 u$

   c $v = iu$

10 a $\dfrac{1}{z} = \dfrac{1}{3}\text{cis}\left(-\dfrac{\pi}{4}\right)$

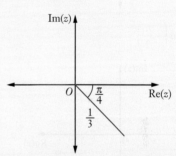

b $\dfrac{1}{z} = \sqrt{2}\,\text{cis}\left(-\dfrac{2\pi}{3}\right)$

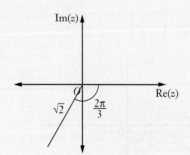

c $\dfrac{1}{z} = \dfrac{1}{4}\operatorname{cis}\dfrac{5\pi}{6}$

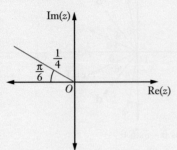

11 a $2\sqrt{2}\operatorname{cis}\dfrac{3\pi}{4}$

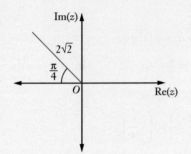

b $2\operatorname{cis}\left(-\dfrac{\pi}{2}\right)$

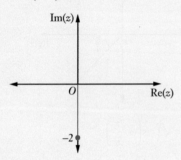

c $\dfrac{1}{16}\operatorname{cis}\left(-\dfrac{2\pi}{3}\right)$

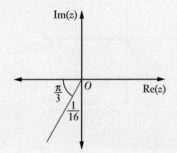

d $\dfrac{1}{3\sqrt{3}}\operatorname{cis}\left(-\dfrac{\pi}{2}\right)$

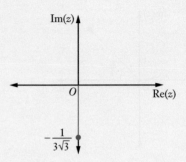

e $16\operatorname{cis}0$

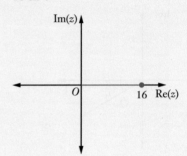

f $8\operatorname{cis}\dfrac{\pi}{2}$

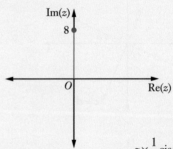

12 a $w = z \times 2\operatorname{cis}\dfrac{\pi}{3}$

b $z \times \dfrac{1}{2}\operatorname{cis}\left(-\dfrac{5\pi}{6}\right)$ or $w = \dfrac{z}{2\operatorname{cis}\left(\dfrac{5\pi}{6}\right)}$

c $w = z^4$ or $w = z \times 8\operatorname{cis}3\theta$

13 a $z_3 = \dfrac{z_1}{i} = -iz_1$

b $z_2 = z_1 + z_3 = z_1 - iz_1$

$\overrightarrow{OB} = z_1 - iz_1$ so B is the point represented by the complex number $z_1 - iz_1$.

c $\dfrac{1}{2}(z_1 - iz_1)$

14 a Proof: See Worked solutions

b $\frac{1}{2}(\delta + \beta)$

15 Proof: See Worked solutions

16 a

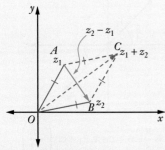

b It is a rhombus.

c It is equilateral.

17 Proof: See Worked solutions

Exercise 4.05

1 a $z_1 = \operatorname{cis} \frac{2\pi}{3}, z_2 = \operatorname{cis}\left(-\frac{2\pi}{3}\right), z_3 = 1$

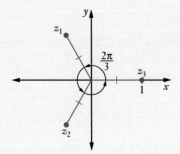

b $z_1 = \operatorname{cis} \frac{2\pi}{5}, z_2 = \operatorname{cis} \frac{4\pi}{5}$,

$z_3 = \operatorname{cis}\left(-\frac{4\pi}{5}\right), z_4 = \operatorname{cis}\left(-\frac{2\pi}{5}\right)$,

$z_5 = 1$

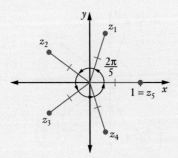

c $z_1 = \operatorname{cis} \frac{\pi}{4}, z_2 = \operatorname{cis} \frac{\pi}{2} = i, z_3 = \operatorname{cis} \frac{3\pi}{4}$,

$z_4 = -1, z_5 = \operatorname{cis}\left(-\frac{3\pi}{4}\right), z_6 = \operatorname{cis}\left(-\frac{\pi}{2}\right) = -i$,

$z_7 = \operatorname{cis}\left(-\frac{\pi}{4}\right)$,

$z_8 = 1$

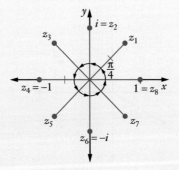

d $z_1 = \operatorname{cis} \frac{2\pi}{9}, z_2 = \operatorname{cis} \frac{4\pi}{9}, z_3 = \operatorname{cis} \frac{2\pi}{3}$,

$z_4 = \operatorname{cis} \frac{8\pi}{9}, z_5 = \operatorname{cis}\left(-\frac{8\pi}{9}\right), z_6 = \operatorname{cis}\left(-\frac{2\pi}{3}\right)$,

$z_7 = \operatorname{cis}\left(-\frac{4\pi}{9}\right), z_8 = \operatorname{cis}\left(-\frac{2\pi}{9}\right)$,

$z_9 = 1$

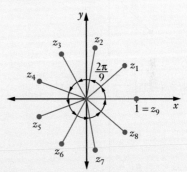

2 Proof: See Worked solutions

3 a $\overline{\alpha} = \alpha^6, \overline{\alpha^2} = \alpha^5, \overline{\alpha^3} = \alpha^4$

b $\overline{\alpha} = \alpha^{10}, \overline{\alpha^2} = \alpha^9, \overline{\alpha^3} = \alpha^8, \overline{\alpha^4} = \alpha^7, \overline{\alpha^5} = \alpha^6$

4 If $z^5 - 1 = 0$ and β is a complex root then

$z^5 - 1 = (z - 1)(z + z^2 + z^3 + z^4 + 1)$

$= (\beta - 1)(\beta + \beta^2 + \beta^3 + \beta^4 + 1)$

$= 0$

then $\beta \neq 1$ so $\beta + \beta^2 + \beta^3 + \beta^4 + 1 = 0$.

Rest is Proof: See Worked solutions

5 a 0   **b** −1   **c** 31

    d 8   **e** −1

6 Proof: See Worked solutions

Exercise 4.06

1 a $z_1 = \sqrt{2} \operatorname{cis} \dfrac{\pi}{6}, z_2 = \sqrt{2} \operatorname{cis}\left(-\dfrac{5\pi}{6}\right)$

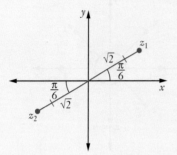

b $z_1 = \operatorname{cis} \dfrac{\pi}{4}, z_2 = \operatorname{cis}\left(-\dfrac{3\pi}{4}\right)$

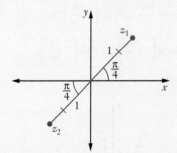

c $z_1 = \sqrt{2} \operatorname{cis}\left(-\dfrac{\pi}{3}\right), z_2 = \sqrt{2} \operatorname{cis} \dfrac{2\pi}{3}$

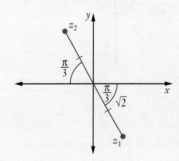

d $z_1 = \operatorname{cis}\left(-\dfrac{\pi}{4}\right), z_2 = \operatorname{cis} \dfrac{3\pi}{4}$

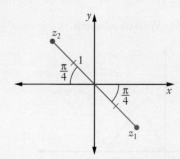

2 a $z_1 = \operatorname{cis} \dfrac{\pi}{3}, z_2 = \operatorname{cis} \pi, z_3 = \operatorname{cis}\left(-\dfrac{\pi}{3}\right)$

    b $z_1 = \operatorname{cis} \dfrac{\pi}{6}, z_2 = \operatorname{cis} \dfrac{5\pi}{6}, z_3 = \operatorname{cis}\left(-\dfrac{\pi}{2}\right)$

    c $z_1 = \operatorname{cis}\left(-\dfrac{\pi}{6}\right), z_2 = \operatorname{cis} \dfrac{\pi}{2}, z_3 = \operatorname{cis}\left(-\dfrac{5\pi}{6}\right)$

3 a $z_1 = 2 \operatorname{cis} \dfrac{\pi}{8}, z_2 = 2 \operatorname{cis} \dfrac{5\pi}{8}, z_3 = 2 \operatorname{cis}\left(-\dfrac{7\pi}{8}\right),$

      $z_4 = 2 \operatorname{cis}\left(-\dfrac{3\pi}{8}\right)$

    b $z_1 = \sqrt[4]{2} \operatorname{cis} \dfrac{\pi}{3}, z_2 = \sqrt[4]{2} \operatorname{cis} \dfrac{5\pi}{6},$

      $z_3 = \sqrt[4]{2} \operatorname{cis}\left(-\dfrac{2\pi}{3}\right), z_4 = \sqrt[4]{2} \operatorname{cis}\left(-\dfrac{\pi}{6}\right)$

    c $z_1 = \operatorname{cis}\left(-\dfrac{\pi}{8}\right), z_2 = \operatorname{cis} \dfrac{3\pi}{8}, z_3 = \operatorname{cis} \dfrac{7\pi}{8},$

      $z_4 = \operatorname{cis}\left(-\dfrac{5\pi}{8}\right)$

4 a, d Proof: See Worked solutions

    b $\operatorname{cis} \dfrac{\pi}{5}, \operatorname{cis} \dfrac{3\pi}{5}, -1, \operatorname{cis}\left(-\dfrac{3\pi}{5}\right), \operatorname{cis}\left(-\dfrac{\pi}{5}\right)$

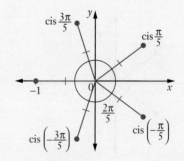

    c $\operatorname{cis} \dfrac{\pi}{5} = \overline{\operatorname{cis}\left(-\dfrac{\pi}{5}\right)}, \operatorname{cis} \dfrac{3\pi}{5} = \overline{\operatorname{cis}\left(-\dfrac{3\pi}{5}\right)}$

5 $z_1 = \operatorname{cis}\dfrac{\pi}{7}, z_2 = \operatorname{cis}\dfrac{3\pi}{7}, z_3 = \operatorname{cis}\dfrac{5\pi}{7}, z_4 = -1,$

$z_5 = \operatorname{cis}\left(-\dfrac{5\pi}{7}\right), z_6 = \operatorname{cis}\left(-\dfrac{3\pi}{7}\right), z_7 = \operatorname{cis}\left(-\dfrac{\pi}{7}\right)$

a $z^7 + 1 = 0$. Sum of roots $= \dfrac{-b}{a} = 0$

b Proof: See Worked solutions

6 a $z_1 = \operatorname{cis}\dfrac{\pi}{6}, z_2 = \operatorname{cis}\dfrac{\pi}{2} = i, z_3 = \operatorname{cis}\dfrac{5\pi}{6},$

$z_4 = \operatorname{cis}\left(-\dfrac{5\pi}{6}\right), z_5 = \operatorname{cis}\left(-\dfrac{\pi}{2}\right) = -i,$

$z_6 = \operatorname{cis}\left(-\dfrac{\pi}{6}\right)$

b $z_1 = \operatorname{cis}\dfrac{\pi}{8}, z_2 = \operatorname{cis}\dfrac{3\pi}{8}, z_3 = \operatorname{cis}\dfrac{5\pi}{8}, z_4 = \operatorname{cis}\dfrac{7\pi}{8},$

$z_5 = \operatorname{cis}\left(-\dfrac{7\pi}{8}\right), z_6 = \operatorname{cis}\left(-\dfrac{5\pi}{8}\right), z_7 = \operatorname{cis}\left(-\dfrac{3\pi}{8}\right),$

$z_8 = \operatorname{cis}\left(-\dfrac{\pi}{8}\right)$

c $z_1 = \operatorname{cis}\dfrac{\pi}{10}, z_2 = \operatorname{cis}\dfrac{\pi}{2} = i, z_3 = \operatorname{cis}\dfrac{9\pi}{10},$

$z_4 = \operatorname{cis}\left(-\dfrac{7\pi}{10}\right), z_5 = \operatorname{cis}\left(-\dfrac{3\pi}{10}\right)$

7 a Proof: See Worked solutions

b $z_2 = \sqrt{2}\operatorname{cis}\left(-\dfrac{7\pi}{12}\right), z_3 = \sqrt{2}\operatorname{cis}\dfrac{\pi}{12}$

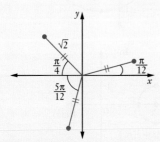

Exercise 4.07

1 a z and w lie on the same line through O (or vector or ray from O) on the same side of O.

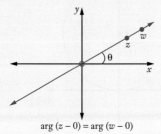

$\arg(z - 0) = \arg(w - 0)$

b z and w lie on the same circle centre O.

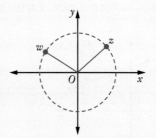

c The line (or vector or ray) OW is a reflection of OZ over the x-axis, but not necessarily the same size.

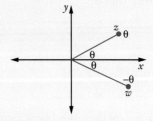

d Vector $z - w \parallel$ vector $u - v$

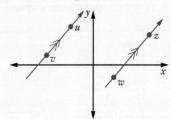

e Lengths of vectors $z - w$ and $u - v$ are equal.

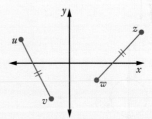

f Vectors $z - w$ and $u - v$ are parallel and equal in length (or collinear with the same argument).

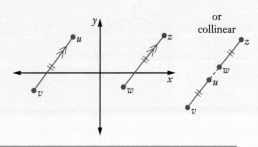

g Lengths of $z - w$ and $z + w$ are equal $\therefore$ O, w, $z + w$ and z form a rectangle (diagonals equal in length).

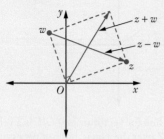

h Either $ZUWV$ form a parallelogram or the vectors are collinear.

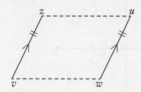

i The diagonals of the quadrilateral formed by O, z, $z + u$ and u are equal and perpendicular; that is, O, z, $z + u$ and u form a square.

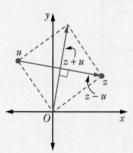

2 a

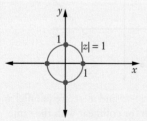

b

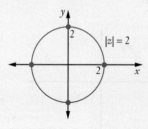

c

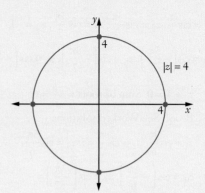

d

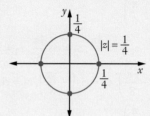

e

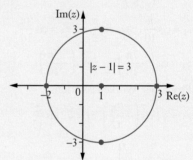

f

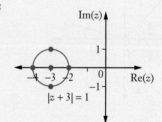

g

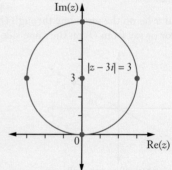

h

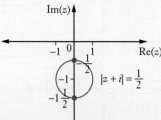

$|z + i| = \frac{1}{2}$

3 a

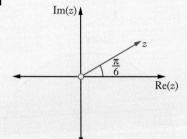

b

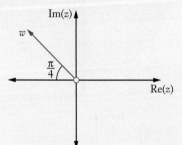

c

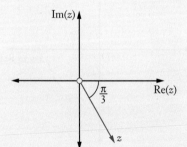

d

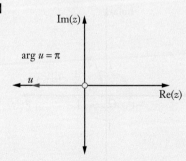

$\arg u = \pi$

4 a

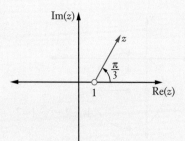

b

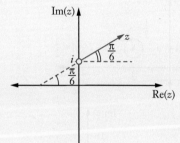

c

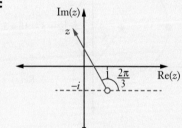

d

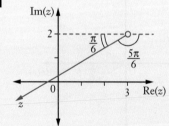

5 a

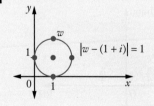

$|w - (1 + i)| = 1$

b

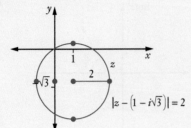

$|z - (1 - i\sqrt{3})| = 2$

c

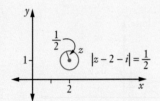

$|z - 2 - i| = \frac{1}{2}$

d

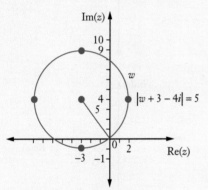

$|w + 3 - 4i| = 5$

6 a

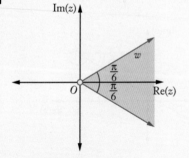

b

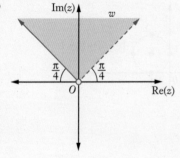

c

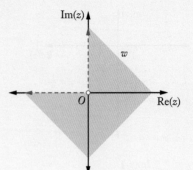

7 a

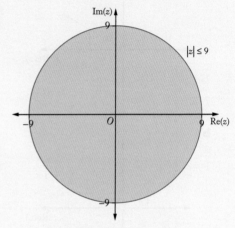

$|z| \leq 9$

b

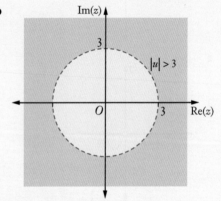

$|u| > 3$

c

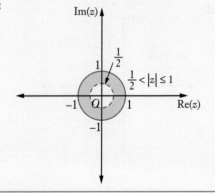

$\frac{1}{2} < |z| \leq 1$

d

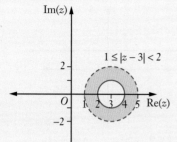

$1 \leq |z - 3| < 2$

8 a

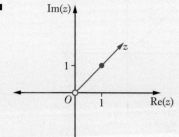

b

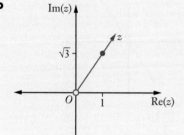

c

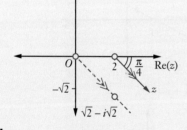

d

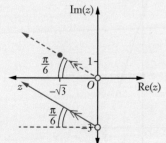

9

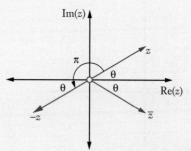

$\arg(-z) = \arg(-1 \times z) = \arg(-1) + \arg z = \pi + \theta$

10 a

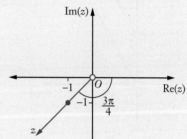

b

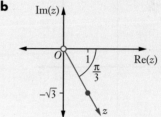

c

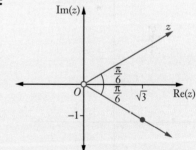

d

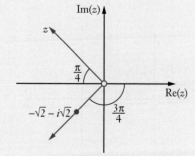

ISBN 9780170413435

e

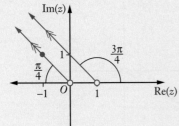

f

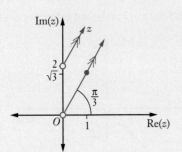

11 a

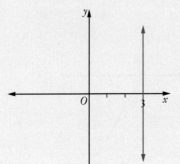

b

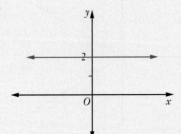

c

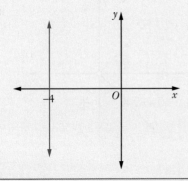

d

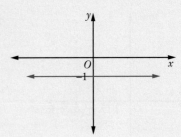

12 a $x + y = 0$

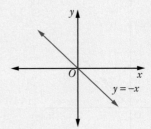

b $y = 2x$

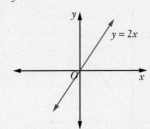

c $y = \dfrac{x+1}{2}$

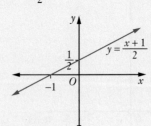

d $3x + y = 6$

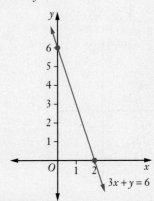

MATHS IN FOCUS 12. Mathematics Extension 2   ISBN 9780170413435

13 a

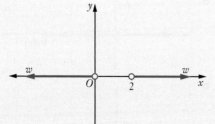

b

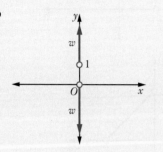

c

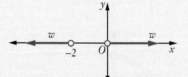

d

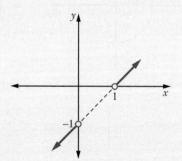

14

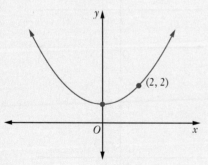

15 a

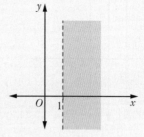

b

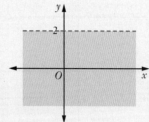

c

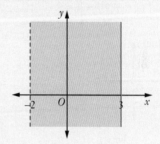

d

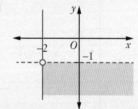

e

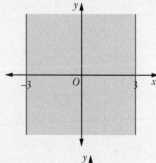

f

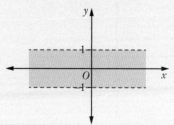

g

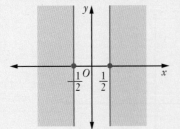

h

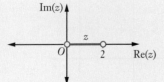

16 a

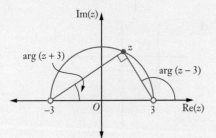

b

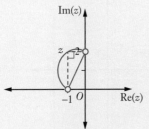

c

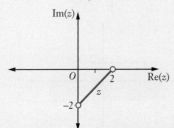

d

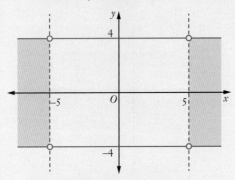

17 a

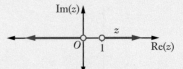

b

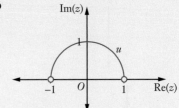

c

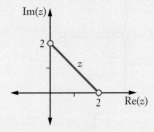

d

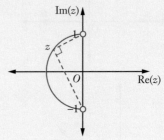

18 perpendicular bisectors of line joining points

a

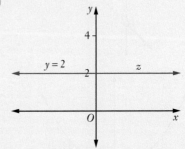

b

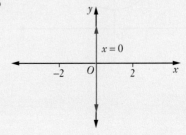

c

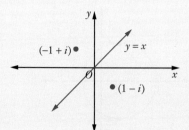

d

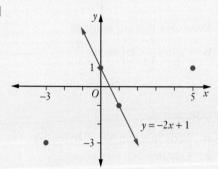

19 a $y = x + 2$ for $x > 2$ or $x < 0$

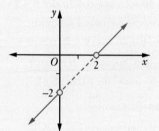

b semicircle centre $(-1, 1)$ radius $\sqrt{2}$

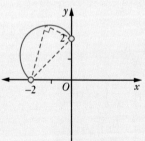

20 a

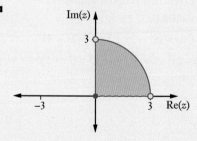

b

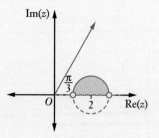

c

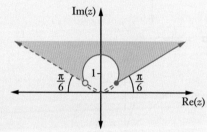

d

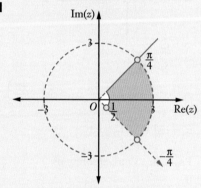

21, 22 Proof: See Worked solutions

23

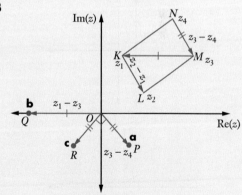

24 Proof: See Worked solutions

25 $b = \bar{a}, c = a \operatorname{cis} \dfrac{\pi}{4}, d = -2a \operatorname{cis} \dfrac{\pi}{4}$

26 a $r = k\dfrac{p}{q}$

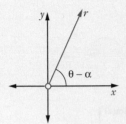

b $r = pq$

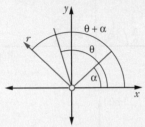

27 a $\alpha = \dfrac{2\pi}{5}$

b $z_1 = \text{cis}\,\dfrac{2\pi}{5}, z_2 = \text{cis}\,\dfrac{4\pi}{5}, z_3 = \text{cis}\left(-\dfrac{4\pi}{5}\right),$

$z_4 = \text{cis}\left(-\dfrac{2\pi}{5}\right), z_5 = 1$

c–e Proof: See Worked solutions

f 0

28 a

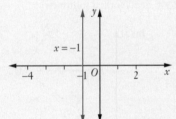

b

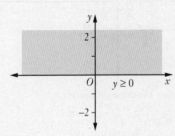

c

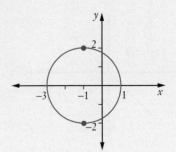

29 Proof: See Worked solutions

b Centre $\left(\dfrac{7}{3}, 0\right)$, radius $\dfrac{8}{3}$

30 $\dfrac{\pi}{6}$

Test yourself 4

1 a $9(\cos 4\delta + i \sin 4\delta)$ **b** $\text{cis}\,\dfrac{\pi}{6}$

 c $3\,\text{cis}\,36°$

2 a i Proof: See Worked solutions

 ii $\sin 5\theta = 16 \sin^5 \theta - 20 \sin^3 \theta + 5 \sin \theta$

 b i Proof: See Worked solutions

 ii $\cos 5\theta = 16 \cos^5 \theta - 20 \cos^3 \theta + 5 \cos \theta$

3 a $\tan 5\theta = \dfrac{\tan^5 \theta - 10 \tan^3 \theta + 5 \tan \theta}{5 \tan^4 \theta - 10 \tan^2 \theta + 1}$

 b $x = \pm \tan \dfrac{\pi}{5}, \pm \tan \dfrac{2\pi}{5}$

4 a Proof: See Worked solutions

 b $-\dfrac{1}{7}\cos 7\theta + \dfrac{7}{5}\cos 5\theta - 7 \cos 3\theta + C$

5 a $z = -3i, i$ **b** $w = 1 - i, 1 - 2i$

 c $x = \pm \dfrac{3(1 - i)}{\sqrt{2}}$

6 a $z = \sqrt{2}\,\text{cis}\left(-\dfrac{\pi}{6}\right), \sqrt{2}\,\text{cis}\,\dfrac{5\pi}{6}$

 b $z = \sqrt{2}\,\text{cis}\left(-\dfrac{3\pi}{8}\right), \sqrt{2}\,\text{cis}\,\dfrac{5\pi}{8}$

7 a Proof: See Worked solutions

 b $x = 1 \pm i\sqrt{2}, 1 \pm i\sqrt{3}$

 c $P(x) = (x^2 - 2x + 3)(x^2 - 2x + 4)$

8 $P(x) = 2x^3 + 9x^2 + 30x + 13 = (2x + 1)(x^2 + 4x + 13)$,

 $x = -\dfrac{1}{2}, -2 \pm 3i$

9 a i $zw = 2\sqrt{2}\,\text{cis}\,\dfrac{3\pi}{4}$

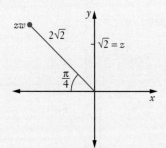

ii $\dfrac{z}{w} = \dfrac{1}{\sqrt{2}}\,\text{cis}\,\dfrac{\pi}{4}$

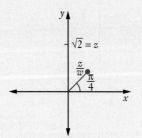

b i $zw = 6\,\text{cis}\,\dfrac{\pi}{6}$

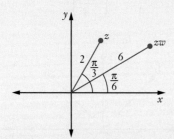

ii $\dfrac{z}{w} = \dfrac{2}{3}\,\text{cis}\,\dfrac{\pi}{2}$

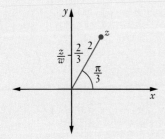

c i $zw = 2\sqrt{2}\,\text{cis}\,\dfrac{\pi}{4}$

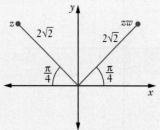

ii $\dfrac{z}{w} = 2\sqrt{2}\,\text{cis}\left(-\dfrac{3\pi}{4}\right)$

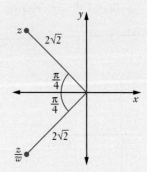

10

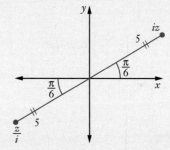

11

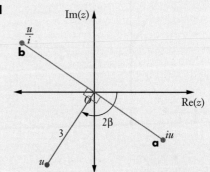

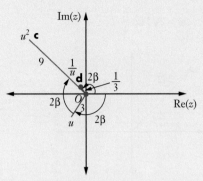

12 a $w = u + v$

b, d Proof: See Worked solutions

c $u - w$ is $v - w$ rotated 90° anticlockwise

e $m = \dfrac{1}{2}(u + iu)$

13 a $z = \pm 1$

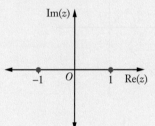

b $z = 1, \text{cis}\dfrac{2\pi}{3}, \text{cis}\left(-\dfrac{2\pi}{3}\right)$; conjugates

$\text{cis}\dfrac{2\pi}{3} = \overline{\text{cis}\left(-\dfrac{2\pi}{3}\right)}$

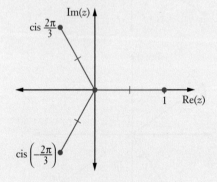

c $z = \text{cis}\dfrac{\pi}{3}, \text{cis}\dfrac{2\pi}{3}, -1, \text{cis}\left(-\dfrac{2\pi}{3}\right), \text{cis}\left(-\dfrac{\pi}{3}\right), 1$;

conjugates $\text{cis}\dfrac{\pi}{3} = \overline{\text{cis}\left(-\dfrac{\pi}{3}\right)}$,

$\text{cis}\dfrac{2\pi}{3} = \overline{\text{cis}\left(-\dfrac{2\pi}{3}\right)}$

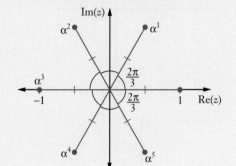

d $z = \text{cis}\dfrac{\pi}{4}, i, \text{cis}\dfrac{3\pi}{4}, -1, \text{cis}\left(-\dfrac{3\pi}{4}\right), -i,$

$\text{cis}\left(-\dfrac{\pi}{4}\right), 1$; conjugates $\text{cis}\dfrac{\pi}{4} = \overline{\text{cis}\left(-\dfrac{\pi}{4}\right)}$,

$i = \overline{-i}, \text{cis}\dfrac{3\pi}{4} = \overline{\text{cis}\left(-\dfrac{3\pi}{4}\right)}$

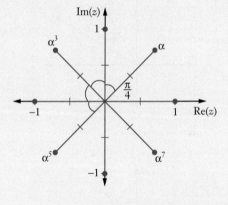

14 a $\alpha = \operatorname{cis} \dfrac{2\pi}{7}, \alpha^2 = \operatorname{cis} \dfrac{4\pi}{7}, \alpha^3 = \operatorname{cis} \dfrac{6\pi}{7},$

$\alpha^4 = \operatorname{cis}\left(-\dfrac{6\pi}{7}\right) = \overline{\alpha^3}, \alpha^5 = \operatorname{cis}\left(-\dfrac{4\pi}{7}\right) = \overline{\alpha^2},$

$\alpha^6 = \operatorname{cis}\left(-\dfrac{2\pi}{7}\right) = \overline{\alpha}, \alpha^7 = 1$

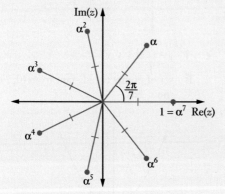

b Sum of roots $= -\dfrac{b}{a} = 0$, or factorise

$(\alpha - 1)(\alpha^6 + \alpha^5 + \alpha^4 + \alpha^3 + \alpha^2 + \alpha + 1) = 0$
and α is not real.

c $z^7 - 1 = (z - 1)(z - \alpha)(z - \overline{\alpha})(z - \alpha^2)(z - \overline{\alpha^2})$

$(z - \alpha^3)(z - \overline{\alpha^3}) = (z - 1)(z^2 - 2z \cos \dfrac{2\pi}{7} + 1)$

$(z^2 - 2z \cos \dfrac{4\pi}{7} + 1)(z^2 - 2z \cos \dfrac{6\pi}{7} + 1)$

d–e Proofs: See Worked solutions

15 a Proof: See Worked solutions

b $x = \cos \dfrac{\pi}{9}, \cos \dfrac{7\pi}{9}, \cos \dfrac{5\pi}{9}$

c Proof: See Worked solutions

16 a -1 **b** 0 **c** 8

17 Proof: See Worked solutions

18 a $\pm(4 - i)$ or only $4 - i$ if $a > 0$

b $\pm e^{\frac{i\pi}{8}}$

c $\pm 2\left(\cos \dfrac{\pi}{12} + i \sin \dfrac{\pi}{12}\right)$

19 a $z_1 = 2 \operatorname{cis} \dfrac{\pi}{4}, z_2 = 2 \operatorname{cis} \dfrac{3\pi}{4}, z_3 = 2 \operatorname{cis}\left(-\dfrac{3\pi}{4}\right),$

$z_4 = 2 \operatorname{cis}\left(-\dfrac{\pi}{4}\right)$

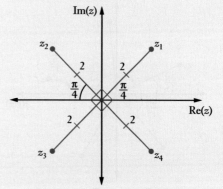

b $z_1 = \sqrt[6]{2} \operatorname{cis}\left(-\dfrac{\pi}{4}\right), z_2 = \sqrt[6]{2} \operatorname{cis} \dfrac{5\pi}{12},$

$z_3 = \sqrt[6]{2} \operatorname{cis}\left(-\dfrac{11\pi}{12}\right)$

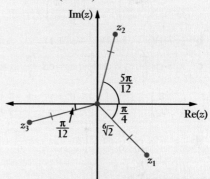

c $z_1 = 2 \operatorname{cis}\left(-\dfrac{\pi}{10}\right), z_2 = -2i, z_3 = 2 \operatorname{cis}\left(-\dfrac{9\pi}{10}\right),$

$z_4 = 2 \operatorname{cis} \dfrac{7\pi}{10}, z_5 = 2 \operatorname{cis} \dfrac{3\pi}{10}$

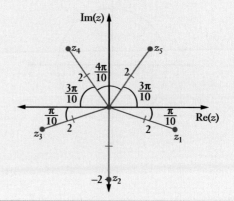

20 $z_1 = \operatorname{cis} \dfrac{\pi}{15}$, $z_2 = \operatorname{cis} \dfrac{7\pi}{15}$, $z_3 = \operatorname{cis} \dfrac{13\pi}{15}$,

$z_4 = \operatorname{cis}\left(-\dfrac{11\pi}{15}\right)$, $z_5 = \operatorname{cis}\left(-\dfrac{\pi}{3}\right)$

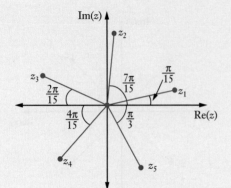

21 a $z_1 = \operatorname{cis} \dfrac{\pi}{9}$, $z_2 = \operatorname{cis} \dfrac{\pi}{3}$, $z_3 = \operatorname{cis} \dfrac{5\pi}{9}$, $z_4 = \operatorname{cis} \dfrac{7\pi}{9}$,

$z_5 = -1$, $z_6 = \operatorname{cis}\left(-\dfrac{7\pi}{9}\right)$, $z_7 = \operatorname{cis}\left(-\dfrac{5\pi}{9}\right)$,

$z_8 = \operatorname{cis}\left(-\dfrac{\pi}{3}\right)$, $z_9 = \operatorname{cis}\left(-\dfrac{\pi}{9}\right)$

b $(z+1)(z^2-z+1)(z^6-z^3+1)$

c $z_1 = \operatorname{cis} \dfrac{\pi}{9}$, $z_3 = \operatorname{cis} \dfrac{5\pi}{9}$, $z_4 = \operatorname{cis} \dfrac{7\pi}{9}$, $z_6 = \operatorname{cis}$

$\left(-\dfrac{7\pi}{9}\right)$, $z_7 = \operatorname{cis}\left(-\dfrac{5\pi}{9}\right)$, $z_9 = \operatorname{cis}\left(-\dfrac{\pi}{9}\right)$

d $\left(z^2 - 2z \cos \dfrac{\pi}{9} + 1\right)$, $\left(z^2 - 2z \cos \dfrac{5\pi}{9} + 1\right)$,

$\left(z^2 - 2z \cos \dfrac{7\pi}{9} + 1\right)$

22 a i $x^2 + y^2 = 36$

  ii set of points 6 units from O

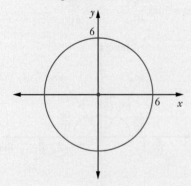

b i $(x-2)^2 + (y-1)^2 = 1$

  ii set of points 1 unit from $(2, 1)$

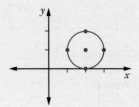

c i $y = \dfrac{1}{4}x^2 + 1$

  ii set of points equidistant from $(0, 2)$ and the x-axis

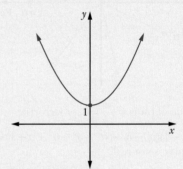

d i $y = 2 - x$

  ii perpendicular bisector of line joining $(2, 2)$ and $(0, 0)$

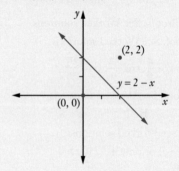

e **i** $3x + 2y + 5 = 0$

ii perpendicular bisector of line joining $(-6, 0)$ and $(0, 4)$

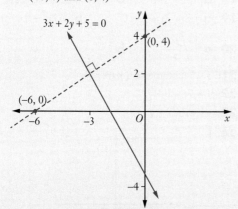

23 a

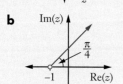

b

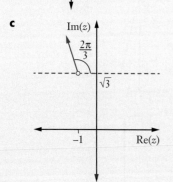

c

24 a

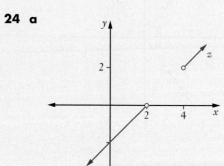

b

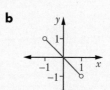

c

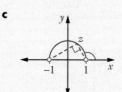

25 a

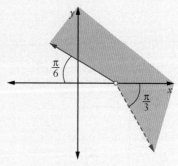

b

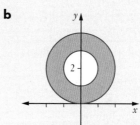

c

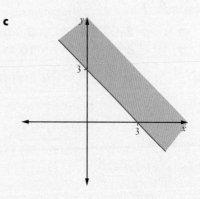

26 a perpendicular bisector

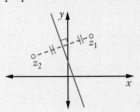

b

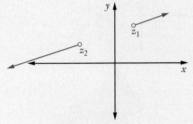

c z is intersection of ray and circle

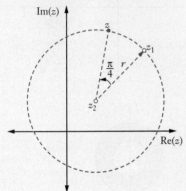

27 Proof: See Worked solutions

Practice set 1

1 C	**2** C	**3** D	**4** C	**5** D
6 B	**7** D	**8** D	**9** A	**10** D

11 a $4i$ **b** $\dfrac{i\sqrt{7}}{2}$ **c** $3 \pm i\sqrt{3}$

12 a 1 **b** $-1 - i$

13 a $x = \pm 8i$ **b** $x = -1 \pm i\sqrt{6}$

c $x = 3 \pm 3i$

14 a $\text{Re}(z) = \dfrac{5}{3}, \text{Im}(z) = -\dfrac{2}{3}$

b $\text{Re}(z) = \dfrac{x+7}{x^2 + y^2}, \text{Im}(z) = \dfrac{2-y}{x^2 + y^2}$

15 a $\bar{z} = 5x + 3iy$

b $\bar{z} = \dfrac{6b - 2a + i(b - a)}{4}$

16 Proof: See Worked solutions

17 a $x = 6, y = -3$ **b** $x = 4, y = 2$

18 a $11 - 19i$ **b** 7 **c** $20i$

19 a $x^2 - 2x + 5 = 0$ **b** $12x^2 + 4x + 1 = 0$

20 a $\dfrac{1 + i\sqrt{3}}{2}$ **b** $\dfrac{2}{9}$ **c** $\dfrac{i}{2}$

21 $5 - i$ (assume real part >0)

22 $1 + 7i, -1 - 7i$

23 $-1 + 3i, 2 - i$

24 i

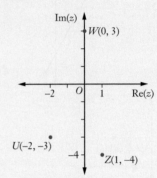

ii

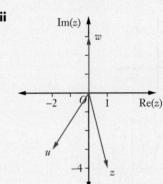

25

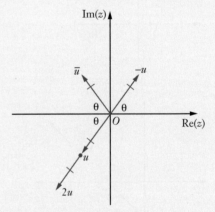

26

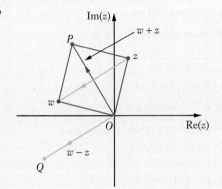

27 a $r = 2, \operatorname{Arg} z = \dfrac{2\pi}{3}$

b $r = 2\sqrt{2}, \operatorname{Arg} z = -\dfrac{\pi}{4}$

c $r = \sqrt{2}, \operatorname{Arg} z = -\dfrac{5\pi}{6}$

28 a $3\left[\cos\left(-\dfrac{\pi}{3}\right) + i\sin\left(-\dfrac{\pi}{3}\right)\right]$

b $\sqrt{2}\left[\cos\left(-\dfrac{\pi}{4}\right) + i\sin\left(-\dfrac{\pi}{4}\right)\right]$

c $5\left(\cos\dfrac{\pi}{2} + i\sin\dfrac{\pi}{2}\right)$

d $4\left(\cos\dfrac{2\pi}{3} + i\sin\dfrac{2\pi}{3}\right)$

e $\dfrac{\sqrt{2}}{3}\left(\cos\dfrac{\pi}{4} + i\sin\dfrac{\pi}{4}\right)$

29 a $u = 2\sqrt{2}\left(\cos\dfrac{3\pi}{4} + i\sin\dfrac{3\pi}{4}\right),$

$v = 2\left[\cos\left(-\dfrac{\pi}{3}\right) + i\sin\left(-\dfrac{\pi}{3}\right)\right]$

b $u = -2 + 2i, v = 1 - i\sqrt{3}$

30 a $r_1 r_2(\cos(\alpha_1 + \alpha_2) + i\sin(\alpha_1 + \alpha_2))$

b $\dfrac{r_1}{r_2}(\cos(\alpha_1 - \alpha_2) + i\sin(\alpha_1 - \alpha_2))$

31 a $n\theta$ **b** $-n\theta$ **c** $-n\theta$ **d** $n\theta$

32 Proof: See Worked solutions

33 a $z_1 z_2 = 6i$ **b** $\dfrac{z_1}{z_2} = \dfrac{3}{2}\operatorname{cis}\left(-\dfrac{5\pi}{6}\right)$

c $(z_2)^3 = -8$ **d** $(z_1)^{-4} = \dfrac{1}{81}\operatorname{cis}\dfrac{2\pi}{3}$

34 a 16   **b** $2\operatorname{cis}\dfrac{11\pi}{12}$

c $\dfrac{1}{16}\operatorname{cis}\left(-\dfrac{2\pi}{3}\right)$

35 $(1 + i)(\sqrt{3} + i) = (\sqrt{3} - 1) + i(1 + \sqrt{3}) = 2\sqrt{2}\operatorname{cis}\dfrac{5\pi}{12};$

$\sin\dfrac{5\pi}{12} = \dfrac{1 + \sqrt{3}}{2\sqrt{2}}$

36 a e^{3i} **b** $4e^{-\frac{i\pi}{5}}$ **c** $2e^{-\frac{5i\pi}{6}}$

37 a $2\operatorname{cis}3\alpha$ **b** $\operatorname{cis}\left(-\dfrac{\pi}{7}\right)$

c $\dfrac{1}{2}\operatorname{cis}\left(-\dfrac{2\pi}{3}\right)$

38 a $6e^{\frac{5i}{2}}$ **b** $e^{-\frac{i\pi}{5}}$

39 Proof: See Worked solutions

40 It rains $\Rightarrow$ the dam is full.

41 If the people are starving then there is not enough food.

42 a If a number is divisible by 2 then it is even. Yes, $P \Leftrightarrow Q$. A number is divisible by 2 iff it is even.

b If its reciprocal is positive then a number is positive. Yes, $P \Leftrightarrow Q$.

c If a quadrilateral it is a rectangle then it has four equal angles. Yes, $P \Leftrightarrow Q$.

d If it eats grass then an animal is a kangaroo. False. $P \not\Leftrightarrow Q$.

43 a The dam is not full.

b The teacher is not good.

c There is at least one cat that is not fluffy.

d There are no smart politicians.

e At least one wine is sweet.

f No sheep are black.

44 a If you do not speed then you do not get a speeding ticket. True.

b If you are not over 65 then you do not get the old-age pension. True.

c If a triangle does not have 3 equal sides then it is not equilateral. True.

d If you do not get wet then you do not go swimming. True.

45 For all natural numbers x there exists a natural number y such that $y = 2x$.

46 $\forall x \in \mathbb{N}: x = 4M$ for some $M \in \mathbb{N},$

$\exists y \in \mathbb{N}: \sqrt{x} = 2\sqrt{y}$

47 C

48 Proof: See Worked solutions

49 Counter-example $f(x) = (x-3)^4$ at $x = 3$

50, 51 Proofs: See Worked solutions

52 $\underset{\sim}{u} \cdot \underset{\sim}{v} = 0$; angle = 90°

53 a $\overrightarrow{AB} = \begin{pmatrix} 2 \\ 3 \\ -2 \end{pmatrix}$ **b** $\left| \overrightarrow{AB} \right| = \sqrt{17}$

   c $\hat{u} = \dfrac{1}{\sqrt{17}} \begin{pmatrix} 2 \\ 3 \\ -2 \end{pmatrix}$

54 $n = 9$

55 a Proof: See Worked solutions

   b $\left| \overrightarrow{AB} \right| = 6, \left| \overrightarrow{CD} \right| = 3$

   c Trapezium

56 136° 13′

57 $\underset{\sim}{r} = \begin{pmatrix} 1 \\ 3 \\ -2 \end{pmatrix} + \lambda \begin{pmatrix} 3 \\ -5 \\ 9 \end{pmatrix}$

58 $x - 3 = \dfrac{y+1}{6} = \dfrac{-z+3}{2}$

59 No

60 $\underset{\sim}{r} = \begin{pmatrix} 2 \\ 1 \\ 4 \end{pmatrix} + \lambda \begin{pmatrix} 6 \\ 2 \\ 3 \end{pmatrix}$

61

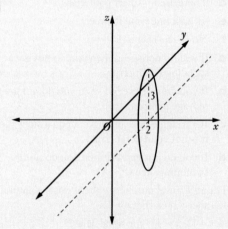

62 Proof: See Worked solutions

63 a $4\sqrt{2}(\cos 15\beta + i \sin 15\beta)$

   b $2(\cos 16° + i \sin 16°)$

   c $\dfrac{1}{256}\left(\cos \dfrac{2\pi}{3} + i \sin \dfrac{2\pi}{3} \right)$

64 a Proof: See Worked solutions

   b $\theta = \dfrac{(2k-1)\pi}{12}, k \in \mathbb{Z}.$

   c Proof: See Worked solutions

65 a Proof: See Worked solutions

   b $\left(z - \dfrac{1}{z} \right)^5 = (2i \sin \theta)^5$ and

    $\left(z - \dfrac{1}{z} \right)^5 = \left(z^5 - \dfrac{1}{z^5} \right) - 5\left(z^3 - \dfrac{1}{z^3} \right) + 10\left(z - \dfrac{1}{z} \right)$

   using the binomial theorem;

   $\sin^5 \theta = \dfrac{1}{16} \sin 5\theta - \dfrac{5}{16} \sin 3\theta + \dfrac{5}{8} \sin \theta;$

   $A = \dfrac{1}{16}, B = -\dfrac{5}{16}, C = \dfrac{5}{8}$

   c $\dfrac{8}{15}$

66 $z = 2i \pm 2\sqrt{2}$

67 $z = \sqrt{2} \operatorname{cis} \dfrac{\pi}{3}, \sqrt{2} \operatorname{cis}\left(-\dfrac{2\pi}{3} \right)$

68 a Degree is odd and coefficients are real so must have 1 real root because complex roots come in conjugate pairs.

   b Proof: See Worked solutions

   c $x = 1, \pm i\sqrt{2}, -\dfrac{1}{2} \pm \dfrac{i\sqrt{3}}{2}$

   d $P(x) = (x - 1)(x^2 + 2)(x^2 + x + 1)$

69

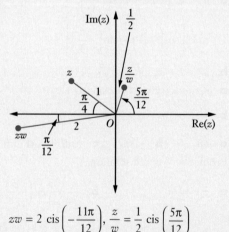

$zw = 2 \operatorname{cis}\left(-\dfrac{11\pi}{12} \right), \dfrac{z}{w} = \dfrac{1}{2} \operatorname{cis}\left(\dfrac{5\pi}{12} \right)$

70 $w = -z = i^2 z, v = \dfrac{z}{i}, u = iz$

71 **a** $d = c + a - b$

   b $\arg\left(\dfrac{d-a}{b-a}\right) = \dfrac{3\pi}{4}$

   c $m = \dfrac{1}{2}(c + a - 2b)$

72 **a** $z_1 = \alpha = \operatorname{cis}\dfrac{2\pi}{5}, z_2 = \alpha^2 = \operatorname{cis}\dfrac{4\pi}{5},$

   $z_3 = \alpha^3 = \overline{\alpha^2} = \operatorname{cis}\left(-\dfrac{4\pi}{5}\right),$

   $z_4 = \alpha^4 = \overline{\alpha} = \operatorname{cis}\left(-\dfrac{2\pi}{5}\right), z_5 = \alpha^5 = 1$

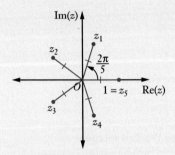

   b Proof: See Worked solutions

   c $z^5 - 1 =$

   $(z-1)\left(z^2 - 2z\cos\dfrac{2\pi}{5} + 1\right)\left(z^2 - 2z\cos\dfrac{4\pi}{5} + 1\right)$

   d Proof: See Worked solutions

73 **a** 0   **b** 0   **c** 3

74 **a** $e^{\frac{i\pi}{4}}, e^{-\frac{3i\pi}{4}}$

   b $3\left(\cos\dfrac{\pi}{6} + i\sin\dfrac{\pi}{6}\right), 3\left[\cos\left(-\dfrac{5\pi}{6}\right) + i\sin\left(-\dfrac{5\pi}{6}\right)\right]$

75 **a** $z = -2, 2\operatorname{cis}\dfrac{\pi}{3}, 2\operatorname{cis}\left(-\dfrac{\pi}{3}\right)$

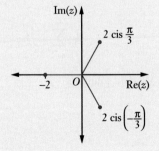

   b $z = \sqrt[4]{2}\operatorname{cis}\dfrac{\pi}{6}, \sqrt[4]{2}\operatorname{cis}\dfrac{2\pi}{3}, \sqrt[4]{2}\operatorname{cis}\left(-\dfrac{5\pi}{6}\right), \sqrt[4]{2}\operatorname{cis}\left(-\dfrac{\pi}{3}\right)$

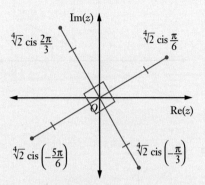

76 **a**

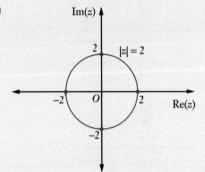

   b

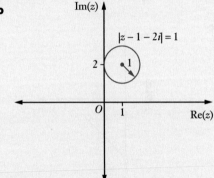

   c

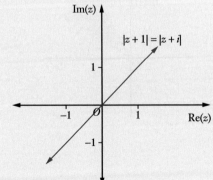

d

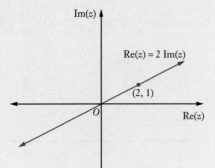

77 a

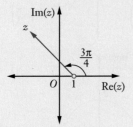

b

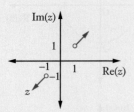

c

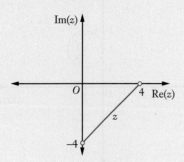

78 a

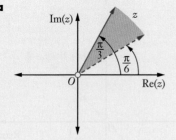

b

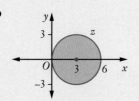

c

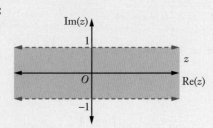

79 Parallelogram

Chapter 5

Exercise 5.01

All proofs: See Worked solutions

Exercise 5.02

All proofs: See Worked solutions

Exercise 5.03

1 a $1^2 + 2^2 + 3^2 + 4^2 + 5^2 + 6^2 + 7^2 + 8^2 + 9^2 + 10^2$

  b $5 + 7 + 9 + ... + (2(n-1) + 3) + (2n + 3)$

  c $\dfrac{1}{1} + \dfrac{1}{2} + \dfrac{1}{3} + ... + \dfrac{1}{M+1}$

  d $-2 + 3 - 4 + ... + 9$

  e $\dfrac{1}{2^0} + \dfrac{1}{2^1} + \dfrac{1}{2^2} + \dfrac{1}{2^3} + ...$

2 a 18   **b** 121

  c 20   **d** 0.0656...

3 a $\displaystyle\sum_{r=1}^{77} (-1)^r r^2$ **b** $\displaystyle\sum_{r=1}^{n-1} \dfrac{1}{r+1}$ or $\displaystyle\sum_{r=2}^{n} \dfrac{1}{r}$

  c $\displaystyle\sum_{r=1}^{99} 3^r$ **d** $\displaystyle\sum_{r=1}^{\infty} \dfrac{(-1)^{r-1}}{2^{r-1}}$

4–6 Proofs: See Worked solutions

Exercise 5.04

1–5 Proof: See Worked solutions

6 $S_n = \dfrac{n^2+n+2}{2}$, proof: see Worked solutions

7 Proof: See Worked solutions

Exercise 5.05

All proofs: See Worked solutions

Exercise 5.06

1 a $x < 1$
    b $-1 \le x \le 3$
  c $-1 < x < 1, x \ne 0$
    d $x \le -3$ or $x \ge 3$
  e $x < 0$ or $x > 1$

2 a $y = |2x|$ lies above $y = |x| - 1$ for all real values of x, so solution is $x \in \mathbb{R}$.

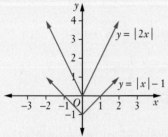

b $y = |x+1|$ never lies below $y = \sqrt{x-1}$ so there is no solution.

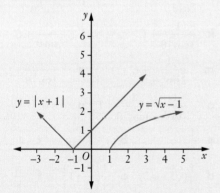

3 $x < -1$ or $x > 2$

4 a Maximum stationary point $\left(1, \dfrac{1}{e}\right)$

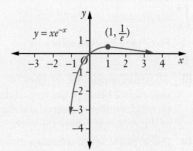

b Proof: See Worked solutions

5 a $m_{\text{secant}} = n \ln\left(\dfrac{n+1}{n}\right)$

b, c Proof: See Worked solutions

d The smaller the compounding interval the higher the rate of return on investment.

6 a The area under the curve is approximated by the sum of the rectangles as $n \to \infty$.

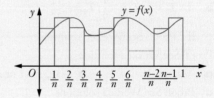

b $\dfrac{2}{\pi}$

7 a Proof: See Worked solutions

b i $\dfrac{1}{2}a, \dfrac{\sqrt{3}}{4}a, \dfrac{3}{8}a, ..., r = \dfrac{\sqrt{3}}{2}$

ii $\dfrac{1}{2}a \times \left(\dfrac{\sqrt{3}}{2}\right)^{n-1}$

iii Proof using part **a**; and let $n \to \infty$. See Worked solutions.

8 a The area under the curve denoted by $\displaystyle\int_{1}^{\sqrt{p}} \dfrac{1}{x}\, dx$ is less than the area of the rectangle, which is $(\sqrt{p}-1) \times 1$.

b, c Proof: See Worked solutions

9 a $\dfrac{d}{dx}(x \ln x) = \ln x + 1$

b $\displaystyle\int_1^n \ln x \, dx. = n \ln n - n + 1$

c **i** Proof: See Worked solutions

    ii $\ln(n!)$

    iii The exact area under the curve is sandwiched between the sum of the rectangles under the curve and the sum of the rectangles above the curve.

    iv Proof: See Worked solutions

Test yourself 5

1–11 Proofs: See Worked solutions

12 a $K^n - L^n$

b, c Proof: See Worked solutions

13 a $\cos(A + B) = \cos A \cos B - \sin A \sin B$

b, c Proof: See Worked solutions

14 a Proof: See Worked solutions

b **i** $T_2 = \dfrac{1}{3}, T_3 = -\dfrac{1}{9}, T_4 = -\dfrac{11}{27}$

    ii Proof: See Worked solutions

15 Proof: See Worked solutions

16 a $-2 \le x \le 2$ **b** $-3 \le x < 1$ or $x \ge 2$

17 a Proof: See Worked solutions

b Approximate the area of the circle using rectangles under the curve and above the curve.

18 a Since curve is concave down then area under curve > area of trapezium.

b Proof: See Worked solutions

Chapter 6

Exercise 6.01

1 a $\ln(e^x + 1) + C$

b $2 \tan^{-1}\left(e^{\frac{x}{2}}\right) + C$ **c** $\dfrac{1}{10}(1 + x^2)^5 + C$

d $-\dfrac{2}{3}\sqrt{1-x}\,(x + 2) + C$

e $2\sqrt{e^x - 1} + C$

f $\dfrac{1}{a} \tan^{-1}\dfrac{x}{a} + C$

g $\dfrac{2}{5}(x-3)^2\sqrt{x-3} + 2(x-3)\sqrt{x-3} + C$

h $\dfrac{2}{3}(x-2)\sqrt{x+1} + C$

i $\dfrac{2}{3}\sqrt{x-1}\,(x+2) + C$

2 a $2e^{\sqrt{x}} + C$ **b** $-\dfrac{1}{2e^{x^2}} + C$

c $\sqrt{x^2 + 1} + C$ **d** $\dfrac{1}{3}(1 + \ln x)^3 + C$

e $\dfrac{1}{3}(1 + \sin^{-1} x)^3 + C$ **f** $\dfrac{1}{2}[\tan^{-1}(x + 1)]^2 + C$

g $-\dfrac{1}{\ln x} + C$ **h** $\ln|\ln x| + C$

3 a $\ln 2$ **b** $\dfrac{1}{2}$

c $\dfrac{3}{8}$ **d** $\dfrac{1}{2}$

e $-\dfrac{1}{5}\sin^5 x + \dfrac{1}{3}\sin^3 x + C$

f $\dfrac{1}{5}\cos^5 x - \dfrac{1}{3}\cos^3 x + C$

g $\dfrac{1}{9}\sin^9 x - \dfrac{2}{7}\sin^7 x + \dfrac{1}{5}\sin^5 x + C$

h $-\dfrac{1}{9}\cos^9 x + \dfrac{2}{7}\cos^7 x - \dfrac{1}{5}\cos^5 x + C$

i $\dfrac{1}{\cos\theta} + C$ **j** $\dfrac{1}{2\cos^2\theta} + C$

k $\dfrac{1}{3\cos^3\theta} + C$ **l** $-\dfrac{1}{\sin\theta} + C$

m $-\dfrac{1}{2\sin^2\theta} + C$ **n** $-\dfrac{1}{3\sin^3\theta} + C$

o $-\dfrac{1}{\cos\theta} + \dfrac{1}{3\cos^3\theta} + C$ **p** $\cos\theta + \dfrac{1}{\cos\theta} + C$

4 a $\dfrac{1}{5}$ **b** $e - 1$ **c** 1

d $\dfrac{1}{2}(\sqrt{3} - \sqrt{2})$ **e** $\dfrac{1}{4}\left(\dfrac{\pi}{2} + 1\right)$ **f** $\dfrac{\pi}{8}$

g $\dfrac{\pi}{8}$

5 a $\dfrac{5}{3}$ **b** $\dfrac{9\pi}{4}$ **c** $\dfrac{\pi}{2}$

d $\dfrac{1}{2} - \dfrac{1}{1+e}$ **e** $2 - \dfrac{\pi}{2}$

6 $\dfrac{9\pi}{4}$

7 a 21   **b** $\dfrac{1}{2}\log_e 3$

  c $-\cot\dfrac{x}{2}+C$

  d, e Proof: See Worked solutions

  f $\dfrac{\sqrt{3}}{2}$

8 a $\dfrac{1}{2}\tan^{-1}\left(\dfrac{x}{2}\right)+C$ **b** $\sin^{-1}\left(\dfrac{x}{3}\right)+C$

  c $\dfrac{1}{6}\tan^{-1}\left(\dfrac{2x}{3}\right)+C$

9 a $\dfrac{\pi}{6}$ **b** $\dfrac{\pi}{3}$ **c** $\dfrac{\pi}{4}$ **d** 2

10 a $32\dfrac{2}{5}$

  b $\log\left|1+\tan\dfrac{x}{2}\right|+C$

  c $\dfrac{1}{3}\sec^3 x+C$ **d** $\dfrac{1}{4}$

  e $\dfrac{1}{2}\log_e(e^{2x}+1)+\tan^{-1}e^x+C$

  f, g Proof: See Worked solutions

Exercise 6.02

1 a $-\dfrac{1}{x+3}+C$ **b** $-\dfrac{1}{x-4}+C$

  c $\dfrac{2}{x+2}+C$ **d** $\tan^{-1}x+C$

  e $\dfrac{1}{3}\tan^{-1}\dfrac{x}{3}+C$ **f** $\dfrac{1}{\sqrt{3}}\tan^{-1}\dfrac{x}{\sqrt{3}}+C$

  g $\dfrac{1}{\sqrt{5}}\tan^{-1}\dfrac{x}{\sqrt{5}}+C$ **h** $-\dfrac{1}{x-2}+C$

2 a $\sin^{-1}x+C$ **b** $\sin^{-1}\dfrac{x}{3}+C$

  c $\sin^{-1}\dfrac{x}{2}+C$ **d** $\dfrac{1}{2}\sin^{-1}\dfrac{2x}{3}+C$

  e $-\sin^{-1}\left(\dfrac{2-x}{2}\right)+C$ or $\sin^{-1}\left(\dfrac{x-2}{2}\right)+C$

  f $-\dfrac{1}{3}\sin^{-1}\left(\dfrac{2-3x}{2}\right)+C$ or $\dfrac{1}{3}\sin^{-1}\left(\dfrac{3x-2}{2}\right)+C$

3 a $\tan^{-1}(x+1)+C$ **b** $\sin^{-1}(x-1)+C$

  c $x+\tan^{-1}x+C$ **d** $\dfrac{1}{2}\tan^{-1}\left(\dfrac{x-1}{2}\right)+C$

4 a $x-\tan^{-1}x+C$ **b** $x-2\tan^{-1}x+C$

  c $x+\dfrac{2}{\sqrt{2}}\tan^{-1}\dfrac{x}{\sqrt{2}}+C$ **d** $x-3\tan^{-1}\dfrac{x}{3}+C$

  e $x+\ln(x^2+1)+C$

  f $x-\ln(x^2+2)-\dfrac{1}{\sqrt{2}}\tan^{-1}\dfrac{x}{\sqrt{2}}+C$

5 a $\ln(x^2+4x+5)-3\tan^{-1}(x+2)+C$

  b $2\ln(x^2+1)+3\tan^{-1}x+C$

  c $\dfrac{1}{2}\ln(x^2+1)+C$

  d $\dfrac{1}{2}\ln(x^2+1)-\tan^{-1}x+C$

  e $\dfrac{1}{2}\ln(x^2-2x+2)+\tan^{-1}(x-1)+C$

  f $\ln|x-1|+C$

6 a $\cos^{-1}x+C$ or $-\sin^{-1}x+C$

  b $\dfrac{1}{2}\sin^{-1}2x+C$

  c $\dfrac{1}{2}\tan^{-1}2x+C$ **d** $\dfrac{1}{2}\tan^{-1}\left(\dfrac{x+2}{2}\right)+C$

  e $\dfrac{1}{6}\tan^{-1}\left(\dfrac{2x+1}{3}\right)+C$ **f** $\tan^{-1}\left(\dfrac{x-3}{2}\right)+C$

7 a π **b** $2\pi\ln 2$ **c** $\dfrac{\pi}{2}\tan^{-1}\dfrac{3}{4}$

Exercise 6.03

1 a $2\ln|x-3|+\ln|x+2|+C$

  b $\ln|x+3|+\dfrac{3}{2}\ln|2x-1|+C$

  c $8\ln|x+3|-5\ln|x+2|+C$

  d $2\ln|x-3|+\ln|x+5|+C$

  e $\dfrac{1}{2}\ln|1+2x|-\dfrac{2}{3}\ln|1-3x|+C$

  f $a\ln|x-a|-b\ln|x-b|+C$

2 a $\ln|x-2|+\ln|x+2|+C=\ln|x^2-4|+C$

  b $\ln|2x-1|-\ln|3x+4|+C$

  c $\dfrac{3}{2}\ln|2x+1|-\ln|x-2|+C$

  d $\ln|x-1|+\ln|x-2|+\ln|x-3|+C$

  e $2\ln|x-1|-\ln|x+1|+\dfrac{3}{2}\ln|2x+3|+C$

  f $\ln|x-6|+C$

3 a $2 \ln |x+1| - \dfrac{3}{x+1} - 2 \ln |3x+2| + C$

b $-\dfrac{1}{4} \ln |x+1| + \dfrac{1}{2(x+1)} + \dfrac{1}{4} \ln |x-1| + C$

c $-\dfrac{2}{x+1} - \dfrac{1}{x-2} - 2 \ln |x-2| + 3 \ln |x+1| + C$

4 a $\dfrac{3}{4} \ln |x^2 + 1| - \dfrac{1}{2} \tan^{-1} x - \dfrac{3}{2} \ln |x+1| + C$

b $-\dfrac{1}{4} \ln |x^2 + 1| + \dfrac{1}{2} \tan^{-1} x + \dfrac{1}{2} \ln |x-1| + C$

c $\tan^{-1} x + C$

5 a $\ln |x+2| - \dfrac{1}{2} \ln |x^2 - 2x + 4| + \sqrt{3} \tan^{-1}\left(\dfrac{x-1}{\sqrt{3}}\right) + C$

b $2 \ln |x-2| - \ln |x^2 + 2x + 4| + \sqrt{3} \tan^{-1}\left(\dfrac{x+1}{\sqrt{3}}\right) + C$

6 a $A = 1, B = -1, C = 1$

b $\ln |x+1| - \dfrac{1}{2} \ln (x^2 + 1) + \tan^{-1} x + C$

7 a $A = -\dfrac{1}{2}, B = \dfrac{1}{2}$ **b** $\dfrac{1}{2} \ln \dfrac{3}{2}$

8 $\dfrac{2x^2 + 5x + 3}{(x-1)^2 (x^2 + 1)} = \dfrac{-\frac{1}{2}}{x-1} + \dfrac{5}{(x-1)^2} + \dfrac{\frac{1}{2}x - \frac{5}{2}}{x^2 + 1};$

$\dfrac{1}{4} \ln (x^2 + 1) - \dfrac{5}{x-1} - \dfrac{1}{2} \ln |x-1| - \dfrac{5}{2} \tan^{-1} x + C$

9 $\dfrac{2x^2 - x - 7}{(x+2)^2 (x^2 + x + 1)} = \dfrac{-2}{x+2} + \dfrac{1}{(x+2)^2} + \dfrac{2x - 1}{x^2 + x + 1};$

$\dfrac{1}{6} - \dfrac{2\pi}{3\sqrt{3}} + \ln \dfrac{4}{3}$

10 a $A = 1, B = 3, C = 3, D = 1$

b $\dfrac{32}{3} + 3 \ln 3$

11 a Proof: See Worked solutions

b $\dfrac{1}{2} \ln |x^2 + 2x + 3| - \dfrac{1}{\sqrt{2}} \tan^{-1}\left(\dfrac{x+1}{\sqrt{2}}\right) + C$

12 a Proof: See Worked solutions

b $\dfrac{\pi}{2ab(a+b)}$

Exercise 6.04

1 a $(x + 1) \ln |x+1| - x + C$

b $x \ln x^2 - 2x + C = 2x(\ln |x| - 1) + C$

c $x \sin x + \cos x + C$

d $-e^{-x}(x + 1) + C$

e $-\dfrac{x}{2} \cos 2x + \dfrac{1}{4} \sin 2x + C$

f $\dfrac{1}{2} e^{2x}\left(x - \dfrac{1}{2}\right) + C$

g $\dfrac{1}{2} x^2 \ln x - \dfrac{1}{4} x^2 + C$

2 a $\dfrac{1}{2} e^x (\sin x - \cos x) + C$

b $\dfrac{1}{2} e^x (\sin x + \cos x) + C$

c $\dfrac{1}{2} e^{x^2} + C$

d $\dfrac{1}{3} x^3 \ln x - \dfrac{1}{9} x^3 + C$

e $-x^2 \cos x + 2x \sin x + 2 \cos x + C$

f $\dfrac{1}{2} x^2 \tan^{-1} x - \dfrac{1}{2} x + \dfrac{1}{2} \tan^{-1} x + C$

g $\dfrac{1}{32} e^{4x}(8x^2 - 4x + 1) + C$

3 a $2 \ln 2 - 1$ **b** 2 **c** 2

d $\dfrac{1}{2}$ **e** -2

f $2(-2 + \ln 5 + \tan^{-1} 2)$ **g** 2π

Exercise 6.05

1 Proofs: See Worked solutions

2 a Proof: See Worked solutions

b $I_5 = \dfrac{5\pi}{32}$

3 Proof: See Worked solutions

4 a $(n - 1) \sin^{n-2} \theta - n \sin^n \theta$

b Proof: See Worked solutions

c $I_4 = \dfrac{3\pi}{16}$

5 Proof: See Worked solutions

6 a, b Proofs: See Worked solutions

c $I_2 = \dfrac{\pi}{4} - \dfrac{2}{3}$

7 a Proof: See Worked solutions

b $I_4 = \dfrac{4}{3}$

Test yourself 6

1 $\dfrac{1}{24} e^{6x^4 + 1} + C$

2 $\dfrac{\pi}{12}$

3 $\dfrac{1}{2}\sec^2\theta + C$

4 Proof: See Worked solutions

5 $\dfrac{3}{10}$

6 $\sin^{-1}\left(\dfrac{3-x}{3}\right) + C$ or $\sin^{-1}\left(\dfrac{x-3}{3}\right) + C$

7 $\tan^{-1}\left(\dfrac{\pi}{4}\right)$

8 $\dfrac{1}{\sqrt{3}}\tan^{-1}\left(\dfrac{x-2}{\sqrt{3}}\right) + C$

9 $\dfrac{1}{2} + \dfrac{1}{2}\ln\dfrac{1}{3}$

10 $A = -1, B = 1; \ln\left|\dfrac{x-1}{x}\right| + C$

11 $A = \dfrac{1}{4}, B = -\dfrac{1}{4}, C = \dfrac{1}{2}; \dfrac{1}{4}\ln\left|\dfrac{x+1}{x-1}\right| - \dfrac{1}{2(x-1)} + C$

12 $7\ln|x+2| - 5\ln|x+1| + C$

13 $3\tan^{-1}x + 2\ln|x-1| + C$

14 0.69

15 $1 - \dfrac{2}{e}$

16 Proof: See Worked solutions

17 $4 - 2\sqrt{e}$

18 $2\ln 2 - 1$

19 Proof: See Worked solutions

Chapter 7

Exercise 7.01

1 $a = x$

2 a Negative
 b $v^2 = 75 - 3x^2$
 c When $x = -5$
 d Positive (towards the origin)
 e Oscillating; greatest speed $5\sqrt{3}$ m s^{-1}

3 a $v^2 = 3600 - 20x$ **b** $v = 60 - 10t$
 c $x = 180$ m   **d** $t = 6$ s
 e $x = 135$ m   **f** $t = 2$ s, 10 s

4 a Proof: See Worked solutions
 b $x = -1$ m, $a = 16$ m s^{-2}; $x = 7$ m, $a = -16$ m s^{-2}
 c $[-1, 7]$ **d** $|\dot{x}| = 8$ m s^{-1}

5 a Positive
 b $v^2 = \sqrt{4x+9} - 3$
 c Proof: See Worked solutions
 d $\dot{x} = \sqrt{2}$ m s^{-1}

6 a $\dot{x} = 5e^{-2x}$ m s^{-1} **b** $x = \dfrac{1}{2}\ln(10t+1)$

7 a $x = e^9$ cm   **b** $v = 3e^9$ cm s^{-1}
 c, d Proof: See Worked solutions

8 a $v^2 = \dfrac{1}{2}\tan^{-1}\dfrac{x}{4}$
 b $v_{\max} = \dfrac{\sqrt{\pi}}{2}$ m s^{-1}

9 $x = t^2 + \sqrt{6}\,t$

10 $x = 2\ln(2t+1)$

11 a $v = 4\sqrt{x+1}$ m s^{-1} **b** $v(3) = 8$ m s^{-1}
 c $x = (2t+1)^2 - 1$ m

12 3012 km

13 $v = -\sqrt{8+e^{2(x-1)}}$ m s^{-1}

14 $a = 512$ m s^{-2}, $x = \dfrac{4}{8t+1}$

Exercise 7.02

1 a $A = 3, n = 2, \alpha = \pi$
 b

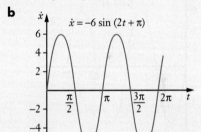

 c

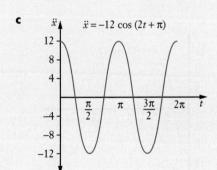

2 a Proof: See Worked solutions

b

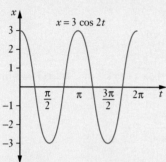

$A = \sqrt{3}, T = 2\pi$

3 a

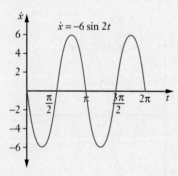

b $x_{max} = 3$ when $t = 0, \pi, 2\pi$ seconds

c $\dot{x} = -6 \sin 2t$

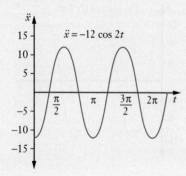

d $\dot{x} = 0$

e $\ddot{x} = -12 \cos 2t$

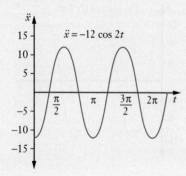

f At $x = 0, \ddot{x} = 0$

4 a Proof: See Worked solutions

b $T = \dfrac{2\pi}{3}$ s   **c** $\left|\dot{x}_{max}\right| = 39$ m s^{-1}

5 $x = \dfrac{3}{2}\cos\left(4t - \dfrac{\pi}{2}\right)$ or $x = \dfrac{3}{2}\sin 4t$

6 a $\ddot{x} = -625x$

b $A = \dfrac{3}{5}, T = \dfrac{2\pi}{25}, \left|\dot{x}_{max}\right| = 15$ m s^{-1}

7 a $x = 0$ m and $x = 6$ m   **b** $x = 3$ m

c $\left|\dot{x}_{max}\right| = 3$ m s^{-1} **d** $\ddot{x} = -(x - 3)$ m s^{-2}

e $T = 2\pi$ s

8 a $\ddot{x} = -4(x + 1)$ **b** $x = -1$

c $A = 4. T = \pi, f = \dfrac{1}{\pi}$

d $x = 4\cos\left(2t - \dfrac{\pi}{2}\right) - 1$

9 a $A = 3, T = \pi, \left|\dot{x}_{max}\right| = 6$ m s^{-1}

b, c Proofs: See Worked solutions

d $\alpha = -\dfrac{\pi}{3}$

10 a Proof: See Worked solutions

b Acceleration is of the form $\ddot{x} = -n^2(x - c)$, where acceleration is proportional to displacement and acts in the opposite direction; $T = \pi$

c $x = -7$ m, $x = 1$ m; $A = 4$

d $\left|\dot{x}_{max}\right| = 8$ m s^{-1}

e Proof: See Worked solutions

11 a $x = 2\sin\left(5t - \dfrac{\pi}{6}\right)$; motion is simple harmonic motion

b $t = \dfrac{\pi}{10}$ s

12 a $T = \dfrac{\pi}{2}$; Proof: See Worked solutions

b $t = \dfrac{\pi}{24}$ s, $\dot{x} = 6\sqrt{3}$ cm s^{-1}

c $\left|\dot{x}_{max}\right| = 12$ cm s^{-1} when $t = 0, \dfrac{\pi}{4}, \dfrac{\pi}{2}, \ldots$ over -3 cm $< x < 3$ cm

13 a $\dot{x} = \pm 2\sqrt{36 - x^2}$

b Proof: See Worked solutions

c $T = \pi, f = \dfrac{1}{\pi}$

d

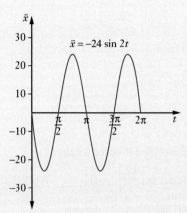

$\ddot{x} = -24 \sin 2t$

14 a $T = 2, A = 2$ **b** $x = 2\cos \pi t$

c 50%

15 a $A = 6$ m, $T = 16$ h   **b** $x = 6.20$ m

c 6 h 43 min

Exercise 7.03

1 61.25 m

2 $10\sqrt{2}$ s

3 a 4.6 s   **b** 103.3 m   **c** 9.2 s

d -33.4 m s^{-1}

4 a Vertically, $\ddot{y} = -10, \dot{y} = -10t + 10\sqrt{2}$,

$y = -5t^2 + 10\sqrt{2}t$

Horizontally, $\ddot{x} = 0, \dot{x} = 10\sqrt{2}, x = 10\sqrt{2}t$

b $2\sqrt{2}$ s   **c** $y = -\dfrac{1}{40}x^2 + x$

5 26.3 m s^{-1}

6 9° or 75°

7 45°

8 Proof: See Worked solutions

9 $30\sqrt{2}$ m s^{-1}

10 a, b Proof: See Worked solutions

11 26.16 m

12 15.53 m s^{-1}

13 $\dfrac{10\sqrt{10}}{3}$ m s^{-1}

14 Proof: See Worked solutions

15 a, b Proof: See Worked solutions

16 a, b Proof: See Worked solutions

Exercise 7.04

1 a Horizontal: 8 N, vertical: $8\sqrt{3}$ N

b Horizontal: $10\sqrt{3}$ N, vertical: 10 N

c Horizontal: $24 \sin 20° \cong 8.2$ N,
vertical: $24 \cos 20° \cong 22.6$ N

d Horizontal: $18 \sin 40° \cong 11.6$ N,
vertical: $18 \cos 40° \cong 13.8$ N

2 a N: 0 newtons along plane, N newtons
perpendicular to plane; mg: $mg \sin 25°$
newtons along plane, $mg \cos 25°$ newtons
perpendicular to plane

b F: F along plane, 0 perpendicular to plane;
mg: $mg \sin 30°$ along plane, $mg \cos 30°$
perpendicular to plane

c N: 0 along plane, N perpendicular to plane;
F: F along plane, 0 perpendicular to plane;
20 N force: $20 \sin 50°$ along plane,
$20 \cos 50°$ perpendicular to plane

3 a $F = \dfrac{5(\cos 40° + \sqrt{3} \sin 40°)}{\sin \theta + \sqrt{3} \cos \theta}$ N

b 4.9 N

4 $\mu = 0.4$

5 $\mu = 1.5$

6 a 1420 N   **b** 1500 N

7 a $T = 0.031$ N   **b** $F = 0.025$ N

8 6 m string: $9g$; 8 m string: $12g$

9 a

N

12 N

45°

$10g$

b $\dfrac{3\sqrt{2}}{5}$ m s^{-2} **c** $10g - 6\sqrt{2}$ N

10 a $\ddot{x} = \dfrac{(m - \mu M)g}{M + m}$ **b** $T = \dfrac{mM(1 + \mu)g}{M + m}$

c Proof: See Worked solutions

11 Proof: See Worked solutions

Exercise 7.05

1 $v = -\dfrac{k}{m}x + u, \quad x = \dfrac{m}{k}(u - v)$

2 $v = ue^{-\frac{k}{m}x}, \quad x = \dfrac{m}{k}\ln\left(\dfrac{u}{v}\right)$

3 Proof: See Worked solutions

4 a Proof: See Worked solutions

   b $t = \dfrac{215M}{F}\ln\left(\dfrac{83}{3}\right)$

5 a, b Proof: See Worked solutions

   c $t = 1.2$ s

6 a $v(t) = \dfrac{e^{2kt} - 1}{e^{2kt} + 1}, \ v_{t \to \infty} = 1$

   b $x = -\dfrac{1}{2k}\ln(1 - v^2)$

7, 8 Proofs: See Worked solutions

Exercise 7.06

1 a Proof: See Worked solutions

   b $v^2 = \dfrac{g}{k}(1 - e^{-2kx})$

   c $v = \sqrt{\dfrac{g}{k}(1 - e^{-2kh})}$

2 a $t = \dfrac{1}{\sqrt{kg}}\tan^{-1}\dfrac{U}{\sqrt{\frac{g}{k}}}$

   b $h_{\max} = \dfrac{1}{2k}\ln\left(1 + \dfrac{k}{g}U^2\right)$

3 a Proof: See Worked solutions

   b $h_{\max} = 160(1 - \ln 2)$ m

   c $t = 4\ln 2$ s

4 a

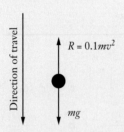

   b Proof: See Worked solutions

   c $v = 10\left(\dfrac{e^{2t} - 1}{e^{2t} + 1}\right)$

   d, e Proofs: See Worked solutions

5 a Explain

   b Proof: See Worked solutions

   c $v_T = \dfrac{g}{k}$

   d $x = \dfrac{g}{k^2}\ln\left(\dfrac{g}{g - kv}\right) - \dfrac{v}{k}$

   e 37.2 m

6 $x = \dfrac{1}{2k}\ln\left(\dfrac{4}{3}\right)$

7 a $t = \dfrac{3}{\sqrt{15}}\tan^{-1}\left(\dfrac{\sqrt{15}}{2}\right)$ s

   b Proof: See Worked solutions

   c $H = 3\ln\left(\dfrac{19}{4}\right)$

8 Proof: See Worked solutions

9 a Proof: See Worked solutions

   b 57.7 m s^{-1}

10 a Proof: See Worked solutions

   b $v_T = \sqrt{\dfrac{g}{k}}$

   c Proof: See Worked solutions

Exercise 7.07

1 a Proof: See Worked solutions

   b -20 m s^{-1} **c** 18.55 m

2 a 570 m   **b** 7.8 s

   c 0.85 s

3 $k \cong 0.158$

4 a 0.4 s   **b** 17.7 m

   c 28.8 m s^{-1}

5 a 0.000 233   **b** 30°

Test yourself 7

1 2 cm s^{-2}

2 $v = \sqrt{10e^{-8x} + 90}$ m s^{-1}.

3 $x(t) = \dfrac{5}{6}\cos\left(6t + \dfrac{\pi}{2}\right)$

4 a Proof: See Worked solutions

   b $x = -1 \pm \sqrt{11}$ m

   c $v_{\max} = 2\sqrt{11}$ m s^{-1}

   d $v^2 = 4[11 - (x + 1)^2]$

5   31.9 m, 5.1 s

6   5.7° or 84.8°

7   30°

8 $\dfrac{250\sqrt{3} - 30g + 75}{100}$ m s^{-1}

9   **a** $\ddot{x} \approx -0.7$ m s^{-2} **b** $T = 52.3$ N

10 $x = \dfrac{m}{k}\left(\dfrac{v_0 - v}{vv_0}\right), v = \dfrac{m}{k}\left(\dfrac{v_0}{v_0 x + \frac{m}{k}}\right)$

11 $x_{40} = 1200\left(1 - \dfrac{1}{e}\right), v_{40} = \dfrac{30}{e}$

12   **a** $v_T = 6mg$ **b** $v_T = \dfrac{mg}{2}$

13 $t = 1.2$ s, $h = 7.9$ m

14, 15 a, b   Proofs: See Worked solutions

Practice set 2

1   C   **2**   A   **3**   D   **4**   C

5   D   **6**   B   **7**   A   **8**   D

9   C   **10**   B   **11**   A   **12**   B

13–22   Proofs: See Worked solutions

23 $-\sin^{-1}\left(\dfrac{2-x}{2\sqrt{2}}\right) + C$

24   1

25 $\dfrac{1}{3}\tan^{-1}\left(\dfrac{x-1}{3}\right) + C$

26 $\dfrac{1}{2}\ln\left|\dfrac{x-2}{x}\right| + C$

27 $\ln\left|\dfrac{(x+2)^5}{(x+1)^3}\right| + C$

28 $\ln\left|\dfrac{(x+3)^3}{(x+2)^2}\right| + C$

29 $\dfrac{1}{2}$

30 $3(\ln 3)^2 - 6\ln 3 + 4$

31 $\dfrac{e^x}{2}(\sin x - \cos x) + C$

32 $x\sin^{-1} x + \sqrt{1 - x^2} + C$

33 $v = \sqrt{\dfrac{2x}{3(x+3)}}, v > 0$

34   4:09 p.m.

35   7.3 m

36   Proof: See Worked solutions

37 $v = 9.8$ m s^{-1}, $h = 4.9$ m

38   **a** $\ddot{x} = -64x$ **b** $v_{max} = 16$ m s^{-1}

39   **a** $a = \dfrac{100 - v}{10}$

     b   Proof: See Worked solutions

     c $a = \dfrac{100 - v^2}{10}$

     d $v_T = 10$ m s^{-1}

40 $\theta = 30°$

41   0.0066 m s^{-2}

42   0.57 s

INDEX

limiting friction 281
linear drag model (resisted projectile motion) 299
locus problems, solving 163–9
locus of z 163

magnitude of a position vector 82, 86
magnitude of a vector 80
Mandelbrot set 147–9
mathematical induction
    applications 202–4, 205–6, 208–9
    proof by 190–2, 194–6, 198–9
mathematical verbs xiii
maximum height (projectiles) 269
    time for (resisted projectile motion) 302
maximum range (projectiles) 269
mechanics 254–309
midpoint of vectors 93
modulus
    of a complex number 23, 24–5
        properties 29–32
    on the complex plane 23–7, 164–6
    of $z - z_1$ 163
modulus–argument form of a complex
    number 25–7
motion
    equations of and forces 277–83
    Newton's laws 277
    projectiles 267–73
    resisted horizontal 286–9
    resisted projectile 299–308
    resisted vertical 291–7
multiplying complex numbers 137–8
    by a constant 21
    by i 141–3
    geometric representation 137

natural numbers 54
negation 51–2, 59
negative power of a complex number,
    property 29, 31
Newton's laws of motion 277
normal force 280
normal vector to a line 110

odd number 64

parallel lines 110–12
parallel vectors 84, 94, 110–12
parallelogram rule for adding and subtracting
    vectors 20
parameters 100
parametric equations
    for the curve 100–1
    for a straight line 108

partial fractions 233–6
    in integration 239–41
    quadratic factors in denominator 237
    repeated linear factors in denominator 236–7
    repeated quadratic factors in
        denominator 238
Pascal's induction 194
Peano axioms 210
period of a pendulum 286
perpendicular lines 110–12
perpendicular vectors 84, 93, 110–12
points, plotting complex numbers as 17–18
polar form of a complex number 25–6
    converting to Cartesian form 27
    properties of moduli and arguments 29–32
polynomial equations 132–4
    complex conjugate root theorem 132–3
    creating 135
    real and complex roots 133
position vector 80
    direction 83
    magnitude 82, 86
positive integers, properties 64–5
power of a complex number 119
    property 29, 30
practice sets 179–87, 312–15
premise 48
principal argument of z (Arg z) 24
product of 2 complex numbers
    proof of property 30
    property 29, 30
product of complex conjugate pairs,
    property 32
product of complex numbers, property 29, 31
product of conjugates of 2 complex numbers,
    property 33
projectile motion 267–73
    maximum height 269
    maximum range 269
    modelling 270
    projection from a height 273–4
    range 269
    resisted 299–308
    resolving horizontal and vertical components
        (projection from the origin) 268
    time of flight 268
    trajectory of projectile 267–8, 269
        in resistive medium 301–5
    velocity components 268
proof(s)
    of binomial theorem 204
    by contradiction 59–60
    by counterexample 61–2

MATHS IN FOCUS 12. Mathematics Extension 2
 ISBN 9780170413435